Charles Seale-Hayne Library

University of Plymouth

(01752) 588 588

LibraryandITenquiries@plymouth.ac.uk

HOLOCENE CLIMATE
VARIABILITY
A MARINE PERSPECTIVE

HOLOCENE CLIMATE VARIABILITY
A MARINE PERSPECTIVE

Edited by

E. JANSEN

University of Bergen
5007 Bergen, Norway

P. DeMENOCAL

Lamont-Doherty Earth Observatory
Palisades, NY 10964, USA

F. GROUSSET

University of Bordeaux 1
F-33405 Talence, France

2004

ELSEVIER

Amsterdam - Boston - Heidelberg - London - New York - Oxford - Paris
San Diego - San Francisco - Singapore - Sydney – Tokyo

ELSEVIER B.V.
Radarweg 29
P.O. Box 211, 1000 AE Amsterdam
The Netherlands

ELSEVIER Inc.
525 B Street, Suite 1900
San Diego, CA 92101-4495
USA

ELSEVIER Ltd
The Boulevard, Langford Lane
Kidlington, Oxford OX5 1GB
UK

ELSEVIER Ltd
84 Theobalds Road
London WC1X 8RR
UK

First edition 2004

Reprinted from Quaternary Science Reviews, Volume 23/20-22, pp. 2061–2268

ISBN: 0080446221

♾ The paper used in this publication meets the requirements of ANSI/NISO Z39.48-1992 (Permanence of Paper).
Printed in Great Britain.

Contents

HOLOCENE CLIMATE VARIABILITY – A MARINE PERSPECTIVE

Edited by E. Jansen, P. deMenocal and F. Grousset

Editorial

Holocene climate variability—a marine perspective

Future climate change will be occurring as a combination of natural climate variability and responses to anthropogenic forcing factors. It is thus necessary to understand the natural element of climate variability to assess our climatic future. This has led to a strong interest in the climate dynamics community to improve our knowledge of the climatic evolution of the Holocene. In the IMAGES program, which is the marine leg of the Past Global Changes (PAGES), this led to the establishment of a specific working group for Holocene climate variability emphasising the marine paleoclimate record, and initiating studies and syntheses on the marine aspects of Holocene climate variability. One of the tasks of the working group is to coordinate meetings and workshops for such syntheses and to facilitate comparisons with results from other paleoclimatic disciplines and climate modelling on Holocene climates.

In this issue we have collected 14 contributions originating from the first workshop of the working group, held in August 2003 in Hafslo, Norway in a setting worthy of the subject. Historical and glaciological evidence from the fjord region nearby tell many stories of hardship encountered due to Holocene climate change.

The 14 contributions in this issue mainly comprise marine proxy reconstructions, but there are also papers summarizing evidence from continental archives, on forcing factors and climate model simulations, as well as on chronological issues. The breadth of the papers in this special edition of *Quaternary Science Reviews* will give the reader a broad perspective on recent research results and approaches, both on the record of climate change in low and high latitudes on regional and/or global scales, and also on the attribution of changes to external and internal forcing factors.

The marine records contain both globally oriented syntheses of the main trends of surface ocean characteristics during the Holocene and the response of the ocean/atmosphere system to the slow orbital forcing (e.g. Kim et al.), as well as a series of papers highlighting major advances obtained by the community in recent years in terms of providing records which document decadal to centennial climate change (e.g. Andersen et al., Giraudeau et al., Knudsen et al., Moros et al.). Papers address high latitudes, latitudinal gradients and the monsoon system. Schmidt et al. and Kim et al. provide modelling perspectives and Cook et al. and Nesje et al. highlight recent advances in continental records. The problem of establishing a chronological basis for correlation of Holocene records are addressed by Eiriksson et al. and Muscheler et al., stressing the need for improvements in an area of key importance. Schulz et al. also highlight an aspect of this by discussing the problems of attributing Holocene climate variability to internal or external mechanisms providing pacing of the variability at certain frequencies.

The collection of papers in this issue will be a good source of knowledge on recent developments in paleoceanography and paleoclimatology, and shows a community in strong progress both in terms of obtaining detailed records, as well as in understanding the scale of changes our planet experienced when human influence played a minor role.

Eystein Jansen
Peter deMenocal
Francis Grousset
University of Bergen, Allegaten 55,
Bergen 5007, Norway
E-mail address: eystein.jansen@geo.uib.no

0277-3791/$ - see front matter © 2004 Published by Elsevier Ltd.
doi:10.1016/j.quascirev.2004.07.019

Quaternary Science Reviews 23 (2004) 2063–2074

Extra-tropical Northern Hemisphere land temperature variability over the past 1000 years

Edward R. Cook[a],*, Jan Esper[b], Rosanne D. D'Arrigo[a]

[a]*Lamont-Doherty Earth Observatory, P.O. Box 1000, Route 9W, Palisades, NY10964, USA*
[b]*Swiss Federal Research Institute WSL, 8903 Birmensdorf, Switzerland*

Abstract

The Northern Hemisphere (NH) temperature reconstruction published by Esper, Cook, and Schweingruber (ECS) in 2002 is revisited in order to strengthen and clarify its interpretation. This reconstruction, based on tree-ring data from 14 temperature-sensitive sites, is best interpreted as a land-only, extra-tropical expression of NH temperature variability. Its strongly expressed multi-centennial variability is highly robust over the AD 1200–1950 interval, with strongly expressed periods of "Little Ice Age" cooling indicated prior to AD 1900. Persistently above-average temperatures in the AD 960–1050 interval also suggest the large-scale occurrence of a "Medieval Warm Period" in the NH extra-tropics. However, declining site availability and low within-chronology tree-ring replication prior to AD 1200 weakens this interpretation considerably.

The temperature signal in the ECS reconstruction is shown to be restricted to periods longer than 20 years in duration. After re-calibration to take this property into account, annual temperatures up to AD 2000 over extra-tropical NH land areas have probably exceeded by about 0.3 °C the warmest previous interval over the past 1162 years. This estimate is based on comparing instrumental temperature data available up to AD 2000 with the reconstruction that ends in AD 1992 and does not take into account the mutual uncertainties in those data sets.

© 2004 Published by Elsevier Ltd.

1. Introduction

A number of efforts have been made to reconstruct the Northern Hemisphere (NH) temperature variability over the past 1000 or more years using well-dated, high-resolution (mostly tree-ring) proxy records (e.g., Jones et al., 1998; Mann et al., 1999; Briffa, 2000; Crowley and Lowery, 2000; Briffa et al., 2001; Esper et al., 2002; Mann and Jones, 2003). Such long records are critically important in attribution studies that seek to determine the causes of the 20th century global warming (e.g., Jones et al., 1998; Crowley, 2000; Hegerl et al., 2003). See also IPCC (2001). Given the progress made to date, significant uncertainty still exists concerning the way in which NH temperatures have varied on multi-centennial time scales.

Briffa and Osborn (2002) highlighted this uncertainty in commentary associated with the publication of the somewhat controversial Esper et al. (2002) extra-tropical NH temperature reconstruction (hereafter referred to as ECS). They compared the ECS reconstruction with six other previously published NH temperature reconstructions (Overpeck et al., 1997; Jones et al., 1998; Mann et al., 1999; Briffa, 2000; Crowley and Lowery, 2000; Briffa et al., 2001). Each of the reconstructions was calibrated against the same observed land-only mean annual NH temperatures for the 20°–90°N latitude band and then smoothed with a 50-year low-pass filter. The result is shown in Fig. 1 (re-plotted from Briffa and Osborn, 2002), with the ECS reconstruction highlighted as the thick-black dashed curve. Over much of the past 1000 years, the expressed multi-centennial variability in the ECS reconstruction stands out from the others. This variability does not in general exceed the estimated uncertainty limits of other

*Corresponding author. Tel.: +1-845-365-8618; fax: +1-845-365-8152.

E-mail address: drdendro@ldeo.columbia.edu (E.R. Cook).

0277-3791/$ - see front matter © 2004 Published by Elsevier Ltd.
doi:10.1016/j.quascirev.2004.08.013

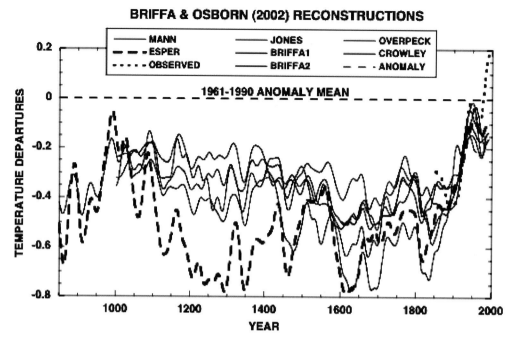

Fig. 1. Seven smoothed NH temperature reconstructions re-plotted from Briffa and Osborn (2002). All of the reconstructions have been re-calibrated identically with NH land-only mean annual temperatures north of 20°N and smoothed with the same 50-year low-pass filter. The ECS reconstruction is highlighted as a dashed curve to illustrate how it departs from the other NH temperature estimates.

large-scale temperature reconstructions (e.g., Fig. 1 in Mann et al., 2003), which suggests that there are few statistically meaningful differences between the NH reconstructions shown in Fig. 1. However, the amplified low-frequency variability in the ECS reconstruction does not appear to be random either. In this sense, it could also represent an emergent large-scale multi-centennial mode of NH temperature variability that is not well understood.

With this latter possibility in mind, we take this opportunity to clarify and tighten up the interpretation of the ECS reconstruction made somewhat incompletely in Esper et al. (2002), examine the robustness of its expressed multi-centennial variability, and address certain criticisms of it. It is also useful to first restate the origin of the Esper et al. (2002) study before proceeding, because the ECS reconstruction was in a sense a by-product of that paper's primary objective. Broecker (2001) argued that the apparent lack of a Medieval Warm Period (MWP) in the Mann et al. (1999) NH temperature reconstruction was due to the inability of long tree-ring records to preserve multi-centennial temperature variability. The Esper et al. (2002) paper sought to dispel that erroneous interpretation by demonstrating how tree-ring records from temperature-sensitive NH locations could preserve multi-centennial variability if the data were processed with that goal in mind. After showing how this could be done, Esper et al. (2002) then interpreted their extra-tropical NH tree-ring record as an alternative expression

of past temperature variability that clearly differs in certain respects from the others shown in Fig. 1.

2. The ECS reconstruction and its interpretation

There are a number of possible reasons why the ECS extra-tropical NH temperature reconstruction differs as much as it does from the others shown in Fig. 1. Each expresses past temperature variability in somewhat different ways due to: (1) differences in the temperature proxies used and their NH locations; (2) differences in the properties of the temperature signals reflected in the proxies; (3) differences in the procedures used to develop the temperature proxies; and (4) differences in how the temperature reconstructions were created. We will address each of these issues with respect to how they relate to the ECS reconstruction, the robustness of its expressed low-frequency signal, and its overall interpretation.

2.1. What temperature proxies were used in the ECS reconstruction and where are they located?

As described in Esper et al. (2002), the ECS reconstruction is based on long, exactly dated, annual tree-ring chronologies from 14 sites in the NH (Fig. 2). The six longest tree-ring chronologies that all extend back to AD 831 are circled. The tree-ring sites are mostly located near the northern or upper-elevation

LOCATIONS OF THE ESPER ET AL. (2002) TREE-RING SITES

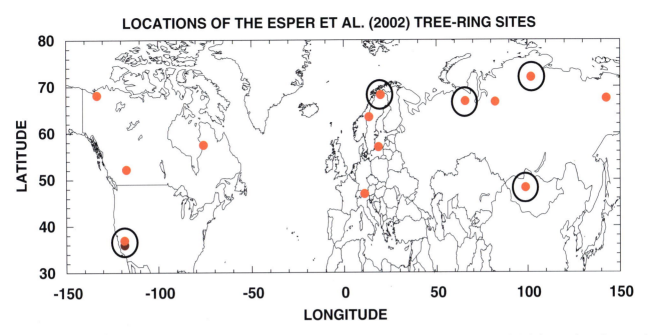

Fig. 2. Map of the Esper et al. (2002) tree-ring sites. Each solid red dot represents one of the 14 sites used. The six circled-sites are those that extend back to AD 831, the beginning of the ECS record.

limits of growth of the sampled tree species. At these locations, temperature variability commonly limits tree growth each year, and this annually changing limitation on growth is reflected in the radial ring-width variations of the tree. This has been shown to be the case for many site-level tree-ring reconstructions and inferred records of past temperatures covering the past millennium in the NH, e.g., northern Fennoscandia (Briffa et al., 1992), northern Polar Urals (Briffa et al., 1995), eastern Taimyr (Naurzbaev et al., 2002), Mongolia (D'Arrigo et al., 2001), and the Columbia Icefield in Alberta (Luckman et al., 1997). Significantly, of the 14 NH tree-ring sites used by Esper et al. (2002) in their study, tree-ring data from each of the five locations just referenced were also used in the ECS reconstruction. So, there is a significant amount of independent support for the expected temperature sensitivity of the tree-ring data at 35% of the sites used by Esper et al. (2002). The remaining sites have not been explicitly subjected to comparisons with local temperature data here and, indeed, some of the more xeric sites (e.g., in California; Lloyd, 1997) may have a mixed temperature/precipitation signal. However as stated earlier, their locations are broadly consistent with the kinds of tree-ring sites that are, from experience known to be sensitive to changing temperatures.

Note that the 14 sites in Fig. 1 are located in the 30–70°N latitude band. For this reason, Esper et al. (2002) also referred to their temperature reconstruction as *extra-tropical*. Interpretations of the ECS reconstruction must take this property into account. It is probably wise to tighten up this spatial interpretation even

further. While it is arguable that some of the temperature information in the tree-ring chronologies extends beyond the continents to the oceans (e.g., Mann et al., 1999), the ECS site locations are almost certainly dominated by regional continentality effects, which would amplify their expressed low-frequency temperature signals over that in tree-ring chronologies from more ocean-dominated sites (e.g., western Tasmania; Cook et al., 2000). For this reason, the ECS reconstruction probably best reflects temperatures over *extra-tropical land areas* of the NH.

3. What temperature signals are in the ECS proxies?

The annual tree-ring chronologies used in the ECS reconstruction all derived from trees with constrained radial growth seasons that typically fall somewhere in the late-spring and summer months. This fact implies that the annual ring-width variations of the trees used in the ECS reconstruction (and in the other NH reconstructions as well) must only reflect warm-season temperatures. Although this would appear to be true, trees are physiologically active over a much broader time period than that associated with the radial growth season alone (Fritts, 1976). Consequently, temperatures outside of the radial growth season can affect longer-term tree growth processes such as foliage production/retention in evergreen conifers and fine root development (Jacoby et al., 1996). In addition, winter climate can have a preconditioning effect on subsequent growing season climate and tree growth through effects on land surface albedo, soil temperatures, and thawing of the root zone in the spring before radial growth commences. Finally, physiological preconditioning from the prior growth season is known to affect the potential for radial growth in the following year (Fritts, 1976). All this serves to broaden the temperature response window and make its interpretation in the ECS reconstruction more complicated and uncertain. Consequently, it is probably best described as a *warm-season-weighted* expression of temperature variability that includes information on temperatures outside of the summer radial growth season as well.

Even if we accept the admittedly nebulous *warm-season-weighted* temperature model in principle, would it have been possible to statistically differentiate it from a simple annual temperature average during the calibration phase of the ECS reconstruction? At the spatial scale that the ECS reconstruction represents and the temporal smoothing of the temperature signal contained within it, probably not. Land-only annual (January–December) and warm-season (April–September) temperatures for the 30–70°N latitude NH extra-tropics have a correlation of 0.87 over the period 1856–2000 without any smoothing and 0.93 after decadal smoothing (see also Esper et al., 2002 online Supplemental Data; http://www.sciencemag.org/cgi/content/full/295/5563/2250/DC1). Given these high correlations and the statistical uncertainty in the smoothed ECS tree-ring chronology used for reconstruction (see

Fig. 3a), an annual temperature model is probably acceptably close to the ill-defined *warm-season-weighted* temperature model, even if the latter is closer to the truth in principle.

3.1. What procedure was used to develop the ECS temperature proxies?

With respect to the procedure used to develop the ECS temperature proxies, the annual tree-ring chronologies were all developed in a uniform way using the Regional Curve Standardization method (RCS; Briffa et al., 1992; Esper et al., 2003). The RCS method of detrending tries to preserve as much multi-centennial time-scale variability as possible in long tree-ring chronologies (Cook et al., 1995). The other NH temperature reconstructions shown in Fig. 1 have also used a variety of RCS tree-ring chronologies, including some used in the ECS reconstruction (e.g., Tornetrask and Polar Urals), but none pooled the site ring-width data together in the way that Esper et al. (2002) did.

In developing the ECS reconstruction, the ring-width data of the 14 sites were first pooled into two ring-width classes according to the shapes of their biological growth trends (non-linear and linear) that needed to be removed as part of the RCS procedure, with three sites mixing between the two trend classes (see Fig. 2 in the Esper et al., 2002 online Supplemental Data). This resulted in the initial development of two reasonably

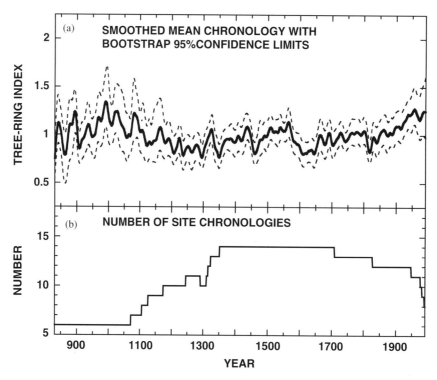

Fig. 3. The ECS mean tree-ring chronology (AD 831–1990) with bootstrap 95% confidence limits (a), all smoothed with a 20-year low-pass filter to highlight inter-decadal and longer fluctuations. The number of site chronologies available each year is shown in (b). They range from 6 to 14.

independent extra-tropical NH RCS tree-ring chronologies, each covering the period AD 831–1990. Over the AD 1200–1990 interval where site and series replication is relatively high in each growth trend class (see Figs. 2b and d in Esper et al., 2002), the two series are significantly correlated ($r = 0.523$; $p < 0.001$ corrected for lag-1 autocorrelation), and the correlation climbs to 0.732 after the series are smoothed with a 50-year low-pass filter. Before that time, the site and series replication decline in each trend class and the correlation between them correspondingly weakens: $r = 0.079$ before smoothing and 0.032 after smoothing.

While this breakdown could indicate a period of more spatially heterogeneous climate variability across the NH (Crowley and Lowery, 2000), a more likely cause is the decline of site and ring-width series replication prior to AD 1200. Further interpretation of this apparent breakdown is hampered by the fact that there is unequal pooled site replication in each trend-class chronology over time, which means that certain sites will dominate certain time periods (e.g., Mongolia, Tornetrask, and Polar Urals in the non-linear class and Taimyr and Upper Wright in the linear class for the AD 900–1100 interval; see Fig. 2 in the Esper et al., 2002 online Supplemental Data). To counteract this uneven weighting effect over time, the grand mean RCS chronology used by Esper et al. (2002) was based on the 14 site RCS chronologies, with each site weighted by the cosine of the latitude band it was located in (centered in either 30°–50°N or 50°–70°N) to provide more realistic regional weighting (again see the Esper et al., 2002 online Supplemental Data). Therefore, the loss of correlation between the non-linear and linear trend classes prior to AD 1200 may not be as severe when the site chronologies are used to calculate the grand mean RCS chronology. Regardless, the pre-1200 interval of the ECS reconstruction needs to be interpreted cautiously, a point that was illustrated by Esper et al. (2002) in their Fig. 2a that showed the linear and non-linear chronologies and their dissimilarities prior to AD 1200.

Fig. 3a shows the grand mean of the 14 RCS tree-ring chronologies for the AD 831–1990 interval with bootstrap 95% confidence limits, each smoothed with a 20-year low-pass filter, *after* the annual values were estimated. The smoothing highlights the inter-decadal and longer fluctuations that are of primary interest here. The number of chronologies available for each year is shown in Fig. 3b. It ranges from 6 to 14. This plot is very similar to the one shown in Fig. 2c in Esper et al. (2002), who regarded it as "compelling evidence" for large-scale low-frequency climate variability over the extra-tropical NH.

Fig. 4 zooms in on the weakly replicated AD 831–1200 interval for easier examination. The number of tree-ring sites available each year ranges from 6 to 10 out of the original total of 14 (see Fig. 3b). The AD

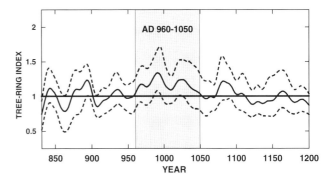

Fig. 4. The weakly replicated AD 831–1200 interval of the ECS mean chronology. The number of site chronologies available each year is shown in Fig. 3b. Note the persistent above-average period in the AD 960–1050 interval that hints at the existence of a spatially coherent temperature signal.

960–1050 interval highlighted in the shaded rectangle is persistently above average in this smoothed record, with two brief periods that also exceed the 95% confidence level. The sites represented are widely scattered, ranging from the western United States to northern Fennoscandia, the northern Polar Urals, the Taimyr Peninsula, and Mongolia (see Fig. 2). This result suggests the existence of a spatially coherent low-frequency signal among these six sites, which Esper et al. (2002) interpreted to be an expression of the MWP, hence their statement that it "supports the large-scale occurrence of the MWP over the NH extratropics". This statement has been strongly criticized by R.S. Bradley (pers. comm.), and indeed it does not fit into the Euro-centric definition of "High Medieval" time (AD 1100–1200) used by both Lamb (1965) and Bradley et al. (2003) in their sensu strictu definition of the MWP. In contrast, the MWP sensu lato is regarded by some as having occurred between the 9th and 14th centuries (e.g., see http://www.ngdc.noaa.gov/paleo/globalwarming/medieval.html), but this is at best an educated guess.

The sensu lato definition of the MWP is more consistent with what the mean RCS chronology shows. However, its suggested 600-year duration also allows for a large range of global climate phenomena to be included, which may not be causally related to the MWP (Bradley et al., 2003). Thus, the MWP sensu lato is susceptible to a variant of "suck-in and smear" (Baillie, 1991), in which a broad time period assigned to an ill-defined climate epoch allows for unrelated climate anomalies to be sucked into it. We do not believe that this is necessarily the case for our mean RCS chronology because of its likely temperature dependency. On the other hand, its weak replication in the AD 831–1200 interval requires more cautious interpretation of MWP warming than that stated by Esper et al. (2002). Little more can be said about the likelihood of a MWP in the ECS data without improving the site chronologies used

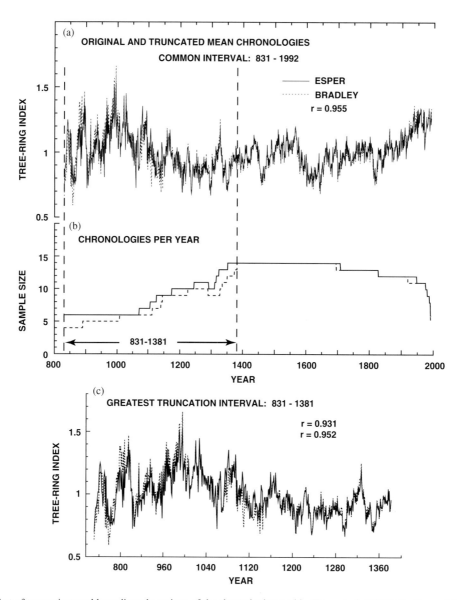

Fig. 5. Test of the effect of truncating weakly replicated portions of the chronologies used by Esper et al. (2002). Portions of chronologies with fewer than five measurements per years were deleted. This resulted in the loss of some chronology replication in the mean series, as indicated by the dashed line in the sample size plot (Fig. 3b). The two correlations in Fig. 3c are for unsmoothed and 40-year low-pass filtered data (in parentheses).

in this and other studies (e.g., better early replication and increased length in some cases) and, more so, in improving the spatial coverage of millennia-long temperature-sensitive tree-ring chronologies (Bradley et al., 2003).

The greater uncertainty of the pre-1200 interval and even the apparent expression of the MWP in the ECS reconstruction could also be related to low replication in the early portions of some of the tree-ring chronologies used (R.S. Bradley, pers. comm.). To investigate this possibility, we truncated the early portions of all chronologies with fewer than five ring-width measurements per year. This cutoff is admittedly somewhat arbitrary, but it serves as a useful demonstration nonetheless. Fig. 5 shows the chronologies before and

after truncation over their full lengths (Fig. 5a) and the change in the number of chronologies per year (Fig. 5b). Most of the truncation occurs in the AD 831–1381 period which causes the number of chronologies to drop to as low as four. Yet, the correlation between the two RCS chronologies remains very high: $r = 0.955$ over the full AD 831–1992 length. Even when we compare only the AD 831–1381 period most affected by truncation (Fig. 5c), the correlation between the mean RCS records remains very high: $r = 0.931$ unfiltered and $r = 0.952$ after 40-year low-pass smoothing. These high correlations, even after losing some sites due to truncation, may be due to the fact that there is still a reasonably good degree of spatial representation in the truncated mean chronology (western United States, northern

NORMALIZED REGIONAL RCS CHRONOLOGIES -- AD 831-1990

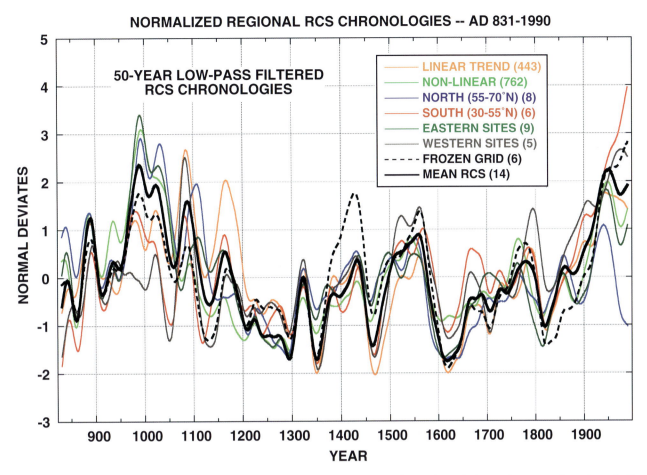

Fig. 6. Testing the regional robustness of the ECS tree-ring data. The "linear" and "non-linear" chronologies are the same as those shown in Esper et al. (2002). The other chronologies have been grouped by geographic location of the sites. The number of sites in each geographic group is provided in parentheses in the plot legend.

Fennoscandia, Taimyr Peninsula, Mongolia). Although the concern of R.S. Bradley (pers. comm.) was well justified, low chronology replication does not appear to have been responsible for creating the persistently above-average period in Fig. 4 that Esper et al. (2002) interpreted as an expression of the MWP sensu lato in their temperature reconstruction.

The robustness of the low-frequency land-only extra-tropical NH signal shown in Fig. 3a is further illustrated in Fig. 6. Besides the "linear" and "non-linear" groupings shown in Esper et al. (2002), the RCS-detrended data were grouped geographically into northern (55–70°N) and southern (30–55°N) bands, by eastern and western hemispheres, and as the mean of only the six longest chronologies (their locations circled in Fig. 2) extending back to AD 831. The latter is a form of "frozen grid" analysis (Jones et al., 1986; Duffy et al., 2001) that tests for the loss of fidelity when data during certain time periods are missing. These chronologies have all been smoothed with a 50-year low-pass filter and normalized to zero mean and unit variance.

Two points are worth making here. First, the low-frequency signal in the mean RCS chronology and the

various subset chronologies is generally robust, particularly over the AD 1200–1950 interval. The average correlation between the mean RCS chronology and the subset chronologies is 0.82 for the full AD 831–1990 period and 0.87 for the AD 1200–1950 period. Prior to AD 1200, there is more separation between the subset chronologies, which is consistent with the significant decline in replication described earlier. After AD 1950, there is clear divergence, particularly between the "North" and "South" subset chronologies. This is probably an expression of the well-documented large-scale loss of growth sensitivity to climate in boreal forest conifers (Briffa et al., 1998a, b), a phenomenon that is not yet fully understood. The "South" chronologies may also show evidence of abnormally accelerated growth in the 20th century (Briffa et al., 1998a, b). So it appears that the data used by Esper et al. (2002) contain a geographically robust low-frequency signal over the period AD 1200–1950 that is almost certainly reflecting large-scale temperature forcing on tree growth. Prior to AD 1200, the evidence for robustness is again weakened by low replication and, perhaps, a period of greater regional climate variability.

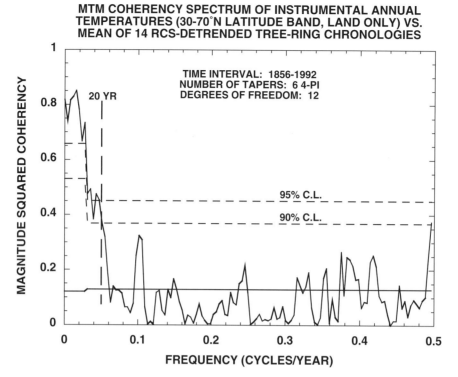

Fig. 7. The magnitude squared coherency spectrum between mean annual extra-tropical NH temperatures and the mean RCS chronology.

Second, the long-network chronology used for the "frozen grid" test correlates well with the mean RCS chronology ($r = 0.85$ over the full AD 831–1990 period and 0.82 over the AD 1200–1950 period). So overall the "frozen grid" chronology based on six sites captures most of the low-frequency variance found in the mean RCS chronology based on 14 sites. Nonetheless, some differences between the two are strongly apparent, especially around AD 1400. While it is tempting to use the "frozen grid" chronology alone as a useful expression of past land-only NH temperature variability, its geographic signal may be spatially biased at times by more geographically restricted effects. This argues again for caution in interpreting the pre-1200 period as a large-scale expression of the MWP and the need for additional long, spatially distributed, tree-ring records covering the past 1000 or more years.

4. How was the ECS temperature reconstruction created?

The ECS reconstruction differs from most of the others shown in Fig. 1 in that it is based on the grand mean of the 14 extra-tropical NH site chronologies *before* it was calibrated with instrumental temperature data. This is not necessarily the optimal way to reconstruct large-scale temperatures. However, the original point of the Esper et al. (2002) paper was to dispel the claim made by Broecker (2001) that long tree-ring series were largely incapable of reconstructing multi-centennial temperature variability. For this reason, the large-scale low-frequency features of tree growth were purposely emphasized in the Esper et al. (2002) paper at the expense of any higher-frequency local temperature signals in the individual site chronologies. Indeed, the way in which Esper et al. (2002) aggregated their data made it difficult, if not impossible, to reconstruct high-frequency temperature changes, which tend to be much more local compared to the multi-decadal and centennial changes (Jones et al., 1997).

To illustrate this point, Fig. 7 shows the magnitude squared coherency (MSC) spectrum between annual temperatures and the mean RCS tree-ring chronology for the common interval 1856–1992. The temperature data are "land-only" in the same 30–70°N latitude band as the tree-ring sites. This differs from the full NH annual average temperature series used by Esper et al. (2002), but that result would be very similar. In this case, the coherency spectrum clearly shows that the only useful temperature signal in the mean RCS chronology occurs at periods greater than 20 years in duration where the MSC exceeds the 90% confidence level. This result fits in nicely with our expectation.

Fig. 8 shows the mean RCS chronology, after 20-year low-pass filtering and calibration in terms of the similarly filtered land-only 30°–70°N annual temperature series used in the coherency spectrum analysis.

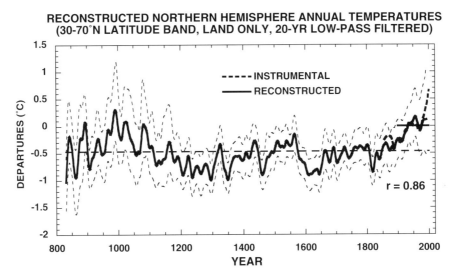

Fig. 8. The temperature-calibrated mean RCS tree-ring chronology, AD 831–1990 (thick solid curve). The thin dashed curves are bootstrap 95% confidence intervals, and the thick dashed curve represents the instrumental data up to 2000. The temperature values are anomalies from the full 20th century instrumental mean anomaly (1900–1999). The difference between the 20th century zero anomaly and the long-term anomaly mean (~0.47 °C) is indicated by the long-term mean (thick dashed horizontal line).

Calibrating on the 20-year low-pass filtered estimates makes sense from a signal processing perspective, a point also made by Osborn and Briffa (2000) and Timm et al. (2004) in their papers on timescale-dependent reconstruction of past climate from tree-ring chronologies. Over the 1856–1992 interval, the instrumental and proxy series have a correlation of 0.86, which based on the coherency spectrum results is statistically significant ($p < 0.05$). The reconstruction is expressed as temperature anomalies from the full 20th century mean of the instrumental record and covers the period AD 831–1992. The zero anomaly period contrasts with the mean anomaly of the entire record, with a difference of −0.47 °C. This difference is clearly related to reconstructed cool periods in the AD 1200–1850 interval that probably represents a large-scale expression of the Little Ice Age (LIA; Grove, 1988) and its various forcings over extra-tropical NH land areas.

The re-calibrated reconstruction in Fig. 8 is effectively the same as the ECS reconstruction shown in Fig. 1 because the same mean RCS chronology was linearly rescaled to temperatures in each case. The main difference is in how the scaling of tree growth in terms of temperature was done. This is a significant issue because it can strongly affect the amplitude of the resulting reconstruction. Briffa and Osborn (2002) performed all of their re-calibrations of the existing reconstructions using annually resolved data. They then low-passed filtered the re-calibrated reconstructions for comparison purposes. In so doing, differing amounts of variance were inevitably lost due to regression, which could explain some of the differences in the expressed amplitudes of the temperature reconstructions in Fig. 1. This procedure also reduced the amplitude of the ECS

reconstruction compared to that published in Esper et al. (2002) because the latter was directly calibrated on 40-year low-pass filtered data, a frequency band with a better signal-to-noise ratio for estimating temperature. This property is illustrated in the MSC spectrum (Fig. 7), which shows that there is little useful temperature information in the mean RCS chronology at periods less than 20 years in duration. For this reason, we argue that the amplitude of the extra-tropical temperature reconstruction produced here, with its approximately 1 °C range over the past 1000 years, is a reasonable estimate of past temperature change within the uncertainty of the data and procedures that we are using here.

5. Other interpretational issues

The re-calibrated mean RCS tree-ring record probably represents the best reconstruction of past land-only, extra-tropical NH annual temperatures that is practical to extract from it at this time. Note that it does very well at tracking the instrumental data on inter-decadal and longer timescales up to about 1982, after which the tree-ring estimates systematically under-estimate the actual warming. This departure probably reflects the loss of climate sensitivity noted earlier in the "North" chronology shown in Fig. 6. Whether or not a similar loss of sensitivity has occurred in the past is unknown with any certainty, but no earlier periods of similar divergence are apparent between the "North" and the other regional chronologies. This result suggests that the large-scale loss of climate sensitivity documented by Briffa et al. (1998a, b) is unique to the 20th century, which argues for an anthropogenic cause. After 1992, smoothed instrumental temperatures (up to 2000, our last available year of data, and scaled to reflect lost variance due to regression in the reconstruction) have increased rapidly and now exceed our warmest estimated past temperature epoch (ca AD 1000) by about 0.3 °C. This estimate is based on comparing instrumental temperature data available up to AD 2000 with the reconstruction that ends in AD 1992 and does not take into account the mutual uncertainties in those data sets.

Based on what we now show here, it appears that late-20th century land-only extra-tropical NH temperatures are warmer than at anytime over the past 1162 years. This conclusion is consistent with other evaluations of recent global warming and its cause(s) (e.g., Jones et al., 1998, 2001; Mann et al., 1999; Crowley, 2000; Mann and Jones, 2003). Even so, our proposed expression of the MWP, with all its caveats, appears to be a period of significant warmth as well. Esper et al. (2002) stated that the MWP in their reconstruction "approached, during certain intervals, the magnitude of 20th century warming at least up to AD 1990". Given the new results in this paper, that earlier statement requires some clarification and correction. Fig. 8 shows that 20th century temperatures since the early 1980s have exceeded those during our proposed expression of the MWP. The separation of the instrumental temperature record from our re-calibrated reconstruction is clearly evident only after 1984 however. Up to that date, the reconstructed warm epoch centered on AD 1000 (with its greater uncertainty) is comparable to that estimated for most of the 20th century. The subsequent divergence between the two records after 1984 up to 1990 may reflect the loss of climate sensitivity in some of the tree-ring records noted earlier. After 1990, the instrumental temperatures steadily increase to the point where they clearly exceed the peak warmth of the earlier warm epoch by about 0.3 °C. Therefore, our present results indicate that our estimate of MWP warming over extra-tropical NH land areas was comparable to that over most of the 20th century, but only up to the early 1980s as best as we can estimate here. Whether or not the MWP was a global phenomenon is still hotly debated (e.g., Hughes and Diaz, 1994; Bradley, 2000; Bradley et al., 2001; Broecker, 2001; Bradley et al., 2003). This question cannot be answered with any certainty either way because, as stated earlier, the global coverage of well-dated, high-resolution, millennial-length temperature proxies is still insufficient.

Given the stated uncertainties in the ECS reconstruction and its re-calibrated version shown in Fig. 8, it is still useful to examine the series in the frequency domain for evidence of band-limited multi-decadal/centennial variability. Fig. 9 shows the power spectrum of the temperature reconstruction shown in Fig. 8. This analysis is necessarily restricted to periods > 20 years in

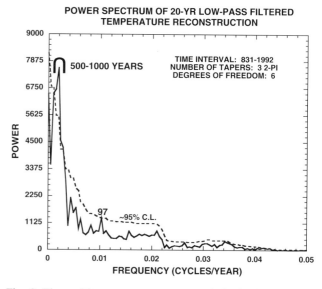

Fig. 9. The multi-taper power spectrum of the land-only, extra-tropical NH annual temperature reconstruction.

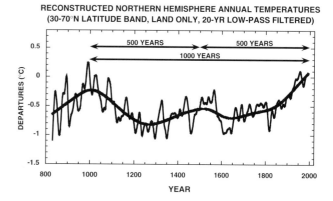

Fig. 10. The extra-tropical NH annual temperature reconstruction showing possible quasi-500 and 1000-year periods, which are visibly apparent in the series.

duration. The spectrum is clearly dominated by significant spectral power in the 500–1000 year band, with a slight indication of additional significant power at around 97 years. The source of this indicated 500–1000 year spectral power is indicated in Fig. 10. Of course, with only 1162 years of data, it is impossible to claim that the variability seen in this reconstruction is representative of longer-term NH temperature variability during the Holocene. At this time it must be viewed only as an estimate of multi-centennial temperature variability over extra-tropical NH land areas for the past 1,162 years. Whether or not this kind of behavior reflects an intrinsic mode of internally or externally forced variability also cannot be determined with any certainty at this time. Variability at similar time scales has been described for the North Atlantic by Bond et al.

(1997, 2001), and also by Chapman and Shackleton (2000). Therefore, it is possible that the suggested quasi-periodic behavior in the temperature reconstruction reflects a mode of long-term variability and forcing that has operated throughout the Holocene. There is, however, a reasonably strong association between certain cold epochs in the reconstruction (e.g., in the AD 1200s and 1600s; see Fig. 8) and explosive volcanism as well (M.E. Mann, pers. comm.). See Briffa et al. (1998c) and Zielinski (2000) for records of past explosive volcanism that support this interpretation, and Shindell et al. (2003) for modeling results that demonstrate the dynamical and radiative effects of volcanic forcing on NH temperatures. So this kind of episodic internal climate forcing must also be considered as an important potential contributor to the indicated multi-decadal variability in our land-only, extra-tropical NH temperature reconstruction.

6. Concluding remarks

The Esper et al. (2002) temperature reconstruction and the data used to develop it have been revisited in this paper in order to clarify and tighten up some of the interpretations made about it. We have argued that this reconstruction is best interpreted as an expression of land-only, extra-tropical NH temperature variability. It probably best reflects warm-season-weighted temperatures, but an annual temperature model can also be used as a reasonable approximation. In the process, we also examined the robustness of its expressed multi-centennial variability and addressed certain criticisms of it. Within the limits of the data and methods used in the original ECS paper, we have determined that the multi-centennial variability evident in the mean RCS chronology used for reconstruction is highly robust over the AD 1200–1950 interval. Prior to AD 1200, site and within-chronology replication weakens considerably, thus making the interpretation of that early period more tenuous. After 1950, a loss of climate sensitivity in the northern boreal zone of the Esper et al. (2002) network is also evident, which weakens the use of these data for climate interpretations.

The temperature signal in the mean RCS tree-ring chronology used by Esper et al. (2002) for reconstruction is largely restricted to periods longer than 20 years in duration. Consequently, the mean RCS chronology was re-calibrated after applying a 20-year low-pass filter to both it and the instrumental record. In so doing, it was determined that annual temperatures over extra-tropical NH land areas have now exceeded earlier reconstructed warm intervals by approximately 0.3 °C. There is also a strongly expressed LIA cooling in the reconstruction, which is the main cause of its separation from the other reconstructions shown in Fig. 1. The

challenge now is to determine why this is so. Three likely candidates are: (1) differences in how the tree-ring chronologies were developed, (2) differences in the methods used to calibrate the tree-ring chronologies for temperature reconstruction, and (3) differences in the regional expressions of the temperature signals recorded in the tree-ring chronologies used for reconstruction. This detailed review and critique of how the ECS reconstruction was developed is a step in that direction.

Acknowledgements

This paper is the outgrowth of an invited talk given by the lead author at the IMAGES—Holocene Working Group Workshop held in Hafslo, Norway on August 27–30, 2003. We thank co-organizers Eystein Jansen and Peter deMenocal for the invitation. We also thank R.S. Bradley for helpful criticisms that resulted in some of the analyses presented here. This work was supported by the NOAA Climate Change Data and Detection program, grant NA16GP2677 (E. Cook and R. D'Arrigo), and the Swiss National Science Foundation, grant 2100-066628 (J. Esper). Lamont-Doherty Earth Observatory Contribution Number 6660.

References

Baillie, M.G.L., 1991. Suck-in and smear: two related chronological problems for the 1990s. Journal of Theoretical Archaeology 2, 12–16.

Bond, G., Showers, W., Cheseby, M., Lotti, R., Almasi, P., deMenocal, P., Priore, P., Cullen, H., Hajdas, I., Bonani, G., 1997. A pervasive millennial-scale cycle in North Atlantic Holocene and glacial climates. Science 278, 1257–1266.

Bond, G., Kromer, B., Beer, J., Muscheler, R., Evans, M.N., Showers, W., Hoffman, S., Lotti-Bond, R., Hajdas, I., Bonani, G., 2001. Persistent solar influence on North Atlantic climate during the Holocene. Science 294, 2130–2136.

Bradley, R.S., 2000. Enhanced: 1000 years of climate change. Science 288, 1353–1355.

Bradley, R.S., Briffa, K.R., Crowley, T.J., Hughes, M.K., Jones, P.D., Mann, M.E., 2001. The scope of the Medieval Warming. Science 292, 2011–2012.

Bradley, R.S., Hughes, M.K., Diaz, H.F., 2003. Climate in medieval time. Science 17 (302), 404–405.

Briffa, K.R., 2000. Annual climate variability in the Holocene: interpreting the message of ancient trees. Quaternary Science Reviews 19, 87–105.

Briffa, K.R., Osborn, T.J., 2002. Blowing hot and cold. Science 295, 2227–2228.

Briffa, K.R., Jones, P.D., Bartholin, T.S., Eckstein, D., Schweingruber, F.H., Karlen, W., Zetterberg, P., Eronen, M., 1992. Fennoscandian summers from AD 500: temperature changes on short and long timescales. Climate Dynamics 7, 111–119.

Briffa, K.R., Jones, P.D., Schweingruber, F.H., Shiyatov, S.G., Cook, E.R., 1995. Unusual twentieth-century summer warmth in a 1000-year temperature record from Siberia. Nature 376, 156–159.

Briffa, K.R., Schweingruber, F.H., Jones, P.D., Osborn, T.J., Shiyatov, S.G., Vaganov, E.A., 1998a. Reduced sensitivity of recent tree growth to temperature at high northern latitudes. Nature 391, 678–682.

Briffa, K.R., Schweingruber, F.H., Jones, P.D., Osborn, T.J., Harris, I.C., Shiyatov, S.G., Vaganov, E.A., Grudd, H., 1998b. Trees tell of past climates: but are they speaking less clearly today? Philosophical Transactions of the Royal Society of London B 353, 65–73.

Briffa, K.R., Jones, P.D., Schweingruber, F.H., Osborn, T.J., 1998c. Influence of volcanic eruptions on Northern Hemisphere summer temperature over the past 600 years. Nature 393, 450–455.

Briffa, K.R., Osborn, T.J., Schweingruber, F.H., Harris, I.C., Jones, P.D., Shiyatov, S.G., Vaganov, E.A., 2001. Low frequency temperature variations from a northern tree-ring density network. Journal of Geophysical Research 106, 2929–2941.

Broecker, W.S., 2001. Was the Medieval Warm Period global? Science 291, 1497–1499.

Chapman, M.R., Shackleton, N.J., 2000. Evidence of 550-year and 1000-year cyclicities in North Atlantic circulation patterns during the Holocene. The Holocene 10 (3), 287–291.

Cook, E.R., Briffa, K.R., Meko, D.M., Graybill, D.S., Funkhouser, G., 1995. The 'segment length curse' in long tree-ring chronology development for paleoclimatic studies. The Holocene 5 (2), 229–237.

Cook, E.R., Buckley, B.M., D'Arrigo, R.D., Peterson, M.J., 2000. Warm-season temperatures since 1600 BC reconstructed from Tasmanian tree rings and their relationship to large-scale sea surface temperature anomalies. Climate Dynamics 16 (2–3), 79–91.

Crowley, T.J., 2000. Causes of climate change over the past 1000 years. Science 289, 270–277.

Crowley, T.J., Lowery, T.S., 2000. How warm was the Medieval Warm Period? Ambio 29, 51–54.

D'Arrigo, R., Jacoby, G., Frank, D., Pederson, N., Cook, E., Buckley, B., Nachin, B., Mijiddorj, R., Dugarjav, C., 2001. 1738 years of Mongolian temperature variability inferred from a tree-ring width chronology of Siberian pine. Geophysical Research Letters 28, 543–546.

Duffy, P.B., Doutriaux, C., Fodor, I.K., Santer, B.D., 2001. Effect of missing data on estimates of near-surface temperature change since 1900. Journal of Climate 14 (13), 2809–2814.

Esper, J., Cook, E.R., Schweingruber, F.H., 2002. Low-frequency signals in long tree-ring chronologies for reconstructing past temperature variability. Science 295, 2250–2253.

Esper, J., Cook, E.R., Krusic, P.J., Peters, K., Schweingruber, F., 2003. Test of the RCS method for preserving low frequency variability in long tree-ring chronologies. Tree-Ring Research 59 (2), 81–98.

Grove, J.M., 1988. The Little Ice Age. Methuen, London.

Hegerl, G.C., Crowley, T.J., Baum, S.K., Kim, K.-Y., Hyde, W.T., 2003. Detection of volcanic, solar and greenhouse gas signals in paleo-reconstructions of Northern Hemispheric temperature. Geophysical Research Letters 30 (5), 1242.

Hughes, M.K., Diaz, H.F., 1994. Was there a 'Medieval Warm Period'? Climatic Change 26, 109–142.

IPCC, 2001. Climate Change: The Scientific Basis. Cambridge University Press, Cambridge 944pp.

Jacoby, G.C., D'Arrigo, R.D., Davaajamts, T., 1996. Mongolian tree rings and 20th century warming. Science 273, 771–773.

Jones, P.D., Raper, S.C.B., Bradley, R.S., Diaz, H.F., Kelly, P.M., Wigley, T.M.L., 1986. Northern Hemisphere surface air temperature variations: 1851–1984. Journal of Climate and Applied Meteorology 25, 161–179.

Jones, P.D., Osborn, T.J., Briffa, K.R., 1997. Estimating sampling errors in large-scale temperature averages. Journal of Climate 10, 2548–2568.

Jones, P.D., Briffa, K.R., Barnett, T.P., Tett, S.F.B., 1998. High-resolution palaeoclimatic records for the last millennium: interpretation, integration and comparison with General Circulation Model control-run temperatures. The Holocene 8 (4), 455–471.

Jones, P., Osborn, T., Briffa, K.R., 2001. The evolution of climate over the last millennium. Science 292, 662–667.

Lamb, H.H., 1965. The early medieval warm epoch and its sequel. Palaeogeography, Palaeoclimatology, Palaeoecology 1, 13–37.

Lloyd, A.H., 1997. Response of tree-line populations of foxtail pine (*Pinus balfouriana*) to climate variation over the last 1000 years. Canadian Journal of Forest Research 27 (6), 936–942.

Luckman, B.H., Briffa, K.R., Jones, P.D., Schweingruber, F.H., 1997. Tree-ring based reconstruction of summer temperatures at the Columbia Icefield, Alberta, Canada, AD 1073–1983. The Holocene 7 (4), 375–389.

Mann, M.E., Bradley, R.S., Hughes, M.K., 1999. Northern Hemisphere temperatures during the past millennium: inferences, uncertainties, and limitations. Geophysical Research Letters 26 (6), 759–762.

Mann, M.E., Jones, P.D., 2003. Global surface temperatures over the past two millennia. Geophysical Research Letters 30 (15), 1820.

Mann, M.E., Ammann, C.M., Bradley, R.S., Briffa, K.R., Crowley, T.J., Hughes, M.K., Jones, P.D., Oppenheimer, M., Osborn, T.J., Overpeck, J.T., Rutherford, S., Trenberth, K.E., Wigley, T.M.L., 2003. On past temperatures and anomalous late 20th century warmth. Eos 84, 256–258.

Naurzbaev, M.M., Vaganov, E.A., Sidorova, O.V., Schweingruber, F.H., 2002. Summer temperatures in eastern Taimyr inferred from a 2427-year late-Holocene tree-ring chronology and earlier floating series. The Holocene 12 (6), 727–736.

Osborn, T.J., Briffa, K.R., 2000. Revisiting timescale-dependent reconstruction of climate from tree-ring chronologies. Dendrochronologia 18, 9–25.

Overpeck, J.T., et al., 1997. Arctic environmental change of the last four centuries. Science 278, 1251–1256.

Shindell, D.T., Schmidt, G.A., Miller, R.L., Mann, M.E., 2003. Volcanic and solar forcing of climate change during the pre-industrial era. Journal of Climate 15, 4094–4107.

Timm, O., Ruprecht, E., Kleppek, S., 2004. Scale-dependent reconstruction of the NAO index. Journal of Climate 17 (11), 2157–2169.

Zielinski, G.A., 2000. Use of paleo-records in determining variability within the volcanism-climate system. Quaternary Science Reviews 19, 417–438.

ELSEVIER

Quaternary Science Reviews 23 (2004) 2075–2088

QSR

Holocene paleoceanography and glacial history of the West Spitsbergen area, Euro-Arctic margin

Morten Hald[a,*], Hanne Ebbesen[a], Matthias Forwick[a], Fred Godtliebsen[b], Liza Khomenko[c], Sergei Korsun[c], Lena Ringstad Olsen[b], Tore O. Vorren[a]

[a]*Department of Geology, University of Tromsoe, Tromsoe, Norway*
[b]*Department of Mathematics and Statistics, University of Tromsoe, Norway*
[c]*Department of Biology, St. Petersburg University, Russia*

Abstract

Two sediment cores from the West Spitsbergen area, Euro-Arctic margin, MD99-2304 and MD99-2305, have been investigated for paleoceanographic proxies, including benthic and planktonic foraminifera, benthic foraminiferal stable isotopes and ice rafted debris. Core MD99-2304 is located on the upper continental margin, reflecting variations in the influx of Atlantic Water in the West Spitsbergen Current. Core MD99-2305 is located in Van Mijenfjord, picturing variations in tidewater glacier activity as well as fjord-ocean circulation changes. Surface water warmer than today, was present on the margin as soon as the Van Mijenfjord was deglaciated by 11,200 cal. years BP. Relatively warm water invaded the fjord bottom almost immediately after the deglaciation. A relatively warm early Holocene was followed by an abrupt cooling at 8800 cal. years BP on the continental margin. Another cooling in the fjord record, 8000–4000 cal. years BP, is documented by an increase in ice rafted debris and an increase in benthic foraminiferal δ^{18}O. The IRD-record indicates that central Spitsbergen never was completely deglaciated during the Holocene. Relatively cool and stable conditions similar to the present were established about 4000 cal. years BP.
© 2004 Elsevier Ltd. All rights reserved.

1. Introduction

The West Spitsbergen continental margin is a climatically sensitive area. The main reason for this is the present oceanographic regime characterized by north-flowing warm Atlantic Water in the West Spitsbergen Current along the continental margin with sharp gradients to cool, partly sea-ice covered Polar Water (Fig. 1). Small geographical displacements in the ocean fronts in this area may impose large climatic changes affecting the society, life habitats and earth processes in the area. Climate models predict that the global warming will have its largest impact in the high northern latitudes (e.g. Hadley Centre (2004)). However, the instrumental record of climate variability is too short and spatially incomplete to reveal the full range of

seasonal to millennial-scale climate variability, or to provide empirical examples of how the climate system responds to large changes in climate forcing. Longer time series may be provided with well dated and high resolution proxy records.

We have investigated two proxy records from the West Spitsbergen area, one from the open ocean upper continental margin and one from a silled fjord setting, Van Mijenfjord. The core from the continental margin is located under the axis of Arctic Water. This core should reflect changes and variability in heat flux to the high northern latitudes, representing the northern limb of the North Atlantic conveyor circulation. The fjord record on the other hand represents more of a continental climate signal, linked to processes such as iceberg rafting from tidewater glaciers and exchange between fjord water and open oceanic water. Our knowledge of the Holocene climatic record of the Svalbard area is sparse due to lack of high resolution and well-dated proxy

*Corresponding author. Tel.: +47-776-44412; fax: +47-776-45600.
E-mail address: morten.hald@ig.uit.no (M. Hald).

0277-3791/$ - see front matter © 2004 Elsevier Ltd. All rights reserved.
doi:10.1016/j.quascirev.2004.08.006

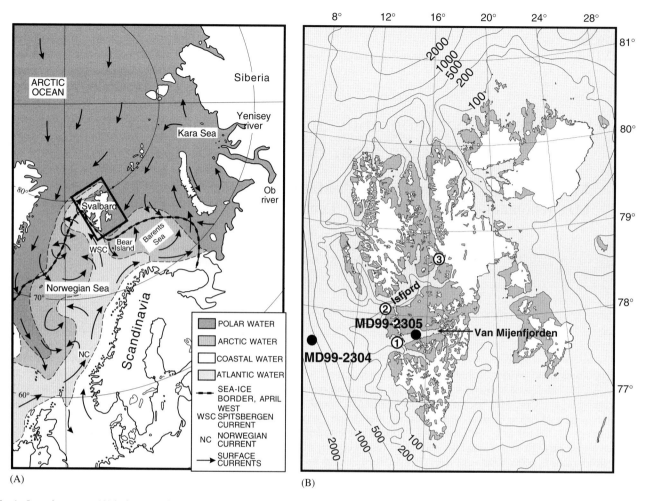

Fig. 1. Location maps: (A) index map showing the surface water masses and surface currents in the Norwegian Sea and adjoining seas; (B) close up of Svalbard and surrounding continental margin. Location of the two investigated sediment cores MD99-2304 and MD99-2305 are shown. 1 = Akseløya; 2 = Linnèvannet and 3 = Billefjorden.

records. The purpose of the present paper is to (1) present well-dated paleoclimate data for the West Spitsbergen area during the Holocene and (2) investigate century to decadal scale climate changes and variability.

2. Physical setting

Spitsbergen is the largest island of the Svalbard archipelago, situated between 76° and 80°N, and bordered by the Arctic Ocean to the north, the Barents Sea to the south and east, and the Norwegian–Greenland Sea to the west (Fig. 1). The archipelago is dominated by Phanerozoic sedimentary rocks (Steel and Worsley, 1984).

The continental margin west of Spitsbergen is characterized by a relatively narrow shelf with a typical glacial morphology represented by shallow banks between glacial troughs, the latter found in the continuation of the east-west trending fjords. The continental slope has a relatively steep gradient of

4–5°. Quaternary sediments rest on Tertiary sediments, and have accumulated in well-defined submarine troughs on the shelf and in submarine fans on the slope (Andersen et al., 1996).

Spitsbergen is cut by several fjords. Van Mijenfjord (Fig. 1) is the second largest of these, 50 km long and ca 10 km broad. It has restricted oceanic communication due to the island Akseløya in the outer part. The fjord is divided into three basins. Maximum depth of the outer basin is 112 m as compared to 74 m in the middle basin and ca 30 m in the inner basin. Most of the fjord is located within Paleogene sedimentary bedrock of various types such as sandstones, siltstones, shales, coals and coals pebbles (Steel and Worsley, 1984).

Early Cretaceous sedimentary rocks are found in the innermost part (Salvigsen and Winsnes, 1989). The inner part of the fjord is dominated by sediments produced during a surge AD 1300 (Punning et al., 1976; Rowan et al., 1982; Hald et al., 2001). The outer basin of the fjord is characterized by a maximum of 25 milli-seconds (ms) of acoustically transparent sediments representing the last deglaciation and the Holocene (Dahlgren, 1998). The sea-floor surface sediments are dominantly silty clay (Hald and Korsun, 1997).

The catchments area of Van Mijenfjord is approximately $2.8 \times 10^3 \, \text{km}^2$, and 50% of this area is covered by glaciers (Hagen et al., 1993). A tidewater glacier (Fridtjofbreen) is located in the northern outermost parts of Van Mijenfjord. The fjord head is occupied by the large tidewater glacier system of Paulabreen (PGS), of which two glaciers (the Paulabreen and Bakininbreen) are the only glaciers that today calve into Van Mijenfjord. Several tributary glaciers feed the main glaciers of the PGS. All the glaciers in the area have experienced a significant loss of volume during the last 100 years (Hagen et al., 1993). The climate of the Van Mijenfjord is cold and dry. Annual temperature of the west coast of Spitsbergen is $-6\,^\circ\text{C}$ and mean annual precipitation varies between 180 and 440 mm (Hanssen-Bauer et al., 1990). The equilibrium line altitudes of the glaciers vary from 200 m above sea level along the coast to > 600 m in the central-eastern parts of Spitsbergen (Hagen et al., 1993).

2.1. Oceanography

The Spitsbergen margin and fjords are influenced by relatively warm Atlantic Water transported from south with the West Spitsbergen Current and cold Polar Water from north (Fig. 1). The sea-ice conditions vary seasonally (Dowdeswell and Dowdeswell, 1989). A cover of land-fast ice forms in the major fjord systems and in other sheltered coastal areas by about late November and is usually retained until late May or June (Wadhams, 1981). Arctic pack ice is usually present around northern Spitsbergen throughout the year, with the lowest flow densities in August and September (Vinje, 1985). Between November and April the pack ice may surround the entire Svalbard archipelago, with minimum densities on the west coast of Spitsbergen. The asymmetric distribution of polar pack ice results from oceanic circulation in which polar waters transport pack ice southward on the eastern side of Svalbard and the West Spitsbergen Current brings warm Atlantic Water northward to the western coast of Svalbard (Wadhams, 1981; Vinje, 1985) (Fig. 1).

The mouth of Van Mijenfjord is almost closed from the open ocean by the island Akseløya and thus allows only for a very modest inflow of Atlantic Water. The water column is dominated by cold local water, with bottom water temperatures of around $-1\,^\circ\text{C}$ and a seasonally warmed surface layer (cf. Hald et al., 2001). Gulliksen et al. (1985) reported a minimum July bottom-water temperature of $-1.5\,^\circ\text{C}$. Bottom water salinity is around 34 PSU in the two outer basins and a pycnocline is observed in the upper 10–15 m. Tidewater ranges are < 1 m.

3. Material and methods

We investigated two sediment cores, core MD99-2304 is located on the upper West Spitsbergen continental margin 77° 37. 26′N and 09° 56. 90′E, at 1315 m water depth, and core MD99-2305 located in Van Mijenfjord 77° 46.87′N and 15° 17.81′E at 110 m water depth. The sampling sites were determined by several high resolution seismic surveys during the 1990s using 3.5 kHz penetration echo sounder and Sparker. Both cores were sampled during the IMAGES-1999 cruise with "R/V Marion Dufresne" using a modified piston coring system, a "Calypso-corer", especially designed to sample long cores. Multi sensor track (MST) logging was applied to the sealed core MD99-2305. The sealed cores, with an inner diameter of 10 cm, were opened by splitting them longitudinally in two equal parts. One half was studied with regard to various geotechnical and sedimentological analyses including MST logging (only MD99-2304), X-radiography, and colour determination by using the Munsell Colour Charts. The whole length of MD99-2305, 18 m, was analysed. MD99-2304 is 23 m long, but only the upper 1.5 m were used in the present study. In core MD99-2305 ice rafted debris (IRD) > 1 mm was counted from X-radiographs.

Preparation of the foraminiferal samples mainly followed the methods of Feyling-Hanssen (1958) and Meldgaard and Knudsen (1979). Between 300 and 400 individuals of benthic and/or planktonic foraminifera were identified from the > 100 μm fraction using a binocular microscope. The planktonic foraminiferal census data were used to reconstruct sea surface summer temperatures (SST) in core MD99-2304. Three different reconstructions were performed. One reconstruction represents quantitative estimations of SST based on the Modern Analogue Technique in the data analysis program C2 version 1.3 (Juggins, 2002). The modern data set used in the estimations is from ATL 947 core-top database of modern planktonic foraminifera in the 150 μm fraction (Pflaumann et al., 2003). The general deviation of SST is about 0.3–1.3 °C. The two other SST reconstructions were based linear regression using smaller modern data sets from the Nordic Seas of respectively 125 μm fraction (Johannessen et al., 1994) and 100 μm fractions (Burhol, 1994).

$\delta^{18}\text{O}$ measurements were carried out at the Norwegian GMS laboratory at the University of Bergen, using Finingan MAT 251 mass spectrometer. The reproducibility at this laboratory is 0.07‰ for $\delta^{18}\text{O}$. The samples were prepared following the procedures described by Shackleton and Opdyke (1973), Shackleton et al. (1983) and Duplessy (1978). In MD99-2305, $\delta^{18}\text{O}$ was analysed on the benthic foraminifer *Cassidulina reniforme*, an infaunal species that appears to precipitate its shell in disequilibrium with an offset of 0.13‰ $\delta^{18}\text{O}$ (Austin and

Kroon, 1996). The plotted $\delta^{18}O$ are corrected for the global ice volume effect (Fairbanks, 1989).

Accelerator Mass Spectrometry (AMS) radiocarbon dates (Table 1) were performed on bivalve tests (MD99-2305) and foraminifera (MD99-2304) in Trondheim, Norway and Uppsala, Sweden (TUa), Arizona, US (AA) and Kiel, Germany (KIA). For the TUa-dates, the targets were prepared at Radiocarbon Laboratory in Trondheim, Norway and were measured at the Svedberg Laboratory in Uppsala, Sweden.

In order to check the statistical reliability and trends of proxy data variations and to identify a lower limit of reliable time resolution we applied the SiZer methodology developed by Chaudhuri and Marron (1999). This method was applied only on the most high resolution data sets in MD99-2305 (i.e. IRD, foraminiferal flux and % *Elphidium. excavatum f. clavatum*) (Fig. 3). The SiZer methodology has not been applied on the lower resolution data sets in MD99-2304 and the $\delta^{18}O$ and

$\delta^{13}C$ data in MD99-2305. For these proxy sets we will only discuss millennial scale changes. The SiZer method is explained in the following.

A realistic statistical model for this situation is given by

$$y_i = m(x_i) + \varepsilon_i, \quad 1 = 1, \ldots, n, \tag{1}$$

where $m(x)$ is the underlying smooth target curve, here representing the "mean behaviour" of the proxy data at time x. The ε_i are independent Gaussian variables with mean 0 and variance σ_i^2, and y_i represents the observed proxy value in data point number i.

In smoothing approaches, one frequently uses a local linear kernel estimator; see Fan (1991), to produce a good estimate of m from the observations $y_1, \ldots, y_n$. To be specific, this means that, at time x, $\hat{m}_h(x)$ equals the fit $\hat{\alpha}_0$ where $\hat{\alpha}_0, \hat{\alpha}_1$ minimizes

$$\sum_{i=1}^{n}(y_i - \alpha_0 - \alpha_1(x_i - x))^2 K_h(x_i - x). \tag{2}$$

Table 1
Radiocarbon ages for MD99-2304 and MD992305, showing the core location, depth in core (cm), reference to the dating laboratory, Radiocarbon dating laboratory of Trondheim and the Tandem Accelerator Laboratory in Uppsala (Tua); Kiel Germany (KIA) and Arizona, US (AA)

Core	Core depth,	Lab ref	Material	14C age	1-sigma	Cal.age	2-sigma	Ranges
MD99-2305	8–9	Tua-4168	*Nucula tenuis*	345	± 50	410	468	294
MD99-2305	100–101	TUa-3249	*N. tenuis*	1115	± 55	1056	1176	944
MD99-2305	156–157	Tua-4169	*Acanthocardia echinata*	1300	± 45	1260	1325	1167
MD99-2305	212–213	Tua-4170	*Thyasira* sp.	1930	± 45	1937	2050	1838
MD99-2305	408–409	TUa-4171	*Yoildia* sp.	3940	± 105	4443	4788	4209
MD99-2305	540–541	TUa-3250	*N. tenuis*	5400	± 55	6208	6316	6316
MD99-2305	625–626B	Tua-4173	*Thyasira equalis*	6475	± 105	7403	7563	7211
MD99-2305	625–626A	Tua-4172	*Nuculana minuta*	6610	± 65	7497	7604	7404
MD99-2305	786–787	Tua-4174	*Yoldiella lenticula*	7780	± 55	8637	8866	8519
MD99-2305	853–854	Tua-4175	*Y. lenticula*	8040	± 65	8929	9059	8784
MD99-2305	926–927	Tua-4176	*Y. lenticula*	8430	± 65	9189	9648	9042
MD99-2305	1106–1107	Tua-4177	*N. tenuis*	8850	± 55	9839	10,278	977
MD99-2305	1132–1135	Tua-3144	*Yoldia hyperborea*	8670	± 65	9788	9950	9417
MD99-2305	1135–1136	TUa-3251	*N. tenuis*	8910	± 65	9929	10,292	9790
MD99-2305	1389–1390	TUa-3252	*Y. hyperborea*	9350	± 65	10,497	10,808	10,248
MD99-2305	1390–1391	TUa-3253	*N. tenuis*	9375	± 70	10,564	10,816	10,274
MD99-2305	1479–1481	TUa-3254	*Y. hyperborea*	9470	± 85	10,617	10,853	10,312
MD99-2305	1480–1481	TUa-3255	*N. tenuis*	9535	± 80	10,801	10,864	10,333
MD99-2305	1535–1536	TUa-3256	*N. tenuis*	9615	± 75	10,990	10,880	10,354
MD99-2305	1569–1570	TUa-3257	*N. tenuis*	9765	± 75	11,118	11,348	10,791
MD99-2305	1580	TUa-4319	*N. tenuis*	9835	± 45	11,141	11,357	10,834
MD99-2305	1595–1596	TUa-4318	*N. tenuis*	9920	± 45	11,171	11,639	11,074
MD99-2305	1763–1767	TUa-3258	2 bivalve fragments	45,285	+3140 −2250			
MD99-2304	2,5	Tua-4421	Plan + bent. forams	580	± 30	560	627	494
MD99-2304	28,5	TUa-3911	Planktonic forams	7855	± 55	8730	8905	8555
MD99-2304	38,5	TUa-3912	Planktonic forams	8105*	± 60	8942*	9080	8804
MD99-2304	56,5	TUa-3913	Planktonic forams	8010	± 65	8855	9023	8688
MD99-2304	80	AA 36609	Shell fragments	8965	± 85	10,044	10,303	9784
MD99-2304	130	KIA9346	Shell fragments	9670	± 55	11,404	11,725	11,082
MD99-2304	156,8	KIA9526	Shell fragments	10,030	± 50	10,607	10,866	10,349
MD99-2304	186	AA 36610	Shell fragments	12,170	± 180	13,945	14,386	13,503
MD99-2304	215	KIA9863	Benthic forams	12,660	± 70	14,835	15,473	14,197

Asterisk denotes age excluded from the age model. The ^{14}C dates were calibrated (cal. ages) using the INTCAL-98 calibration data set (Stuiver et al., 1998).

In Eq. (2), $K_h(\cdot) = (1/h) \cdot K(\cdot / h)$, where K is a kernel function symmetric around zero. Typically, we have

$$K(x) = \frac{1}{\sqrt{2\pi}} \exp\left(-\frac{x^2}{2}\right). \tag{3}$$

The degree of smoothness in the estimate $\hat{m}_h$ is controlled through the bandwidth h. Note that whereas traditional smoothing methods seek an "optimal" value of h, SiZer exploits "all" values of h to find significant features. This is crucial in climatological applications since important features typically change as a function of scale. Hence, the whole scale-space is needed in order to find all significant features in a data set. By the SiZer approach one also avoids the bias difficulty that appears in traditional smoothing. This is caused by the fact that $\hat{m}_h$ is an unbiased estimator of the true curve at the resolution corresponding to h.

4. Chronology

The chronology of MD99-2305 is based on 21 AMS radiocarbon dates performed on mollusc bivalves (Table 1). Most of these bivalves were identified to species level. All the dates were corrected for a reservoir effect of 464 years (including a ΔR value of 64 +/− 35 years) (Mangerud and Gulliksen, 1975) and show increasing ages versus depth in core (Fig. 2). An age model was produced by converting the dates to calendar years following the calibration model of Stuiver et al. (1998) (Fig. 4F). Sedimentation rates (mm/cal. years) were calculated by applying a fourth order polynomial fit through the dates. The age model implies that sedimentation rates vary from 66 cm/1000 cal. year to >500 cm/1000 cal. years. The highest rates occurred during the early Holocene, the lowest during mid-Holocene and then a slight increase the last 1700 years. The sedimentation rates in the core allow a high stratigraphic resolution in the various proxies. Average resolution for the IRD counts is 6.8 years; foraminiferal counts 42 years and stable isotope analyses 175 years.

The age model for MD99-2304 is based on 8 out of 9 AMS dates performed on mixed benthic/planktonic foraminifera; planktonic foraminifera only, or unidentified fragments of bivalve shells (Fig. 5H,F, Table 1). One date, TUa 3912 (8105 +/−60 yrs BP) appears too old and was excluded from the age model. This may be due to reworking. However, based on the low relief of the core location (Labeyrie et al., 2003), the fine sediment texture, and lack of structures indicating sediment disturbances we assume reworking of foraminifera generally to be low. A reservoir age of 464 years was applied also for these dates, although we are aware that the reservoir age may be higher for the benthic foraminifera at sea bottom (1315 m), compared to the planktonic foraminifera reflecting the upper water

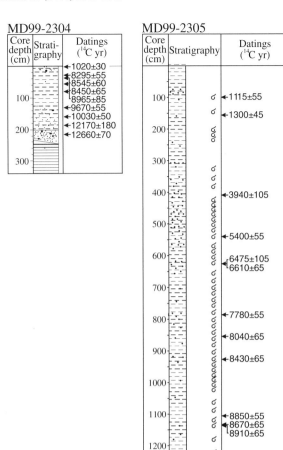

Fig. 2. Lithostratigraphy and radiocarbon dates in Core MD99-2304 on the continental margin (left) and MD99-2305 in Van Mijenfjord (right), West Spitsbergen. The legend at bottom right applies to both records.

masses. The reservoir age may also have varied through time (Waelbroeck et al., 2001). However, at present there is no data from the study area to confirm this. The dates were calibrated to calendar years (Stuiver et al., 1998) and sedimentation rates were calculated by linear interpolation between the dated levels. The results reveal relatively high sedimentation during early Holocene (40 to >200 cm/1000 cal. years) dropping abruptly around 8800 cal. years BP to 3.2 cm/1000 cal. years for the remaining part of the Holocene. Due to low carbonate content it was only possible to retrieve one date from the uppermost part of the Holocene. Stratigraphical analyses were performed on sediment samples from each cm in the core. This implies a mean time resolution of 27 years prior to 8800 cal. years BP and 314 years after this date.

5. Results

5.1. The SiZer analysis

The results of the SiZer analysis for core MD99-2305 are shown in Fig. 3. In the top panels of Fig. 3 A–C, a family of smoothes (h) is given. The curves plotted have smooth ranges as given in the figure. The thick solid line corresponds to a choice of h one typically would use if only one scale was to be used. This "optimal" h is a data driven bandwidth and a best choice of the bandwidth from a purely mathematical point of view (Ruppert et al., 1995). In the lower panels of Fig. 3A–C, SiZer plots are given. The plot shows a feature map as a function of scale (controlled by h) and location (given by x) for the signal. For each (x,h) position, SiZer tests whether, $\alpha_1 \neq 0$, i.e. whether the curve at this level of resolution has a derivative different from zero which means that there is a true increase or decrease of the curve. Readers interested in more details about this are referred to Chaudhuri and Marron (1999).

A significantly positive derivative in the SiZer map is flagged as black, while a significantly negative derivative is flagged as light grey. This means that black shows areas where there is a significant increase, while light grey shows where there is a significant decrease. At locations where the derivative is not found to be significantly different from zero, the colour grey is used. Hence grey means that there is no significant change. The horizontal black line (around h = 100) corresponds to the smoothing obtained by the solid black line in the upper panel of Fig. 3A. Dark grey is used to indicate that too few data points are available to do inference. Typically, dark grey colour occurs at very small scales for SiZer. It should be noted that when the SiZer map show an, e.g. positive derivative of the curves, this corresponds to a rise in the actual proxy value, and vice versa for negative derivatives. In the following we describe from too (modern) to bottom (Younger Dryas) and explain the variability and trends in the curves as shown by the SiZer map.

The results show that for the IRD variations the "optimal" h is 109 years (Fig. 3A). Following the horizontal line for this h, there was a marked decrease followed by an increase between 800 and 1000 years ago. Further a marked increase is shown around 3000 cal. years BP and a significant increase/decrease about

7500 cal. years BP. Finally the large amount of IRD in the older part of the record is depicted by two marked increases older than 10,000 cal. years BP.

For the foraminiferal flux the "optimal" time horizon is 311 years (Fig. 3B). It shows a short decrease about 2000 cal. years BP and an increase between 3500 and 4000 cal. years BP. In addition, significant decreases are depicted at ca 7000 cal. years BP and 8500 cal. years BP. A reduction in foraminiferal flux is evidenced for the lowermost part of the stratigraphy, older that 10,500 cal. years BP.

The "optimal" h for % E. excavatum is 161 years (Fig. 3C). At this scale the only significant changes are increases at about and 11,000, 10,200 and 8500 cal. years BP. For the period from 8400 to 6200 cal. years BP there are too few data to do any analysis. Both foraminiferal flux and E. excavatum have too few observations to give any information on variability on time scales less than 100 years throughout most of the Holocene (Figs. 3B and C).

5.2. Proxy data variations in MD99-2305

The lithology of MD99-2305 is characterized by two units (Fig. 2). The lower 2 m of the core comprise a massive and firm diamicton with a high content of clasts and a low content of fossils. One AMS date performed on shell fragments from this unit gave a close to infinite age of 45385 +/− 3140/2250 ^{14}C years (Table 1), assumed to result from reworking. The diamicton has a sharp boundary to the upper unit characterized by a soft mud with scattered clasts. The acoustic data show that both units are widespread in the fjord; Dahlgren (1998) and Hald et al. (2001). We interpret the diamicton to represent a basal till and the upper unit to result from glaciomarine fjord sedimentation. The boundary between the two units is dated to 11,170 cal. years BP. We consider this to be a minimum age for the deglaciation of Van Mijenfjord.

The proxy data for IRD, foraminiferal flux and E. excavatum are represented by the "optimal" curve as given in Fig. 3. The early deglaciation of Van Mijenfjord around 11,200 cal. years BP is characterized by high iceberg rafting decreasing rapidly reaching relatively low values from ca 10,000 cal. years BP (Fig. 4E). A peak in IRD occurred around 7500 cal. years BP, followed by high flux values between 7000 and 4000 cal. years BP.

As IRD declined there was a rise in the flux of foraminifera showing relatively high values during early Holocene (11,000–7000 cal. years BP). The same time interval was characterized by a high abundance of bivalve shells. The benthic foraminiferal species E. excavatum shows peak values during the earliest part of the deglaciation, followed by a marked decline until ca 8600 cal. years BP. There was a marked rise of this species after 7000 cal. years BP, reaching peak values

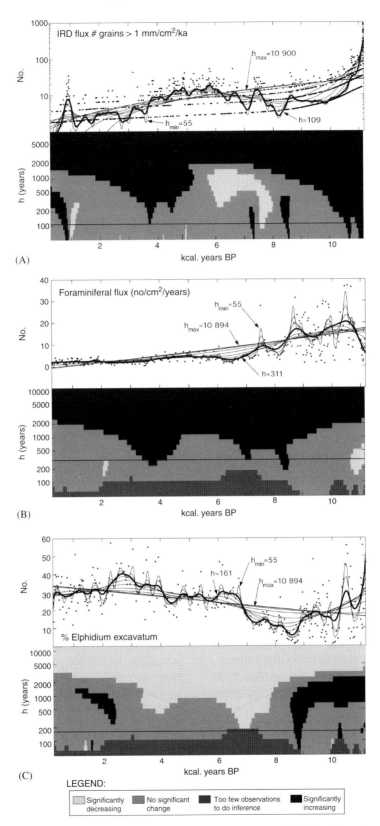

Fig. 3. SiZer analysis of proxy data from core MD99-2305, Van Mijenfjord, West Spitsbergen (A) IRD grains > 1 mm (B) Foraminiferal flux and (C) % *E. excavatum*. In the top panel of each figure a family of smoothes (*h*-values) is given. The dots shows all data, the thick line shows the "optimal" smooth (cf. discussion in text) and the thin lines show various smooth values, with maximum and minimum values given. In the lower panel a SiZer plot is shown depicting at which time the proxy show a significant increase, decrease, no change or where there are too few observations.

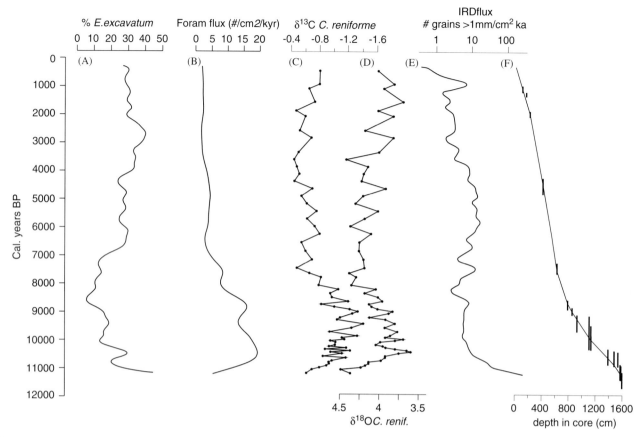

Fig. 4. Proxy record for sediment core MD99-2305, Van Mijenfjord, West Spitsbergen: (A) % *E. excavatum* smoothed at value $h = 161$ years; (B) foraminiferal flux, smoothed by $h = 311$ years; (C) $\delta^{13}C$ measured on the benthic foraminifer *C. reniforme*; (D) $\delta^{18}O$ measured on the benthic foraminifer *C. reniforme*; (E) ice rafted debris (IRD) smoothed at $h = 109$ years; (F) age (cal. years BP) vs. cm depth in core plotted as a fourth polynomial fit through the dated levels in the core. Cf. discussion in text of smoothing technique in (A), (B) and (E).

around 2800 cal. years BP. *E. excavatum* shows a increasing trend (Fig. 4A) and foraminiferal flux (Fig. 4B) a decreasing trend throughout the Holocene. The $\delta^{13}C$ (Fig. 4C) and $\delta^{18}O$ (Fig. 4D), measured on the benthic foraminifer *C. reniforme*, show a parallel trend throughout most of the Holocene. Both show a marked depletion during the earliest part of the deglaciation leading to peak low values between 10,500 and 8300 cal. years BP. Maximum values occurred between ca 8000 and 4000 cal. years BP followed by a slight depletion during the last 3000 years.

5.3. Proxy data variations in MD99-2304

The lithology of the upper 150 cm of MD99-2304 is characterized by a brownish-grey homogenous mud with scattered clasts. The planktonic foraminifera are dominated by the polar species *Neogloboquadrina pachyderma* (sinistral) (Fig. 5A), followed by the subpolar species *Globigerina quinqueloba* (Fig. 5B). In addition, the following subpolar species are recorded: *N. pachyderma* (dextral) (Fig. 5C), *Globigerinita glutinata* (not shown), *G. bulloides* (Fig. 5D), *G. scitula* and

G. uvula (not shown). The SST reconstruction shows highly varying temperatures during earliest Holocene until ca 10,800 cal. years BP, followed by stable, relatively warm temperatures until a rapid cooling of ca 4 °C around 8800 cal. years BP. This cooling step is bracketed by two AMS dates (Table 1, Fig. 5F). The remaining part of the Holocene appears to be characterized by stable, cool sea surface temperatures. However, possible short-lived changes may be obscured by the low stratigraphic resolution for this part of the core. Reconstruction of SST for the core-top sample (ca 4 °C) is in agreement with instrumental observations of summer temperatures in the area.

6. Paleoceanographic implications

6.1. The proxy indicators

Planktonic foraminifera are reliable tracers of surface and sub-surface ocean temperatures (e.g. Bé and Tolderlund, 1971) and quantitative reconstructions of SST based on various statistical approaches have been

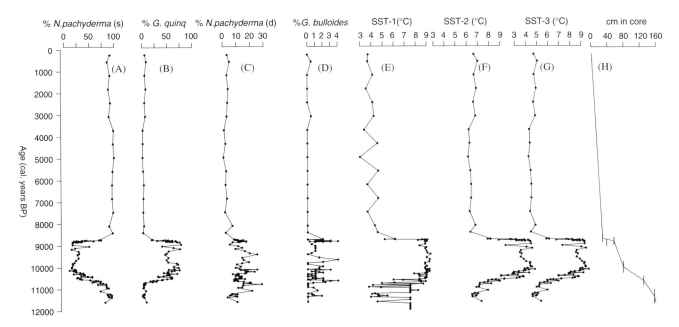

Fig. 5. Proxy record for sediment core MD99-2304, western Svalbard continental margin. (A–D) Percent frequencies for the most important planktonic foraminiferal species. (E) Sea surface temperature SST-1 reconstructed by the modern analogue technique, using the modern database of Pflaumann et al. (2003). (F) Sea surface temperature (SST-2) reconstructed by linear regression applying a modern data set from Johannesen et al. (1994). (G) Sea surface temperature (SST-2) reconstructed by linear regression applying a modern data set from Burhol (1994). (H) Age (cal. years BP) vs. cm depth in core plotted by linear interpolation between the dated levels in the core.

applied over the last three decades. In the present study the planktonic foraminiferal fauna was counted in the 100 μm fraction, were as the modern data set of Pflaumann et al. (1996, 2003) used for the SST-1 reconstruction (Fig. 5E) represents the 150 μm fraction. This adds an uncertainty to the reconstruction, since one of the subpolar species, *G. quinqueloba*, frequently is smaller than 150 μm (Carstens and Wefer, 1992). Thus, this species may be overrepresented in the proxy record compared to the modern data set and lead to too high reconstructed SSTs. Thus, in order to elucidate this uncertainty, we additionally reconstructed SST by applying two smaller modern data sets in the 125 μm fraction (Johannessen et al., 1994) (Fig. 5F) and the 100 μm fraction (Burhol Fig. 5G), respectively. The planktonic foraminiferal fauna in the coldest water masses (0–3 °C) is often completely dominated by the single species *N. pachyderma* (sinistral). Estimates in this temperature interval are therefore considered to be uncertain (Pflaumann et al., 2003). In general, the main structure of three SST reconstructions compare well. All show a marked SST reduction at 8800 cal. years BP and fairly stable cool temperatures for the remaining part of the record. The SST-2 appears to be offset towards warmer temperatures by 1–2 °C compared to SST-1. The oxygen and carbon isotopes measured on benthic foraminifera reflect both the ambient bottom water temperature as well as the water mass isotopic composition during the time of precipitation of the carbonate test. The benthic foraminiferal species *E. excavatum* is

common in Arctic fjords and occurs with high frequencies close to tide-water glacier termini (Hald and Korsun, 1997; Korsun and Hald 1998).

6.2. Early Holocene

The SST-reconstruction from the continental margin off West Spitsbergen (MD99-2304, Figs. 5E–G) show that significantly warmer water than at present was starting to develop around 11,000 cal. years BP. Stable warm conditions prevailed from 10,800 cal. years BP until 8800 cal. years BP. This relatively warm period suggests that Atlantic Water heat flux in the West Spitsbergen Current was stronger than at present. This warm period is reflected in the Van Mijenfjord by depletion in benthic $\delta^{18}O$ of 0.8‰ between 10,900 and 10,300 cal. years BP (Fig. 4D), suggesting a warming of the bottom water in the fjord. Further, warming is supported by the rise in foraminiferal flux, depletion in benthic $\delta^{13}C$ (Fig. 4C) and high abundance of macrofossils (Fig. 2), indicative of more productive waters than today. Influx of relatively warm oceanic water from the continental margin across the fjord sill probably caused this warming.

6.3. Middle and Late Holocene

The rise in $\delta^{18}O$, $\delta^{13}C$ and *E. excavatum* and depletion in foraminiferal flux in the fjord record (Fig. 4) indicates a gradual cooling starting around 8800 cal. years BP.

This gradual cooling in the fjord record differs from the abrupt cooling observed on the continental margin at this time (Fig. 5). We assume that the abrupt cooling on the margin may partly be an indication of the slow sedimentation rates and hence low stratigraphic resolution after 8800 cal. years BP as documented for this site (Fig. 5). The fall in SST on the margin indicates reduced influence of warm Atlantic Water and increased influence of Polar Water and/or runoff from land. This may be due to a change in the drift pattern and/or intensity of the West Spitsbergen Current. The conditions on the continental margin apparently remained stable and cool for the remaining part of the Holocene. However, neither dating nor time resolution for this part of the record does not permit any detailed reconstructions.

We suggest that the following three factors may have contributed to the marked reduction in sedimentation rates on the continental margin around 8800 cal. years BP: (1) Early Holocene sedimentation rates on continental margins that were glaciated during the last glacial, are often found to be high due to a large access to easily erodable glaciogenic sediments from recently deglaciated areas (Hald et al., 1999); (2) increase in bottom currents due to a glacio-eustatic fall in sea level (Landvik et al., 1987) and (3) increased bottom water erosion due to a deeper flow of the Atlantic Water during cooler conditions. Today, Atlantic Water attains a gradually deeper flow as it flows northward along the Svalbard margin, due to cooling and increase of density (Hopkins, 1991).

In Van Mijenfjord, the period 7500–4000 cal. years BP seems to represent a general cooling. This is supported by high benthic $\delta^{18}O$ indicating cool bottom water, a reduction in the biogenic production, rise in *E. excavatum* and enhanced IRD (Fig. 4). This general cooling may partly be influenced by flux of cool waters from the continental margin to the fjord, and partly due to a more restricted fjord/margin water-mass exchange across the fjord sill due to glacio-isostatic rebound. As the isostatic rise continued towards the present time (Landvik et al., 1987), the sill of Van Mijenfjord gradually became shallower, and the fjord would likely develop more of a local climatic signal with reduced connection to the open ocean.

The proxy data in the fjord record suggest that cool conditions prevailed for the last 4000 years, as indicated by continued high frequencies of *E. excavatum* and low foraminiferal flux (Fig. 4). The slight depletions in both $\delta^{18}O$ and $\delta^{13}C$ indicate changes of the source of the bottom water, rather than a rise temperature since modern bottom water temperatures in Van Mijenfjord today are already at a minimum between -1 and $-1.5\,^{\circ}C$. One possible mechanism for depleting the bottom water with respect to $\delta^{18}O$, without changing the temperature, is increased brine formation. When sea-ice

freezes during the fall, brine, characterized by very low temperatures and high salinity, sinks to a deeper level. This probably explains the very cold bottom water in Van Mijenfjord today (Skardhamar, 1998). As brine formation takes place in the surface water, it also attains a light $\delta^{18}O$ composition. The brine rapidly sinks through the water column and thereby brings $\delta^{18}O$-depleted water to the bottom. Current investigations recently revealed $\delta^{18}O$-depleted and very cold bottom water in silled fjords of Svalbard, including Van Mijenfjord (Mikalsen and Hald, unpublished). The reduction in iceberg rafting over the last 4000 years, except for the surge event around 800 cal. years BP (Hald et al., 2001), may be a response to a more continental climate of the region. This may have developed during an increasing isolation of Van Mijenfjord due to glacio-isostatic rebound (cf. Landvik et al., 1987).

7. Holocene glacial history of Spitsbergen

The content of grains >1 mm in the MD99-2305 (Fig. 4) may reflect both rafting from icebergs and sea-ice. In addition, coarser grains may be induced by storm-induced sediment suspensions in shallow water (Rumohr et al., 2001). However, lack of turbidites or other bottom current structures in the sediments, indicate that this latter process is of less importance in van Mijenfjord. To our knowledge, no systematic studies have been undertaken with respect to the content of debris in sea-ice in the Svalbard fjords. However, we assume that most of the IRD reflects rafting from icebergs based on the following. At present, the fjord is covered by sea-ice 6–9 months per year, and has three glaciers that calve into the fjord, but there is a very low content of IRD both at present and during the last 1000 years. This is well documented both in the present study (Fig. 4E) and by Hald et al. (2001). For two of the intervals with high IRD flux, there is very good evidence for increased ice berg rafting. One of these intervals is the marked IRD-peak at ca 800 cal. years BP (Fig. 4E) that is correlated to a surge event that is well documented both on land and in submarine sediments in the fjord (Punning et al., 1976; Rowan et al., 1982; Hald et al., 2001). Another interval with high IRD flux was the last deglaciation of Van Mijenfjord following the Younger Dryas. This deglaciation is well documented both in the present core and from studies on land (Landvik et al., 1987; Mangerud et al., 1992).

Van Mijenfjord was deglaciated by 11,200 cal. years BP (corresponding to 9900 ^{14}C years). The lowest date is located just above a diamicton, interpreted to be a till (Fig. 2). This age is assumed to represent a minimum age for the deglaciation of the fjord and is relatively young compared to the oldest dates of the deglaciation of van

Mijenfjord obtained by Mangerud et al. (1992). They retrieved a number of radiocarbon dates from raised marine deposits underlain by till east of the sill in the Van Mijenfjord area, ranging between 10,650 and 9500 ^{14}C years BP. We cannot exclude that the oldest deglacial sediments in MD99-2305 is missing, thus implying a younger minimum age for the deglaciation compared to oldest dates retrieved by Mangerud et al. (1992).

The deglaciation phase of the fjord was characterized by a very high IRD flux (Fig. 4E). A marked reduction of IRD occurred between 11,200 and 11,000 cal. years BP (Figs. 3A and 4E). We assume that this reduction reflects that a larger part of the glacier withdrew on land. Again, the present data gives somewhat younger ages for this deglaciation phase compared to results from Mangerud et al. (1992), inferring that the Van Mijenfjord glacier broke up very fast, between 10,600 and 10,300 ^{14}C years BP. However, the chronology of the IRD curve from 11,200 cal. years BP (9900 ^{14}C years BP) and onwards is well supported in the present study. This we believe is a good indicator that an intense break-up of the glacier prevailed at least until 11,000 cal. years BP.

The moderate to low IRD flux between 10,500 and 8000 cal. years BP (Fig. 4E) corresponds partly to a period during which the other proxy data in MD99-2305 (Figs. 4A–D) indicate a climatic optimum. Further, increased rafting peaking around 7500, as well as between 7000 and 4000 cal. years BP, correlates to a cooling of the bottom water of the fjord, and to low SSTs on the continental margin. Thus, increased rafting may be a response to a reduction of the equilibrium line altitude causing increased tide-water glaciation in Van Mijenfjord. After 4000 cal. years BP, ice rating was reduced. Today, 50% of the catchment area of Van Mijenfjord is glaciated, and only three glaciers calve directly into the fjord (Hagen et al., 1993). The significant IRD-flux at ca 800 cal. years BP is, as mentioned above, correlated to a marked surge event, namely that of the Paula Glacier System (PGS) in the inner part of the fjord (Hald et al., 2001; Dahlgren, in prep.). During this event the PGS advanced 35 km beyond its present glacier front, leaving an IRD spike that is evident as a marker horizon in high resolution acoustic profiles as well as in a number of sediment cores throughout the fjord. We find no evidence in the acoustic record of any older surges during, e.g. early and middle Holocene. The presence of IRD throughout the entire Holocene indicates that central Spitsbergen was never completely deglaciated during this interglacial period.

Holocene glacial fluctuations on Svalbard have been investigated in the Isfjord area (Elverhøi et al., 1995; Svendsen and Mangerud, 1997). These studies infer reduced glacial activity during early/middle Holocene

and increased glaciation for the last 4000 years in Billefjord in the inner Isfjord, and the last 2000 years at Linnévannet, outer Isfjord. Reduced glaciation during early Holocene correlates well to the present study. However for the mid and late Holocene the Isfjord reconstructions and the present study differ. The IRD record in Van Mijenfjord indicates enhanced tidewater glaciation during mid-Holocene and somewhat reduced during late Holocene. Svendsen and Mangerud (1997) present good evidences that the Linnévannet glacier, in the outer Isfjord area, started to form 4000–5000 years ago reaching a maximum during the Little Ice Age. Thus, we cannot exclude that there are local differences in glacial history on Svalbard, although this is not what we would expect since the areas are only some 50–100 km apart. The Holocene glacier variations in inner Isfjord are based largely on sedimentation rates (Elverhøi et al., 1995). In terms of sedimentation rates all records show a similar pattern. However, we consider IRD to be a more reliable indicator of tidewater glaciation, rather than sedimentation rates. Changes in sedimentation rate are not only dependent on relative glaciation in the catchment area, but also the conditions in the accumulation basin.

8. Correlations

We compare the proxy records from the Svalbard margin and Van Mijenfjord with a high resolution marine SST proxy record from the Western Barents Sea, ca 75°N (Sarnthein et al., 2003) (Fig. 6). Both SST-records are reconstructed by applying the modern analogue technique to planktonic foraminiferal faunal data, and they are interpreted to reflect latitudinal temperature gradients of Atlantic Water in the Spitsbergen Current during the Holocene. These SST-records are compared to the Van Mijenfjord IRD record and summer insolation at 78°N (Berger and Loutre, 1991). All records have individual, independent, radiocarbon-based time scales. The absolute temperatures estimates on the Svalbard margin are somewhat higher compared to those from the Barents Sea. We assume this can be attributed to the difference in size fraction used for the foraminiferal analysis. In the present study we used the 100 μm fraction that will include more of the small subpolar species, implying higher SST, compared to the study of Sarnthein et al. (2003) using the 150 μm fraction. However, both SST-records show a remarkable similarity during the early Holocene, including the abrupt cooling around 8800 cal. years BP. The early Holocene appears to represent the Holocene climatic optimum. This supports modelling experiments by Liu et al. (2003) predicting increased SST with increasing northern latitude as a function of stronger insolation during the early Holocene (Fig. 6D). The middle Holocene SST

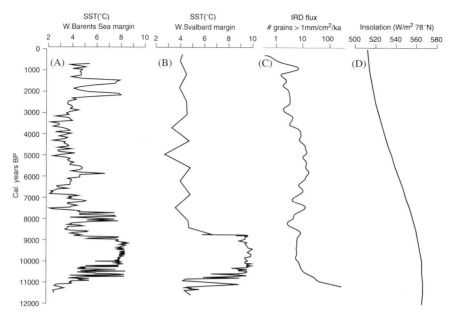

Fig. 6. Correlation between SST records from (A) Western Barents Sea (Sarnthein et al., 2003), (B) W. Svalbard margin (SST-1, present study), (C) Van Mijenfjord IRD (present study) and (D) insolation 78°N (Berger and Loutre, 1991).

depicts a decreasing trend in the Barents Sea. Super-imposed on this trend are millennial–centennial scale fluctuations. These fluctuations were not identified in the Svalbard record, probably due to low stratigraphic resolution. The overall cooling trend during the middle Holocene in the Barents Sea correlates well to increased tidewater glaciation in Van Mijenfjord. The Barents Sea record is located at 75°N in a sensitive boundary area (Fig. 1A), and abrupt coolings/warmings can be linked to relatively small displacements of the Arctic oceanic front system, separating cool Arctic Water from warmer Atlantic Water (Sarnthein et al., 2003).

An early Holocene warm period is well documented in a number of proxy records from the Svalbard and western Barents Sea region. In addition to high SSTs of the West Spitsbergen Current (Hald and Aspeli, 1997; Sarnthein et al., 2003), warmer surface waters in the fjords are indicated by thermophilic molluscs (Salvigsen et al., 1992). The conclusion of Svendsen and Mangerud (1997) that the glacier Linnébreen in outer Isfjord did not exist during early Holocene, may imply atmospheric temperatures of at least 1.5–2.5 °C higher than at present, if winter precipitation was similar to the present. In a study of lacustrine sediments on the island Bear Island, (74°N, western Barents Sea), Wohlfarth et al. (1995) concluded that a climatic optimum occurred between 9500 and 8000 ^{14}C years. This is in fairly good agreement with the present study (Fig. 6). However, there are indications of a relatively warm middle Holocene which contradict the present study. This includes both reduced glaciation of the glacier Linnébreen, as well as presence of thermophilic bivalves

(Salvigsen et al., 1992). However, the frequency of the thermophilic bivalves appears to decline during the Holocene, implying a cooling trend in line with the present study. In a study of plant macrofossils in a sediment core from a lake in the outer coastal Isfjord area Birks (1991) suggests that mean July temperatures were 2 °C warmer than today between 8000 and 5000 ^{14}C years BP. But the period prior to 8000 years BP was not investigated in that study.

9. Conclusions

- Sea surface temperatures, significantly warmer than at present, were recorded on the continental margin off West Spitsbergen from ca 11,200–8800 cal. years BP.
- As the Van Mijenfjord was deglaciated after the Younger Dryas, ca 11,200 cal. years BP, water warmer than present almost immediately invaded the fjord bottom.
- The presence of IRD throughout the entire Holocene indicates that central Spitsbergen was never completely deglaciated during this interglacial period. After the glaciation during the Younger Dryas, there was an intense deglaciation phase until ca 11,000 cal. years BP. Moderate to low tidewater glaciation characterized the early Holocene, 11,000–7500 cal. years BP, followed by enhanced glaciation around 7500 cal. years BP and between 7000 and 5000 cal. years BP.
- The glacial history is supported by the quantitative reconstructions of SST on the continental margin and

relative bottom water temperature in the Van Mijenfjord record, indicating that the warm early Holocene, 11,200–8800 cal. years BP, was followed by a cooling, 8800–4000 cal. BP and relatively stable, cool conditions the last 4000 years.

Acknowledgements

Funding was provided by the VISTA programme, the Research Council of Norway to the Strategic University Programme SPONCOM, and by the Roald Amundsen Centre, University of Tromsø. Technical assistance was provided by the crews onboard R/V Jan Mayen and R/V Marion Dufresne, Jan P. Holm (computer drawings), Edel Ellingsen, Trine Dahl, Odd Aasheim (laboratory work), Elsebeth Thomsen Hanken (macrofossil identification). Carin Andersson Dahl and Kari-Lise Rørvik helped with the C2-SST reconstructions. Stable isotopes were measured at the University of Bergen, Norway, Radiocarbon dates were run by the Radiocarbon Laboratory in Trondheim, Norway, Uppsala, Sweden, Kiel, Germany and Arizona, USA. To all these persons and institutions we offer our sincere thanks.

References

Andersen, E.S., Dokken, T.M., Elverhøi, A., Solheim, A., Fossen, I., 1996. Late Quaternary sedimentation and glacial history of the western Svalbard continental margin. Marine Geology 133, 123–156.

Austin, W.E.N., Kroon, D., 1996. Late glacial sedimentology, foraminifera and stable isotope stratigraphy of the Hebridean continental shelf, northwest Scotland. Late Quaternary Paleoceanography of the North Atlantic Margins, vol. 111. Geological Society Special Publication, London, pp. 187–213.

Bé, A.W.H., Tolderlund, D.S., 1971. Distribution and ecology of living planktonic foraminifera in surface waters of the Atlantic and Indian Oceans. In: Funnel, B.M., Riedel, W.R. (Eds.), The Micropaleontology of Oceans. Cambridge University Press, London.

Berger, A., Loutre, M.F., 1991. Insolation values for the climate of the last 10 million years. Quaternary Sciences Review 10 (4), 297–317.

Birks, H.H., 1991. Holocene vegetational history and climate change in west Spitsbergen—plant macrofossils from Skardtjørna, an Arctic lake. The Holocene 1 (3), 209–218.

Burhol, A., 1994. Resent planktonic foraminifera in the NE Norwegian Sea and western Barents Sea. Unpublished Master Thesis, University of Tromsø (in Norwegian).

Carstens, J., Wefer, G., 1992. Recent distribution of planktonic foraminifera in the Nansen Basin, Arctic Ocean. Deep-Sea Research 39, S507–S524.

Chaudhuri, P., Marron, S., 1999. SiZer for exploration of sturctures in curves. Journal of the Amerikan Statistical Association 94, 807–823.

Dahlgren, T., 1998. Holocene Sedimentation And Glacial history of inner Van Mijenfjorden, Spitsbergen, University of Tromsø, 86pp.

Dowdeswell, J., Dowdeswell, E.-K., 1989. Debris in icebergs and rates of glaci-marine sedimentation; observations from Spitsbergen and a simple model. Journal of Geology 92 (2), 221–231.

Duplessy, J.-C., 1978. Isotope studies. In: Gribbin, J.R. (Ed.), Climatic Change. Cambridge University Press, Cambridge, pp. 46–67.

Elverhøi, A., Svendsen, J. I., Solheim, A., Andersen, E. S., Milliman, J., Mangerud, J., Hooke, R.L., 1995. Late Quaternary Sediment Yield from the High Actic Svalbard Area: The Journal of Geology, vol. 103, pp. 1–17.

Fairbanks, R.G., 1989. A 17000-year glacio-eustatic sea level record: influence of glacial melting rates on the Younger Dryas event and deep-ocean circulation. Nature 342, 637–642.

Fan, J., 1991. Design-adaptive nonparametric regression. Journal of the American Statistical Association 87, 998–1004.

Feyling-Hanssen, R.W., 1958. Mikropaleontologiens teknikk. Norwegian Geotechnical Institute 29, 1–14.

Gulliksen, B., Holte, B., Jakola, K.-J., 1985. The soft bottom fauna in van Mijenfjord and Raudfjord, Svalbard. In: Gray, J.S., Christiansen, M.E. (Eds.), Marine Biology of Polar Regions and Effects of Stress on Marine Organisms, pp. 199–215

Hadley Centre, 2004. http://www.met-office.gov.uk/research/hadleycentre/models/modeldata.html.

Hagen, J.O., Liestøl, O., Sollid, J.L., Wold, B., Østrem, G., 1993. Glacier atlas of Svalbard and Jan Mayen. Norsk Polarinstitutt Meddelelser(129), 141pp.

Hald, M., Aspeli, R., 1997. Rapid climatic shifts of the northern Norwegian Sea during the last deglaciation and the Holocene. Boreas 26, 15–28.

Hald, M., Korsun, S., 1997. Distribution of modern Arctic benthic foraminifera from fjords of Svalbard. Journal of Foraminiferal Research 27, 101–122.

Hald, M., Kolstad, V., Polyak, L., Forman, L., Herlihy, F.A., Ivanov, G., Nescheretov, A., 1999. Late-glacial and Holocene paleoceanography and sedimentary environments in the Saint Anna Trough, Eurasian Arctic Ocean Margin. Palaeogeography, Palaeoclimatology, Palaeoecology 146, 229–249.

Hald, M., Dahlgren, T., Olsen, T.-E., Lebesbye, E., 2001. Late Holocene paleoceanography in Van Mijenfjorden, Svalbard. Polar Research 20, 23–35.

Hanssen-Bauer, I., Kristensen, M.S., Steffensen, E.L., 1990. The Climate of Spitsbergen. DNMI, Oslo.

Hopkins, T.S., 1991. The GIN Sea—a synthesis of its physical oceanography and literature review 1972–1985. Earth-Science Reviews 30, 175–318.

Johannessen, T., Jansen, E., et al., 1994. The relationship between surface water masses, oceanographic fronts, and paleoclimatic proxies in surface sediments of the Greenland, Iceland, Norwegian Seas. Nato ASI Series 117, 61–85.

Juggins, S., 2002. C2 1.3 version. http://www.staff.ncl.ac.uk/stephen.juggins.

Korsun, S., Hald, M., 1998. Modern benthic foraminifera off tide water glaciers, Novaja Semlja, Russian Arctic. Arctic and Alpine Research 30 (1), 61–77.

Labeyrie, L., Jansen, E., Cortijo, 2003. Les rapports de campagnes à la mer. MD 114/IMAGES V OCE/2003/02. Brest, Institut Polaire Francais-Paul Emile Victor, 850pp.

Landvik, J.Y., Mangerud, J., Salvigsen, O., 1987. The late Weichselian and Holocene shoreline displacement on the west-central coast of Svalbard. Polar Research 5, 29–44.

Liu, Z., Brady, E., Lynch-Stieglitz, J., 2003. Global ocean response to orbital forcing in the Holocene. Paleoceanography 18, 1041, doi:10.1029/2002PA000819.

Mangerud, J., Gulliksen, S., 1975. Apparent Radiocarbon Ages of Recent Marine Shells from Norway, Spitsbergen and Arctic Canada. Quaternary Research 5, 263–273.

Mangerud, J., Bolstad, M., Elgersma, A., Helliksen, D., Landvik, J.Y., Lønne, I., Lycke, A.K., Salvigsen, O., Sandahl, T., Svendsen, J.I.,

1992. The Last Glacial Maximum on Spitsbergen, Svalbard. Boreas 38, 1–31.

Meldgaard, S., Knudsen, K.L., 1979. Metoder til indsamling og oparbejding af prøver til foraminifer-analyser. Dansk natur Dansk skole.

Pflaumann, U., Duprat, J., Pujol, C., Labeyrie, L.D., 1996. SIMMAX: A modern analog technique to deduce Atlantic sea surface temperatures from planktonic foraminifera in deep-sea sediments. Paleoceanography 11, 15–35.

Pflaumann, U., Sarnthein, M., Chapman, M., Ld'AZbreu, L., Funnell, B., Huels, M., Kiefer, T., Maslin, M., Schultz, H., Swallow, J., van Kreveld, S., Vautravers, M., Vogelsang, E., Weinelt, M., 2003. Glacial North Atlantic Sea-surface conditions reconstructed by GLAMAP 2000. Paleoceanography 18 (3) 10-1–10-21.

Punning, J.-M., Troitsky, L., Rajamae, R., 1976. The genesis and age of the Quaternary deposits in the eastern part of Van Mijenfjorden, West Spitsbergen. Geologiska Föreningens Förhandlingar 98, 343–347.

Rowan, D.E., Pewe, T.L., Pewe, R.H., 1982. Holocene glacial geology of the Svea lowland, Spitsbergen, Svalbard. Geografiske Annaler 64A, 35–51.

Rumohr, J., Blaume, F., Erlenkeuser, H., Fohrmann, H., Hollender, F.-J., Mienert, J., Schäfer-Neth, C., 2001. Records and processes of near-bottom sediment transport along the Norwegian–Greenland Sea margins during the Holocene and Late Weichselian (Termination I) times. In: Schäfer, P., Ritzrau, W., Schlüter, M., Thiede, J. (Eds.), The Northern North Atlantic, a Changing Environment. Springer, Berlin, Heidelberg, New York, 500pp.

Ruppert, D., Sheather, S.J., Wand, M.P., 1995. An effective bandwidth selector for local least square selections. Journal of the American Statistical Association 90, 1257–1270.

Salvigsen, O., Winsnes, T.O., 1989. Geological map of Svalbard 1: 100,000. Sheet C10G Braganzavågen (Map). Norsk Polarinstitutt Temakart 4.

Salvigsen, O., Forman, S.L., Miller, G.H., 1992. Thermophilous molluscs on Svalbard during the Holocene and their paleoclimatic implications. Polar Research 11 (1), 1–10.

Sarnthein, M., Van Kreveld, S., Erlenkeuser, H., Grootes, P.M., Kucera, M., Pflaumann, U., Schulz, M., 2003. Centennial-to-millennial-scale periodicities of Holocene climate and sediment injections off the western Baretns shelf, 75°N. Boreas 32, 447–461.

Shackleton, N.J., Opdyke, N.D., 1973. Oxygen Isotope and Palaeomagnetic Stratigraphy of Equatorial Pacific Core V28-238: Oxygen Isotope Temperatures and Ice Volume on a 10^5 Year and 10^6 Year Scale. Quaternary Research 3, 39–55.

Shackleton, N.J., Imbrie, J., Hall, M.A., 1983. Oxygen and carbon isotope record of The East Pacific core V19-30: implications for the formation of deep water in the late Pleistocene North Atlantic. Earth Planetary Science Letter 65, 233–244.

Skardhamar, J., 1998. Sirkulasjonen i Van Mijenfjorden, en arktisk fjord (Circulation in Van Mijenfjorden, an Arctic fiord). Master Thesis, University of Bergen, Bergen, 79pp.

Steel, R.J., Worsley, D., 1984. Svalbards post-Caledonian strata—an atlas of seidmentological patterns and paleogeographic evolution. In: Spencer, A.M.e.a.e. (Ed.), Petroleum Geology of the Northern European Margin. Graham and Trotman, London, pp. 109–135.

Stuiver, M., et al., 1998. INTCAL98 Radiocarbon age calibration, 24,000-0 cal BP. Radiocarbon 40 (3), 1041–1083.

Svendsen, J.I., Mangerud, J., 1997. Holocene glacial and climatic variations on Spitsbergen, Svalbard. The Holocene 7, 45–57.

Vinje, T., 1985. The physical environment, the western Barents Sea. Drift, composition, morphology and distribution of the sea ice fields in the Barents Sea. Norsk Polarinstitutt Skrifter 179c p. 26.

Wadhams, P., 1981. The ice cover in the Greenland and Norwegian Seas. Review Geophysical Space Physics 19, 345–393.

Waelbroeck, C., Duplessy, J.-C., Miches, E., Labeyrie, L., Paillard, D., Duprat, J., 2001. The timing of the last deglaciation in North Atlantic climate records. Nature 412, 724–727.

Wohlfarth, B., et al., 1995. Early Holocene environment on Bjørnøya (Svalbard) inferred from multidisciplinary lake sediment studies. Polar Research 14, 253–275.

ELSEVIER

Quaternary Science Reviews 23 (2004) 2089–2099

Mid Holocene origin of the sea-surface salinity low in the subarctic North Pacific

M. Sarnthein[a,*], H. Gebhardt[a], T. Kiefer[b], M. Kucera[c], M. Cook[b], H. Erlenkeuser[d]

[a]*Institut für Geowissenschaften, University of Kiel, Olshausenstr. 40, D-24098 Kiel, Germany*
[b]*Department of Earth Sciences, University of Cambridge, Cambridge CB2 3EQ, UK*
[c]*Department of Geology, Royal Holloway University of London, Egham, TW20 OEX, UK*
[d]*Leibniz Laboratory, University of Kiel, D-24098 Kiel, Germany*

Abstract

IMAGES core MD01-2416 (51°N, 168°E) provides the first centennial-scale multiproxy record of Holocene variation in North Pacific sea-surface temperature (SST), salinity, and biogenic productivity. Our results reveal a gradual decrease in subarctic SST by 3–5 °C from 11.1 to 4.2 ka and a stepwise long-term decrease in sea surface salinity (SSS) by 2–3 p.s.u. Early Holocene SSS were as high as in the modern subtropical Pacific. The steep halocline and stratification that is characteristic of the present-day subarctic North Pacific surface ocean is a fairly recent feature, developed as a product of mid-Holocene environmental change. High SSS matched a salient productivity maximum of biogenic opal during Bølling-to-Early Holocene times, reaching levels similar to those observed during preglacial times in the warm mid-Pliocene prior to 2.73 Ma. Similar productivity spikes marked every preceding glacial termination of the last 800 ka, indicating recurrent short-term events of mid-Pliocene-style intense upwelling of nutrient-rich Pacific Deepwater in the Pleistocene. Such events led to a repeated exposure of CO_2-rich deepwater at the ocean surface facilitating a transient CO_2 release to the atmosphere, but the timing and duration of these events repudiate a long-term influence of the subarctic North Pacific on global atmospheric CO_2 concentration.
© 2004 Elsevier Ltd. All rights reserved.

1. Introduction

As summarized by Haug et al. (1999), the surface waters of the modern subarctic North Pacific Ocean are isolated from the highly nutrient-rich waters below by a steep vertical gradient in salinity, which leads to a robust stratification of the surface ocean in this region (Reid, 1969; Warren, 1983; Clarke et al., 2001). Large and Nurser (2001) depicted modern sea surface salinity (SSS) values in the North Pacific subarctic gyre, which are approximately 2 p.s.u. lower than in the subtropical gyre (Fig. 1). In terms of sea-surface density (SSD) the strong decrease in SSS almost compensates for the marked northward decrease in sea surface temperature (SST) during winter (down to approx. 3 °C). The physical

processes which maintain this halocline and, in turn, its physical, biological, and geochemical effects have long been subjects of intense investigation (Chisholm and Morel, 1991). Haug et al. (1999) and Sigman et al. (2004) concluded from the major reduction of opal accumulation rates and $\delta^{15}N$ ratios that the stable stratification of the northern North Pacific was a self-sustained, quasi-permanent feature since the onset of Northern Hemisphere glaciation at 2.73 Ma. Since then a decreased biological pump may have led to lowered CO_2 evasion and a significant decrease in atmospheric CO_2 concentration and greenhouse warming.

However, the actual details of Pleistocene and late deglacial-to-Holocene ocean variability in the far northwestern Pacific have remained little known and poorly understood for various reasons: (1) Modest sedimentation rates and increased $CaCO_3$ dissolution below 2500 m water depth have largely hampered the reconstruction of high-resolution records; (2) severe dating

*Corresponding author. Tel.: +49-431-880-2882; fax: +49-431-880-4376.

E-mail address: ms@gpi.uni-kiel.de (M. Sarnthein).

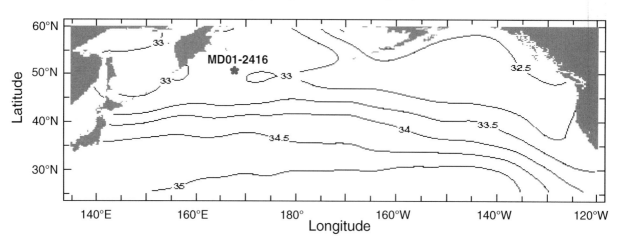

Fig. 1. Location of Marion Dufresne (MD) Site MD01-2416 on the Detroit Seamount, near the center of the subarctic gyre in the northwest Pacific. Distribution pattern of modern sea surface salinities for 50 m water depth after Levitus et al. (1994).

problems occur on high-resolution time scales because the paleo-^{14}C reservoir ages of surface water are poorly known and possibly highly variable in this region (estimates for modern surface water ages range from 700 to 950 yr; Keigwin et al. (1992) and Keigwin, (1998). Moreover, (3) different General Circulation Models have produced controversial results on the millennial-scale phase relationship between North Pacific and North Atlantic climate variability (e.g., Mikolajewicz et al., 1997, vs. Schmittner et al., 2003), depending on whether the atmosphere or the ocean are dominating the interhemispheric signal transfer in the different models.

We present new data for IMAGES core MD01-2416 (Detroit Seamount, 51°N, 168°E; 2317 m; Fig. 1), including planktic δ^{18}O and two independent SST proxies. The records cover an interval in the Holocene between 12,000 and 4000 yr B.P. with 50–200 yr resolution thus providing the first well-dated high-resolution Holocene paleoenvironmental signal from the subarctic North Pacific, allowing comparison with climatic records from Greenland ice cores (Johnsen et al., 2001). This new data base enabled us to establish first Holocene paleosalinity records for the subarctic North Pacific that provide a conservative estimate of short-term intra-Holocene variations in SSS and SSD, showing that the modern pronounced pycnocline is a relatively recent feature in this high-latitude sea region.

2. Methods and age control

Stable oxygen and carbon isotopes were measured in the Leibniz Laboratory of Kiel University. Each analysis is based on 30 specimens of subpolar *N. pachyderma* (sin.) (Nps) picked from the 150 to 250 μm size fraction. Sample preparation and analytical procedures were recently described in Kiefer et al. (2001). SST records were derived from two independent

proxies. (1) SST were deduced from Mg/Ca ratios measured on ca. 40 shells of Nps per sample, picked from the size fraction 150–250 μm. Foraminifera tests were gently crushed to open chambers, cleaned to remove clays (ultrasonicating in water and methanol) and organic matter (hydrogen peroxide treatment), and polished with dilute acid (Barker et al., 2003). Samples were analyzed with a Varian Vista inductively coupled plasma-atomic emission spectrometer (de Villiers et al., 2002) at the Department of Earth Sciences, Cambridge University. Calcification temperatures T (°C) were estimated from Mg/Ca ratios (mmol/mol) using the *N. pachyderma* calibration Mg/Ca = 0.50 exp (0.10 T) of Elderfield and Ganssen (2000). Analytical precision (1σ s.d.) was 0.004 mmol/mol for a standard solution of Mg/Ca = 1.29 mmol/mol. Replicate analyses on 8 samples (including re-picking) resulted in a reproducibility of 0.07 mmol/mol, equivalent to approximately 0.8 °C.

(2) Two SST records were reconstructed from planktic-foraminifer counts with both the SIMMAX (Pflaumann et al., 2003) and ANN techniques (Malmgren et al., 2001). The Atlantic database of GLA-MAP-2000 (Pflaumann et al., 2003) was employed to calibrate faunal compositions to caloric summer (July–August–September average in Northern Hemisphere) SST at 10 m water depth. We used the same set of trained neural networks as in Sarnthein et al. (2003).

We are aware that many of the key morphospecies of planktic foraminifera in the North Pacific are genetically different from their North Atlantic counterparts (Darling et al., 2003). If the ecological preferences of these genetic types also differ between the two oceans, Pacific SST estimates produced by transfer functions trained on North Atlantic faunas may be in error (Kucera and Darling, 2002). However, no appropriate Pacific coretop database is available with a sufficient coverage of the northern high-latitude regions and with adequate age control for individual coretop samples.

Nevertheless, in an attempt to determine the extent of differences in patterns and magnitude of SST reconstructions caused by the use of the more distant calibration dataset, we have experimented with the best available Pacific dataset, recently developed within the MARGO project (Kucera et al., 2004). The resulting SST gradients were very similar but steeper than those produced by neural networks trained on the Atlantic database. In terms of absolute values, the Pacific-based ANN temperature estimates reached 12–13 °C for the Younger Dryas, that is approximately 4 °C warmer than the already fairly high Atlantic-based values. These Pacific-based SST reconstructions would result in very unrealistic SSS values of 38 p.s.u. for the Younger Dryas, if employed for calculating SSS values from coeval planktic $\delta^{18}O$.

There is as yet no clear indication that the genetic differences between the North Pacific and North Atlantic populations translate into significant differences in ecology. On the other hand, the caveat of potential genetic differences may also apply to the purely Pacific-based dataset employing populations from the far northern and far southern Pacific. Moreover, SST estimates derived from the Atlantic database for the last 12 ky parallel reasonably well the somewhat lower Mg/Ca-based estimates. The differences rarely exceed 2–2.5 °C and in part result from the calcification depth of Nps close to the thermocline (Simstich et al., 2002). Here temperatures are 1–2° lower (Reid, 1997) than faunal SST values which are calibrated to 10 m water depth. In addition, partial dissolution of Nps tests (many of which are particularly thin-walled) has induced a preferential loss of Mg and hence an apparent SST reduction.

In summary, some caution must be exercised when interpreting the results from planktic foraminifera counts. Whilst the general trend of the reconstructed SST curve is robust, the absolute values of the inferred SSTs may be slightly affected by the choice of calibration database. However, temporal shifts in absolute SST, which exceed 1–2°C, should be identifiable by comparison with independently established planktic $\delta^{18}O$ and the general trend of Mg/Ca-based SST estimates, the noise of which appears largely

induced by cleaning deficiency, since it is not paralleled by similar noise in $\delta^{18}O$ of Nps.

Records of SSS were reconstructed from planktic $\delta^{18}O$ values corrected for Holocene ice volume/ sea level changes (Geyh et al., 1979) and normalized for SST changes, following the techniques proposed by Duplessy et al. (1991). We assume 1‰ $\delta^{18}O$ equal to 3.57 °C near the "cold end" and/or 2.75 p.s.u. in local SSS (Keigwin, 1998; Duplessy et al., 2002). Moreover, the modern SSS level of 33 p.s.u. which was reported for the center of the surface layer at 50 m water depth (Keigwin, 1998) is used as reference value for the 4 ky old core top at Site MD01-2416.

Biogenic opal concentrations were measured at GEOMAR in Kiel following the automated leaching method of Müller and Schneider (1993) on 2 mg of freeze-dried and crushed sediment. Values are reproduced within ±0.5 wt%.

Five AMS radiocarbon ages in the Holocene (Table 1; Fig. 2) and numerous ^{14}C ages farther below (Kiefer et al., 2004) were corrected for an apparently persistent ^{14}C reservoir age of 800 years for modern surface water (Southon et al., 1990) and converted to calendar ages using the CALIB-98 program for terrestrial ^{14}C values (Stuiver et al., 1998). The age of 7.98 calendar (cal.) ka was ignored as "too young" because it lies on top of an ash layer that barred the otherwise continuous upward bioturbational mixing of "old" Nps tests from deeper sediment sections. In this text all ages are presented as calendar years B.P., if not specified otherwise. The estimate of 800 yr reservoir age was largely corroborated for the early Holocene right after glacial Termination Ib by ^{14}C dating a significant geomagnetic intensity minimum at 50 cm core depth to 10.3 cal ka, in line with a reference age of about 10 ka from the SAPIS stacked geomagnetic record (Stoner et al., 2002). Further upcore, a major ash layer at 44±4 cm nominal depth (=48–58 cm c.d. in Fig. 2) has an age of 9.2–10 ka (Keigwin, 1998). On the basis of this stratigraphy, apparent Holocene sedimentation rates range from 6 to 14 cm/ka. The time resolution of the planktic $\delta^{18}O$ record is 70–100 yr, that of the SST and SSS records 100–200 yr (Fig. 2).

Table 1
AMS radiocarbon ages measured at *N. pachyderma* (sin.) from trigger-weight (TW) and piston cores (PC) MD01-2416

Laboratory number	Core	Nominal depth (cm bsf)	Composite depth (cm bsf)	^{14}C ages (corr.) (years)	±1σ (years)	Calendar age (yr B.P.)
KIA17084	TW	0.25	0.25	3745	50	4217
KIA20147	PC	0.75	8.75	4884	30	5632
KIA20148	PC	16	24	5945	33	6790
KIA20149	PC	39	47	7069	36	7928[a]
KIA20150	PC	51	59	9232	44	10,306

KIA numbers refer to the Leibniz Laboratory at Kiel University. Midpoint ^{14}C ages of 1 cm thick samples were corrected for 800 yr oceanic reservoir age and converted into calendar ages after Stuiver et al. (1998).
[a] The age of 7.93 ka is ignored as "too young" (see text for explanation).

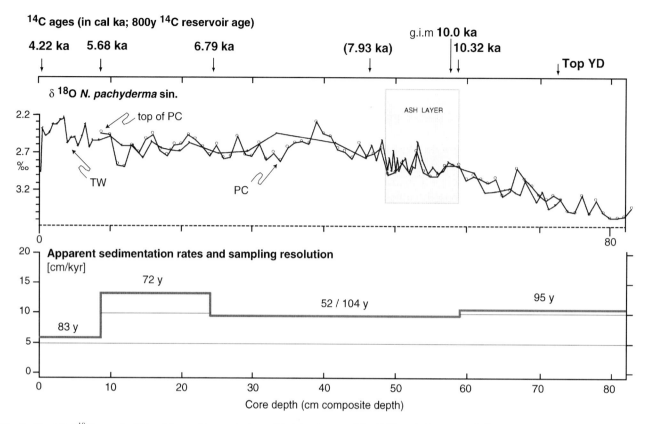

Fig. 2. Planktic δ^{18}O record of *Neogloboquadrina pachyderma* (sin.) in trigger weight (TW) and piston core (PC) MD01-2416 versus composite core depth, ^{14}C ages plotted as cal. ka, apparent sedimentation rates, and sampling resolution (in years). YD = Younger Dryas. The top of the YD was defined by interpolation between neighboring ^{14}C ages and the onset of decreasing planktic δ^{18}O values recording of Termination Ib. g.i.m. = geomagnetic intensity minimum. Ash layer is shaded.

All data are available at http://www.pangaea.de/PangaVista?query = NW-PacHolo

3. Temperature and salinity trends

In general, the SIMMAX-based reconstruction of summer SST is slightly more muted than the ANN-based SST curve and in particular, than the noisier Mg/Ca-based record (Fig. 3). The range of absolute temperature estimates is closely compatible in all three records, especially between ANN- and SIMMAX-based estimates where relative deviations amount to < ±0.5 °C, rarely rising to ±1.2 °C, well within the limit of the expected prediction errors of the two techniques (Pflaumann et al., 2003; Sarnthein et al., 2003). All three proxy records show a clear temperature maximum of 7–9 °C during the late Younger Dryas and early Preboreal (12.0–11.1 ka), followed by an abrupt drop of approximately 1 °C. Later, temperatures drop more gradually towards a long-term mid-Holocene minimum starting at 9.0/9.3 ka with temperatures of 4.0–7.3 °C, a level much lower than during the Younger Dryas. This low SST level continued with short-term oscillations until 4 ka. ANN- and Mg/Ca-based SST

depict a further drop by 1–2 °C near 5.7/5.3 ka, Mg/Ca-based SST also shows an apparent rise by 2.5 °C after 4.8 ka. No younger sediments are preserved at Site MD01-2416.

These SST gradients contrast with temperature trends found on Greenland (Johnsen et al., 2001) and for the West Spitsbergen Current (Sarnthein et al., 2003), that is the final derivate of warm poleward heat flow from the North Atlantic into the Norwegian Sea. In those records, SST markedly increased during the Preboreal, from 11.1 to 10.7 ka, and reached a distinct Early Holocene SST maximum from 10.7 to 8.8 ka and near 8.3 to 7.7 ka (Fig. 3). The different SST trends observed in the Norwegian Sea and northwest Pacific records cannot be explained by simple joint atmospheric forcing of northern hemisphere climate. The opposed temperature trends of the two regions agree with GCM results by Schmittner et al. (2003), who showed a mainly ocean-controlled antiphase evolution of SST changes on millennial timescales. Similar shifts in far North Pacific SST, which are opposed to North Atlantic SST changes, were reported for preceding Marine Isotope Stages (MIS) 2 and 3 (Kiefer et al., 2001, 2004).

Likewise, the extreme δ^{18}O minimum in core MD01-2416 near 8.3–8.6 ka (Fig. 3) may represent a slight

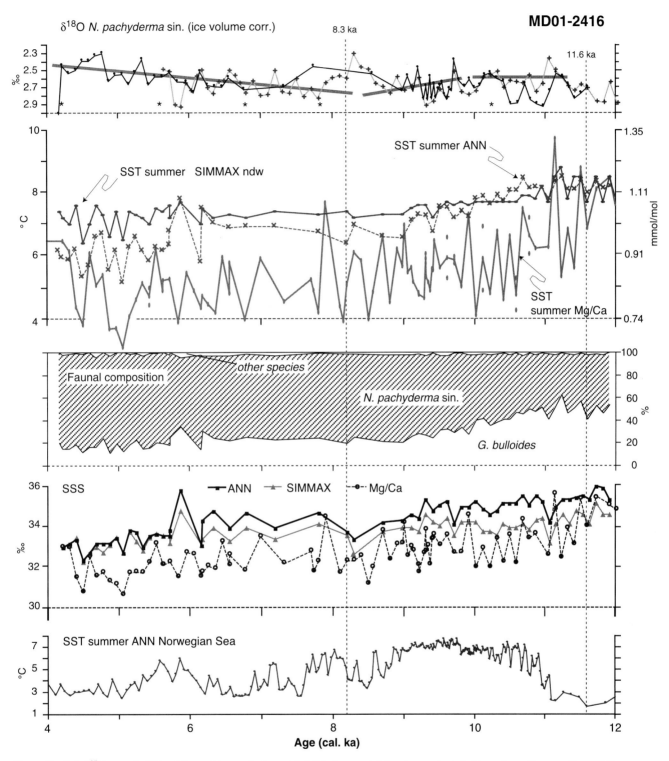

Fig. 3. Planktic $\delta^{18}O$ record of *Neogloboquadrina pachyderma* (sin.) (Nps) corrected for changes in global ice volume (top panel), (SST) for summer (panel 2) as deduced from Mg/Ca ratios of Nps tests (with dots indicating replicate analyses) and from census counts of planktic foraminifera species using the (ndw = non-distance weighted) SIMMAX and ANN transfer functions (Pflaumann et al., 2003; Malmgren et al., 2001), (SSS) curves as deduced from planktic $\delta^{18}O$ and the various SST records (panel 4), and percentages of Nps, *Globigerina bulloides*, and sum of other planktic foraminifera species for the Holocene interval 12–4 ka (panel 3), compared to ANN-based SST record of core M23258 from the northern Norwegian Sea, 75 °N (Sarnthein et al., 2003). $\delta^{18}O$ record of trigger weight (TW) core MD01-2416 is marked by full dots, piston core (PC) data are marked by + signs.

warming, being an immediate response to the ice outbreak from the Hudson Bay and coeval widespread cooling in the North Atlantic at this time (Barber et al., 1999; Sarnthein et al., 2003), a signal that unfortunately is not resolved in our SST records.

Similar to SST, SSS shows an overall significant decrease from 35 to 36 p.s.u. in the Younger Dryas and earliest Holocene to 33 p.s.u. and less in the Late Holocene (Fig. 3). Similarly high SSS values may be deduced for the Younger Dryas from interpreting the $\delta^{18}O$ and *Cycladophora davisiana* records from the nearby core RAMA 44PC (Keigwin et al., 1992; although poorly resolved). In total, this decrease amounts to 3.0–4.0 p.s.u. (depending on the SST record employed for SSS reconstruction), with the absolute SSS value at 4 ka arbitrarily set at the modern local SSS level of 33 p.s.u.. The long-term salinity loss is restricted to three distinct phases of major SSS reduction, from 11.7–11.1, 9.3–8.5, and 6.0–4.4 ka. An early extreme but short-lasting SSS minimum occurs in all three proxy records at 11.1 ka, with a further minimum near 8.4 ka. A slight recovery of SSS ends abruptly after 5.9 ka. The maximum in SSS (and $\delta^{18}O$) at 5.9 ka may be an artifact of sediment distortion at the core-top of the piston corer (see Fig. 2).

The ANN-based SSS record suggests an overall stronger SSS reduction over the Holocene than SIM-MAX-based SSS. Most Mg/Ca-based SSS estimates appear noisier because of cleaning problems and are 1–2 °C lower because of both Nps calcifying near the thermocline and calcite dissolution. High Mg/Ca-based SSS estimates come closer to the SIMMAX-based SSS, except for the extreme SSS values high near the end of the Younger Dryas, which precisely match the ANN-based estimates.

4. Discussion

In part, the SSS records in Fig. 3 depend on the calcification depths of *N. pachyderma* (sin.) (Nps), whose tests served for obtaining the $\delta^{18}O$ and Mg/Ca records, at which we base the SSS records. For comparison, calcification of this morphospecies in the northern North Atlantic is centered between 30 and 100/200 m water depth (Simstich et al., 2003). As the stratification of the surface ocean increases, these calcification depths will also increase. Signals of low-salinity surface water in tests of Nps will thus be increasingly smoothed. Modern hydrographic data near Site MD01-2416 (Keigwin, 1998) show that salinity increases by 0.8 p.s.u. from the sea surface down to the base of the pycnocline at 150 m water depth, where most tests of Nps may form today per analogy to the North Atlantic. Accordingly, the overall decrease in SSS depicted in Fig. 3 may be regarded as a record of

subsurface water composition and thus only as conservative estimate of an actually much stronger salinity reduction of North Pacific sea surface waters during the Holocene.

In summary, the widely cited (Haug et al., 1999) robust halocline which today characterizes the total subarctic North Pacific (Large and Nurser, 2001) appears to be a product of relatively recent environmental change. This halocline only developed over the last 11.1/9.3 ky, in particular over the last 6 ky through a major, probably stepwise salinity reduction by > 2–3-p.s.u., which resulted in a significant decrease of SSD (Figs. 3 and 4). Unlike today, the surface ocean was well mixed in the Early Holocene and Younger Dryas, when subarctic salinity was as high as in the modern low-latitude subtropical Pacific (> 35 p.s.u.). SSD was sufficient to allow for seasonal convection of Upper Pacific Deepwater by mere cooling of surface water (arrows in Fig. 4), since SSD was similar to the density of the deepwater which today spills the Detroit Seamount at 2300–4000 m water depth (Fig. 4) (Keigwin, 1998).

An explanation is still required for the origin of the extensive Holocene freshwater supply which has changed so fundamentally the thermohaline structure of the northern North Pacific in Holocene times. The surrounding continents neither provide major fluvial runoff nor carry major glacial ice sheets for continuing meltwater discharge. Various potential freshwater sources may be considered: (1) Enhanced Late Holocene precipitation from the Westerlies; (2) increased freshwater advection with the low-salinity Aleutian Current which collects the runoff from the Rocky Mountains to the north of Portland, WA, and carries it up to the eastern Bering Sea (Dodimead et al., 1963); (3) advection of meltwater from Arctic sea ice through the Bering Strait and Oyashio Current subsequent to the major opening/deepening of this strait around 9 ka (Dalton, 2003), although the modern net transport of water today goes from the Pacific into the Arctic; or (4) a combination of these sources. At present, no tracer evidence such as isotopic proxy data is available for any of these hypotheses. On the other hand, the phase of high salinity was apparently linked to a unique phase of intense global ocean thermohaline circulation and North Pacific upwelling, conditions obviously tied to Pleistocene glacial terminations.

Extremely short, approximately 100-year long oscillations in planktic $\delta^{18}O$ and SSS near 9.2–9.4 ka (55–45 cm c.d.) are not considered as analytical artifact but appear to represent real oceanographic signals, since they occur in the upper part and on top of a major ash layer where bioturbational mixing was largely suppressed. This has also affected a ^{14}C age that is far too low for this core depth, as outlined above.

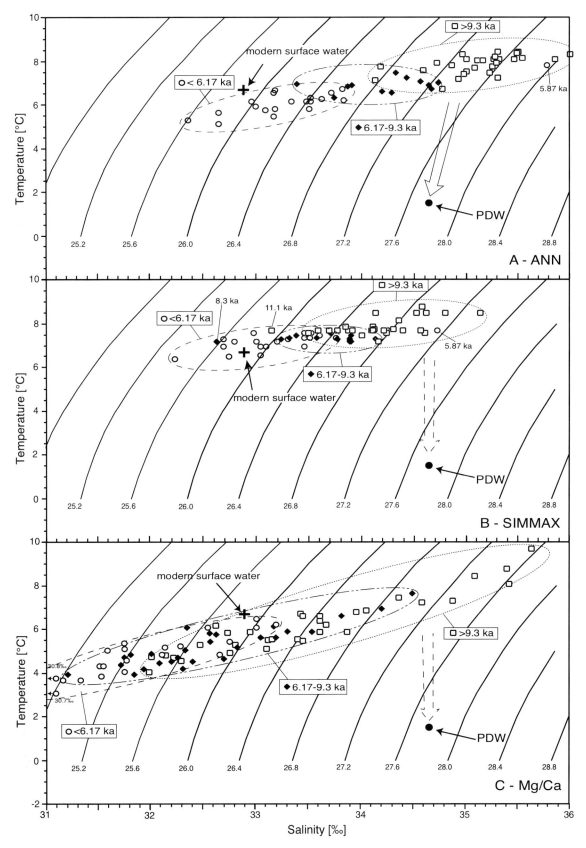

Fig. 4. Temperature, salinity, and density of Holocene surface water masses in the subarctic North Pacific, compared to modern surface water and upper Pacific Deepwater (PDW; temperature and salinity after Keigwin, 1998). Density lines are following Cox et al. (1970) for σ_0 level. Sea surface temperatures are calibrated to 10 m water depth, SSS are regarded as average of top 100 m water depth. SSS estimates are calculated from planktonic $\delta^{18}O$ corrected for global ice volume and SST based on (A) ANN transfer function; (B) SIMMAX transfer function; and (C) Mg/Ca ratios. Elliptic circles envelop sample groups from the Early, Middle, and late Holocene. Arrows indicate potential local deepwater formation by cooling of surface water. Extremely low salinity values in (C) probably result from carbonate dissolution.

5. Implications

A well-mixed surface ocean during the Younger Dryas and preceding Termination-Ia times (Kiefer et al., 2003) will lead to a vast surface exposure of extremely nutrient enriched Pacific Deepwater in northern high latitudes. This nutrient concentration will be similar to that of preglacial times in the mid-Pliocene, prior to 2.73 Ma (Haug et al., 1999; Sigman et al., 2004) and to the modern (interglacial) scenario in the Southern Ocean. In contrast, the modern low-salinity lid in the far northern Pacific is suppressing major advection of upwelled deepwater from below (Reid, 1962). Accordingly, we may expect a fundamental intra-Holocene decrease in nutrient supply to the sea surface and coeval increase in nutrient utilization, which should be clearly recognized in fertility and paleoproductivity records, i.e. in the efficiency of the biological pump at Site MD01-2416.

Indeed, high concentrations (>25%) and accumulation rates (1.2–4.4 g cm^{-2} ky^{-1}) of biogenic opal reveal a prominent maximum in upwelling-induced phytoplankton productivity during large parts of glacial Termination I. It started shortly after 15 ka (Kiefer et al., 2004; Keigwin et al., 1992), culminated from 14.7 to 12.9 ka, and lasted until 10.8 ka (Fig. 5), as long as SSS was high. Later, opal productivity decreased rapidly to much lower values in the Middle and Late Holocene after 9.5 ka, largely coeval with the decrease in SSS. This trend further confirms the establishment of the low-salinity stratification of surface water in the far northwestern Pacific only during the Holocene (Fig. 4). The

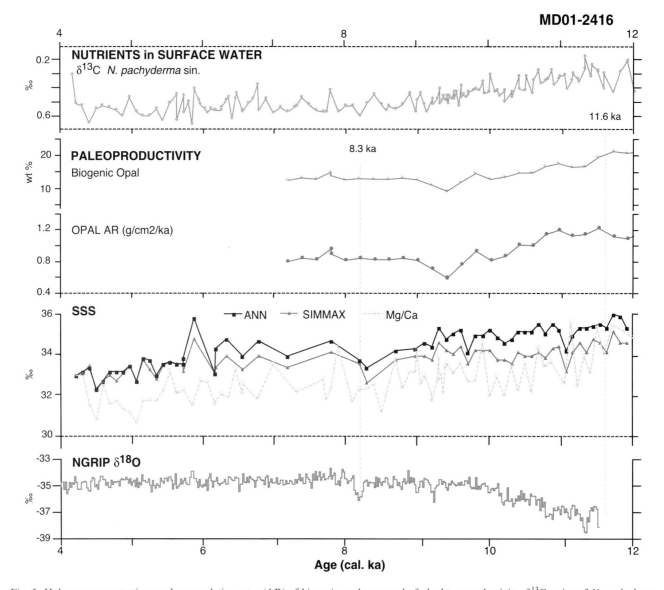

Fig. 5. Holocene concentrations and accumulation rates (AR) of biogenic opal as record of plankton productivity, δ^{13}C ratios of *N. pachyderma* (sin.) as (inverse) proxy of nutrient concentrations in the surface ocean, and SSS at Site MD01-2416, compared to NGRIP δ^{18}O temperature record from Greenland (Johnson et al., 2001).

productivity high during the Bølling-Allerød, Younger Dryas, and Earliest Holocene is paralleled by extremely low $\delta^{13}C$ values of *N. pachyderma* (sin.) (Fig. 5), a proxy of high nutrient concentration in the depth habitat of this species, moreover, by a coeval pronounced benthic $\delta^{13}C$ minimum in upper Pacific Deepwater at the Detroit Seamount (Keigwin, 1998), and thus may record a short-term but pronounced pulse of deepwater upwelling. Further support for this event comes from a salient peak in $CaCO_3$ concentration during the Boelling-Allerød in the otherwise carbonate-poor North Pacific sediment sections (Kiefer et al., 2004). A new high-resolution XRF record of core MD01-2416 shows similar extreme and short-term but well dated $CaCO_3$-based maximima of paleoproductivity as characteristics for each glacial termination over the last 800 ky (unpubl. Kiel-Bremen data).

The extreme rates of biogenic-opal production at 14.9–10.8 ka and especially, the rates at 14.6–12.9 ka closely match the upwelling-induced high productivity values reported for the preglacial mid-Pliocene prior to 2.73 Ma (2–6 g opal cm^{-2} ky^{-1}; Haug et al., 1999). This holds true, even if the Late Pleistocene values are corrected for a slight overestimation of sedimentation rates because of core length extension by piston coring (Skinner and McCave, 2003). The close match of productivity rates demonstrates that intervals with intense upwelling of nutrient-rich deepwater were a truly recurrent feature throughout the Pleistocene and that low SSS and stable stratification with complete nutrient utilization do not present the sole Pleistocene mode of the subarctic Pacific surface ocean. During a major portion of all glacial terminations, the extensive exposure of nutrient- and CO_2-rich deep water at the ocean surface with CO_2 release to the atmosphere presents a distinct third, Pliocene-style mode of enhanced global thermohaline circulation, in addition to the two modes delineated for the Pleistocene in the scheme of Haug et al. (1999) and Sigman et al. (2004). This finding somewhat downplays the central importance of the subarctic Pacific for long-term global atmospheric CO_2 changes, previously deduced from the carbon cycle experiments of Haug et al. (1999). However, the deglacial CO_2 release to the atmosphere has probably contributed to rapid deglacial warming, but never led to a full restoration of preglacial atmospheric pCO_2 (Petit et al., 1999), for reasons yet unknown.

6. Conclusions

Modern wide-spread low SSS (32–33 p.s.u.) leads to stable upper-ocean stratification of the subarctic North Pacific, thus suppressing the upwelling of nutrient-rich Pacific Deepwater from below and reducing biological productivity and CO_2 evasion to the atmosphere. The stratification has first developed 2.73 Ma, coeval with the onset of extensive glaciation of the Northern Hemisphere, and is considered largely self-sustained since this time (Haug et al., 1999). This concept of a long-term persistent structure has now been tested for the Holocene time span on the basis of new centennial-scale paleoceanographic records, which lead us to the following conclusions:

1. SST have decreased by 3–5°C from the latest Younger Dryas and earliest Holocene until 5000 yr B.P. This trend is in contrast to atmospheric temperatures and SST in Greenland and the northern Norwegian Sea, which have strongly increased only after the Younger Dryas and during the earliest Holocene, the Preboreal.
2. In the Younger Dryas, SSS of subarctic North Pacific surface water were as high as in the modern subtropical gyre (up to > 35.5 p.s.u.). They stepwise dropped to the present level in the Early Holocene and early Late Holocene. Accordingly, the present prominent low-salinity anomaly of the North Pacific is a product of early Holocene environmental change.
3. The precise source of extensive Holocene freshwater supply, or conversely, the origin of the freshwater deficit during the last glacial terminations remain unclear.
4. The high salinity in the Younger Dryas and Earliest Holocene is closely related to outstanding maxima in nutrients and biogenic-opal and carbonate productivity in the surface ocean. During the preceding Bølling-Allerød this productivity has even reached the extreme of the preglacial mid-Pliocene. Accordingly, upwelling-induced CO_2 evasion from the North Pacific to the atmosphere may have also reached a mid-Pliocene level. Interesting to note, similar mid-Pliocene-style productivity spikes mark each glacial termination of the last 800 ka and thus record recurrent short-term events of upwelling intensity, possibly linked to unique states in the intensity of global ocean thermohaline circulation, which characterize glacial terminations around the globe.

Acknowledgments

We thank the crew on board of RV *Marion Dufresne* (MD) for retrieving an undisturbed, 45-m long, large-diameter piston core MD01-2416. Sincere thanks goes to G. Haug for valuable scientific discussion and to U. Pflaumann, who helped us in handling the SIMMAX tranfer function to produce sound SST records. We thank K. Kissling for her efficient laboratory assistance on board of MD, to D. Gudehus for her indefatigable assistance in preparing foraminifera samples, and to

Claudia Sieler for extensive support and skill with data management and processing. We acknowledge P. Grootes and his team at the Kiel Leibniz Laboratory, who provided great expertise in measuring and interpreting AMS-^{14}C ages, and H. Cordt and H. Heckt who took care of the stable-isotope mass spectrometry. We thank the Deutsche Forschungsgemeinschaft (DFG) for generously funding the retrieval of cores on IMAGES cruise VII and the subsequent scientific evaluation of extensive core material.

References

Barber, D.C., Dyke, A., Hillaire-Marcel, C., Jennings, A.E., Andrews, J.T., Kerwin, M.W., Bilodeau, G., McNeely, R., Southon, J., Morehead, M.D., Gagnon, J.-M., 1999. Forcing of the cold event of 8,200 years ago by catastrophic drainage of Laurentide lakes. Nature 400, 344–348.

Barker, S., Greaves, M., Elderfield, H., 2003. A study of cleaning procedures used for foraminiferal Mg/Ca paleothermometry. Geochemistry, Geophysics, Geosystems, doi:10.1029/2003GC000559.

Chisholm, S.W., Morel, F.M.M. (Eds.), 1991. What controls phytoplankton production in nutrient-rich areas in the open sea? Limnology and Oceanography 36, 1507–1970.

Clarke, A., Church, J., Gould, J., 2001. Ocean Processes and Climate Phenomena. In: Siedler, G., Church, J., Gould, J. (Eds.), Ocean Circulation and Climate. International Geophysics Series 77, 11–30.

Cox, R.A., McCartney, M.J., Culkin, F., 1970. The specific gravity/salinity/temperature relationship in natural seawater. Deep-Sea Research 17, 679–689.

Dalton, R., 2003. The coast road. Nature 421, 10–12.

Darling, K.F., Kucera, M., Wade, C., von Langen, P., Pak, D., 2003. Seasonal distribution of genetic types of planktonic foraminifer morphospecies in the Santa Barbara Channel and its paleoceanographic implications. Paleoceanography 18 (2), 1032 doi:10.1029/2001PA000723.

De Villiers, S., Greaves, M., Elderfield, H., 2002. An intensity ratio calibration method for the accurate determination of Mg/Ca and Sr/Ca of marine carbonates by ICP-AES. Geochemistry, Geophysics, Geosystems 3, doi: 10.1029/2001GC000160.

Dodimead, A.J., Favorite, F., Hirano, T., 1963. Salmon of the North Pacific Ocean, Part II: Review of oceanography of the Subarctic Pacific region. International North Pacific Fisheries Comission Bulletin 13, 1–195.

Duplessy, J.C., Labeyrie, L., Juillet-Leclerc, A., Maitre, F., Duprat, J., Sarnthein, M., 1991. Surface salinity reconstruction of the North Atlantic Ocean during the Last Glacial Maximum. Oceanologica Acta 14, 311–324.

Duplessy, J.-C., Labeyrie, L., Waelbroek, C., 2002. Constraints on the ocean oxygen isotopic enrichment between the last glacial maximum and the Holocene; Paleoceanographic implications. Quaternary Science Reviews 21, 315–330.

Elderfield, H., Ganssen, G., 2000. Past temperature and δ^{18}O of ocean surface waters inferred from foraminiferal Mg/Ca ratios. Nature 405, 442–445.

Geyh, M.A., Kudrass, H.-R., Streif, H., 1979. Sea-level changes during the late Pleistocene and Holocene in the Strait of Malacca. Nature 278, 441–443.

Haug, G.H., Sigman, D.M., Tiedemann, R., Pedersen, T.F., Sarnthein, M., 1999. Onset of permanent stratification in the subarctic Pacific Ocean. Nature 401, 779–782.

Johnsen, S.J., Dahl-Jensen, D., Gundestrup, N., Steffensen, J.P., Clausen, H.B., Miller, H., Masson-Delmotte, V., Sveinbjörnsdottir, A.E., White, J., 2001. Oxygen isotope and paelotemperature records

from six Greenland ice-core stations: Camp Century, Dye-3, GRIP, GISP2, Renland, and North GRIP. Journal of Quaternary Science 16, 299–307.

Keigwin, L.D., 1998. Glacial-age hydrography of the far northwest Pacific Ocean. Paleoceanography 13, 323–339.

Keigwin, L.D., Jones, G.A., Froelich, P.N., 1992. A 15,000 year paleoenvironmental record from Meiji Seamount, far northwestern Pacific. Earth and Planetary Science Letters 111, 425–440.

Kiefer, T., Sarnthein, M., Erlenkeuser, H., Grootes, P.M., Roberts, A.P., 2001. North Pacific response to millennial-scale changes in ocean circulation over the last 60 kyr. Paleoceanography 16, 179–189.

Kiefer, T., Sarnthein, M., Elderfield, H., Erlenkeuser, H., Grootes, P.M., 2004. Warmings in the far northwestern Pacific support possible pre-Clovis immigration to America. Geology, submitted for publication.

Kucera, M., Darling, K.F., 2002. Genetic diversity among modern planktonic foraminifer species: its effect on paleoceanographic reconstructions. Philosophical Transactions of the Royal Society of London A 360, 695–718.

Kucera., M., Weinelt, M., Kiefer, T., Pflaumann, U., Hayes, A., Weinelt, M., Chen, M.-T., Mix, A.C., Cortijo, E., Duprat, J., Juggins, S., Waelbroeck, C., 2004. Reconstruction of sea-surface temperatures from assemblages of planktonic foraminifera: multi-technique approach based on geographically constrained calibration datasets and its application to glacial Atlantic and Pacific Oceans. Quaternary Science Reviews, in press (doi:10.1016/j.quasci-rev.2004.07.014).

Large, W.G., Nurser, A.J.G., 2001. Ocean Surface Water Mass Transformation. In: Siedler, G., Church, J., Gould, J. (Eds.), Ocean Circulation and Climate. International Geophysics Series 77, 317–336.

Levitus, S., Boyer, T., 1994. World Ocean Atlas Volume 4: Temperature NOAA Atlas NESDIS 4, U.S. Department of Commerce, Washington, DC.

Malmgren, B.A., Kucera, M., Nyberg, J., Waelbroeck, C., 2001. Comparison of statistical and artificial neural network techniques for estimating past sea surface temperatures from planktonic foraminifer census data. Paleoceanography 16, 520–530.

Mikolajewicz, U., Crowley, T.J., Schiller, A., Voss, R., 1997. Modelling teleconnections between the North Atlantic and North Pacific during the Younger Dryas. Nature 387, 384–387.

Müller, P.J., Schneider, R., 1993. An automated leaching method for the determination of opal in sediments and particulate matter. Deep-Sea Research I 40, 425–444.

Petit, J.R., Jouzel, J., Raynnaud, D., Barkov, N.I., et al., 1999. Climate and atmospheric history of the past 420,000 years from the VOSTOK ice core, Antarctica. Nature 399, 429–436.

Pflaumann, U., Sarnthein, M., Chapman, M., Funnell, B., Huels, M., Kiefer, T., Maslin, M., Schulz, H., Swallow, J., van Kreveld, S., Vautravers, M., Vogelsang, E., Weinelt, M., 2003. The glacial North Atlantic: Sea surface conditions reconstructed by GLAMAP-2000. Paleoceanography 18 (3), doi:10.1029/2002PA000774, 10, 1–28.

Reid, J.L., 1962. On circulation, phosphate-phosphorus content, and zooplankton volumes in the upper part of the Pacific Ocean. Limnology and Oceanography 7, 287–306.

Reid, J.L., 1969. Sea surface temperature, salinity, and density of the Pacific Ocean in summer and in winter. Deep-Sea Research 16 (suppl.), 215–224.

Reid, J.L., 1997. On the total geostrophic circulation of the Pacific ocean: Flow patterns, tracers and transports. Progress in Oceanography 39, 263–352.

Sarnthein, M., van Kreveld, S., Erlenkeuser, H., Grootes, P.M., Kucera, M., Pflaumann, U., Schulz, M., 2003. Centennial-to-millennial-scale periodicities of Holocene climate and sediment injections off the western Barents shelf, 75°N. Boreas 32, 447–461.

Schmittner, A., Saenko, O.A., Weaver, A.J., 2003. Coupling of the hemispheres in observations and simulations of glacial climate change. Quaternary Science Reviews 22, 659–671.

Sigman, D.M., Jaccard, S.L., Haug, G.H., 2004. Polar ocean stratification in a cold climate. Nature 428, 59–63.

Simstich, J., Sarnthein, M., Erlenkeuser, H., 2002. Paired δ¹⁸O signals of *N. pachyderma* (s) and *T. quinqueloba* show thermal stratification structure in the Nordic Seas. Marine Micropaleontology 48, 107–125.

Skinner, L.C., McCave, I.N., 2003. Analysis and modelling of gravity- and piston coring based on soil mechanics. Marine Geology 199, 181–204.

Southon, J.R., Nelson, D.E., Vogel, J.S., 1990. A record of past ocean-atmosphere radiocarbon differences from the Northeast Pacific. Paleoceanography 5, 197–206.

Stoner, J.S., Laj, C., Channell, J.E.T., Kissel, C., 2002. South Atlantic and North Atlantic geomagnetic paleointensity stacks (0-80 ka) implications for interhemispheric correlation. Quaternary Science Reviews 21, 1141–1152.

Stuiver, M., Reimer, P.J., Bard, E., Beck, J.W., Burr, G.S., Hughen, K.A., Kromer, B., McCormac, F.G., van der Plicht, J., Spurk, M., 1998. INTCAL98 radiocarbon age calibration, 24,000-0 cal BP. Radiocarbon 40, 1041–1083.

Warren, B., 1983. Why is no deepwater formed in the North Pacific? Journal of Marine Research 41, 327–347.

Quaternary Science Reviews 23 (2004) 2101–2111

Causes and timing of the 8200 yr BP event inferred from the comparison of the GRIP ^{10}Be and the tree ring Δ^{14}C record

Raimund Muscheler[a,*], Jürg Beer[b], Maura Vonmoos[b]

[a]*GeoBiosphere Science Centre, Department of Geology, Quaternary Sciences, Lund University, Sölvegatan 12, SE-22362 Lund, Sweden*
[b]*Swiss Federal Institute of Environmental Science and Technology (EAWAG), Ueberlandstrasse 133, CH-8600, Dübendorf, Switzerland*

Abstract

We analyse the 8200 yr BP cold event by comparing the high-resolution ^{10}Be record from the GRIP ice core from Central Greenland with the well-known tree ring Δ^{14}C record. By transferring the absolute dated tree ring chronology to the ice core time scale, we show that the coldest phase in the GRIP record occurred around 8150 yr BP. Furthermore, this method allows us to disentangle production and climate effects on ^{10}Be and ^{14}C with important implications for the reconstruction of past solar activity, and changes in the carbon cycle and ^{10}Be transport. We show that, in principle, it is possible to infer changes in ocean circulation by comparing ^{10}Be and ^{14}C records. However, the duration of the 8200 yr BP event is too short to assign unambiguously a significant change in atmospheric ^{14}C concentration to changes in the global ocean circulation. Based on the comparison of ^{10}Be with climate records, one could argue that the 8200 yr BP cold event is triggered by a change towards lower solar activity. However, this link is questioned by the fact that around this period there are other similar and even stronger changes in solar activity that have no apparent connection to climate changes.
© 2004 Elsevier Ltd. All rights reserved.

1. Introduction

The cosmogenic radionuclides (for example, ^{10}Be, ^{36}Cl, ^{14}C) are produced in interactions of galactic cosmic ray particles with the atoms of the atmosphere (Lal and Peters, 1967). Since solar cosmic rays have not sufficient energy to produce significant amounts of ^{10}Be and ^{14}C, the variable solar activity has only an indirect influence on the production rate of these radionuclides. Solar wind emitted from coronal holes on the sun carries magnetic fields which deflect the galactic cosmic rays depending on the field intensity. High solar activity causes strong solar wind emission and, therefore, leads to a stronger deflection of galactic cosmic rays and a decreased production of cosmogenic radionuclides (Masarik and Beer, 1999). Changes in the geomagnetic field intensity influence the radio-

nuclide production in a similar way. High geomagnetic field intensity causes stronger galactic cosmic ray deflection and accordingly a lower radionuclide production. Since the production of ^{10}Be and ^{14}C is well understood and can be modelled (Masarik and Beer, 1999), it is possible to compare radionuclide records quantitatively. In all probability common signals are caused by the common production process and differences between such records can be attributed to the completely different geo-chemical behaviour of ^{10}Be and ^{14}C. After production, ^{14}C oxidises to $^{14}CO_2$ and becomes a part of the global carbon cycle (Lal and Peters, 1967) whereas ^{10}Be becomes attached to aerosols and is removed from the atmosphere within 1–2 years (Raisbeck et al., 1981). Due to its comparatively long atmospheric residence time, atmospheric ^{14}C can be regarded as a global signal. However, it can be influenced by changes in the carbon cycle (Siegenthaler et al., 1980). By contrast, ^{10}Be records could include a local component depending on the atmospheric transport and ^{10}Be deposition.

---

*Corresponding author. Tel.: +46-46-222-3956; fax: +46-46-222-4830.
E-mail address: raimund.muscheler@geol.lu.se (R. Muscheler).

0277-3791/$ - see front matter © 2004 Elsevier Ltd. All rights reserved.
doi:10.1016/j.quascirev.2004.08.007

In the following, we will present the comparison of tree ring $\Delta^{14}C$ and ^{10}Be from the GRIP ice core for the period around 8200 yr BP. We show that the common variability can be used to date the ice cores with a very high precision (Beer et al., 1988). Furthermore, the large similarity indicates that the radionuclide records are dominated by the common production rate variations due to geomagnetic field and solar activity changes. Nevertheless, there are also differences between the records that can be attributed to changes in the ^{10}Be transport and/or to changes in the carbon cycle. This is of particular interest around 8200 yr BP where the strongest Holocene climate deterioration is recorded in the Summit ice cores. The causes of this event are still a matter of debate especially also due to uncertainties in the dating of different climate records (Alley et al., 1997; Snowball et al., 2002). It has been proposed that this cold phase is connected to changes in the ocean circulation in the North Atlantic possibly caused by increased meltwater influx from the Laurentide ice sheet (Klitgaard-Kristensen et al., 1998; Barber et al., 1999). The comparison of ^{10}Be and ^{14}C records has the potential to gain additional information about this event since changes in ocean circulation influence the ^{14}C distribution within the oceans and therefore potentially also the atmospheric ^{14}C concentration.

2. Dating using radionuclide records

There is increasing evidence that regional and temporal variability plays an important role in Holocene climate change. Very accurate dating is crucial for the understanding of such climate changes. Regional variability might even be masked if climate events in different records are matched because of a lack of accurate dating. Also, the comparison with changes in solar forcing as, for example, indicated by radionuclide records might feign a solar influence on climate after wiggle matching which is often justified by the un-certainties in the time scales. The best-established radionuclide dating technique is the radiocarbon dating. This method is based on the decay of ^{14}C which has a half-life of 5730 years (Stuiver and Polach, 1977). However, ^{14}C dating is complicated due to the variable atmospheric ^{14}C concentration caused by changes in the ^{14}C production and the carbon cycle. The knowledge of the initial ^{14}C concentration is a prerequisite to calculate an accurate age based on ^{14}C. Therefore, a lot of effort has been put into the reconstruction of a calibration curve to be able to account for these variations (Stuiver et al., 1998). For the last 11,500 years this calibration curve is based on ^{14}C measurements in tree rings that can be independently dated via dendrochronology (Friedrich et al., 1999). Unfortunately, due to low CO_2 content and high in situ production of ^{14}C, radiocarbon

dating of ice cores is not feasible. However, ^{10}Be and ^{36}Cl can be measured in ice cores and the high degree of similarity between ^{10}Be, ^{36}Cl and ^{14}C records can be used to date ice cores by matching common features (Finkel and Nishiizumi, 1997; Muscheler et al., 2000). In this way, the absolutely dated tree ring time scale can be transferred to the ice core.

To compare ^{10}Be and ^{14}C records, it is indispensable to account for the different geo-chemical behaviour of these radionuclides. Because of its relatively short atmospheric residence time, ^{10}Be provides a relatively direct measure for past production changes. For example, the solar 11-year cycle is detectable in the high-resolution ice core record from Dye3 in Greenland (Beer et al., 1990). However, it has always to be kept in mind that incomplete atmospheric mixing and potential changes in ^{10}Be deposition might lead to a climatic and/or local component in the ^{10}Be signal. By contrast, ^{14}C has a relatively long mean atmospheric residence time that allows ^{14}C to mix well within the atmosphere. However, because of the large reservoirs (atmosphere, biosphere and ocean) and the constant exchange between these reservoirs, the short-term changes in the ^{14}C production rate are strongly dampened in the atmosphere. For example, the 11-years cycle appears in annual tree ring $\Delta^{14}C$ records with an amplitude in the order of 1‰ (Stuiver and Braziunas, 1993) which compares to variations in the ^{14}C production rate of approximately 20%. For dating it is especially impor-tant to account for the delayed response of atmospheric ^{14}C variations in relation to the original production variations. For example, a solar-induced production variation with a periodicity of 200 years leads to corresponding changes in the atmospheric ^{14}C concen-tration which are delayed by approximately 20 years (Siegenthaler and Beer, 1988).

Different studies indicate that the ^{10}Be flux to Summit is a very good proxy for past changes in the cosmogenic radionuclide production rate. For example, it is possible to reconstruct geomagnetic field changes during the last ice age on time scales longer than approximately 2000–3000 years under the assumption that the ^{10}Be flux to Summit is proportional to the global ^{10}Be production rate (Beer et al., 2002; Wagner et al., 2000). In addition, throughout the last 50,000 years typical solar cycles are detectable in the GRIP ^{10}Be record (Yiou et al., 1997; Wagner et al., 2001a). Fig. 1a shows the ^{10}Be concentration from 7800 to 8500 yr BP from the GRIP ice core (Muscheler et al., 2004). The data are shown on the ss08c time scale which is based on layer counting and which is the best estimate of the GRIP time scale for the Holocene period (Johnsen et al., 1999). For the last ice age it is indispensable to take the variable accumulation rate into account to estimate past variations in ^{10}Be production (Alley et al., 1995; Finkel and Nishiizumi, 1997; Yiou et al., 1997; Wagner et al.,

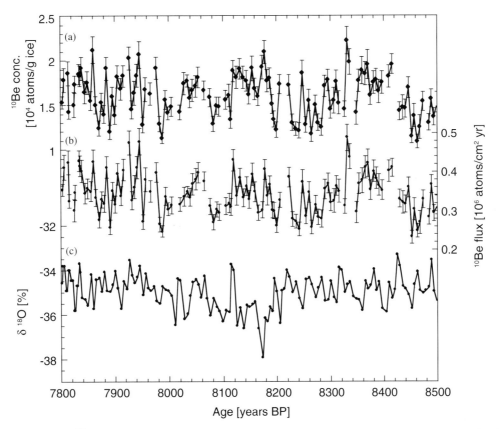

Fig. 1. ^{10}Be concentration (a) and ^{10}Be flux (b) measured in the GRIP ice core (Summit, Central Greenland). Panel c shows the climate parameter δ^{18}O measured in the GRIP ice core (Johnsen et al., 1997).

2001b). It is not a priori clear if this also applies for the Holocene period when only small variations in the accumulation rate are observed (Johnsen et al., 1995). Therefore, all the following calculations and conclusions are based on both ^{10}Be concentration and ^{10}Be flux that is shown in Fig. 1b. The ^{10}Be flux was calculated based on the accumulation rate published for the GRIP ice core (Johnsen et al., 1995). The data consist of 114 individual ^{10}Be measurements resulting in an average time resolution of 6.14 years. Few ^{10}Be measurements are missing mainly because of the lack of ice available for ^{10}Be measurements. Missing data are indicated by the absence of a line connecting neighbouring data points. Assuming that the ^{10}Be concentration or the ^{10}Be flux is proportional to the global ^{10}Be production, it is possible to calculate the past ^{14}C production around 8200 yr BP. This ^{14}C production rate allows us to model the atmospheric ^{14}C concentration using a carbon cycle model (Siegenthaler, 1983; Muscheler et al., 2004).

Fig. 2 shows the comparison of tree ring and ^{10}Be-based atmospheric ^{14}C concentration. The atmospheric ^{14}C concentration is expressed in terms of Δ^{14}C which is defined as the per mil deviation from the National Institute of Standards and Technology ^{14}C standard after correction for decay and fractionation (Stuiver and Polach, 1977). All curves in Fig. 2 are detrended by

removing the linear trend from the data from 8500 to 7800 yr BP. Panel a shows Δ^{14}C measured in tree rings. The time scale of these data is based on dendrochronology which has errors of less than 1 year during the Holocene (Friedrich et al., 1999). Panels b and c show the modelled Δ^{14}C which are based on the ^{10}Be concentration (b) and the ^{10}Be flux (c) assuming a stable carbon cycle. The indicated error bands are based on the uncertainties of the individual ^{10}Be measurements. For each modelled curve we applied 100 Monte-Carlo runs to infer the errors of the modelled Δ^{14}C.

The tree ring Δ^{14}C shows three main maxima during the period from 8500 to 7800 yr BP centered around 7900, 8100 and 8300 yr BP. The maximum around 8100 yr BP has a double-peak structure. The modelled Δ^{14}C basically exhibits the same structure. Even the double peak is clearly visible in the modelled Δ^{14}C in panel b. The differences between modelled Δ^{14}C based on the ^{10}Be concentration and the ^{10}Be flux are small. Only the maximum around 8100 yr BP is not so prominent in the modelled Δ^{14}C based on the ^{10}Be flux which is caused by the lower accumulation rate around 8200 yr BP. It is possible to identify the three main peaks in all curves. Because of the different geo-chemical behaviour of ^{14}C and ^{10}Be only the similar production can be responsible for these common changes. This

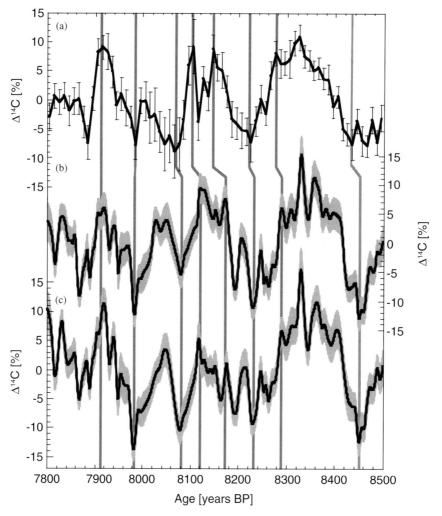

Fig. 2. Measured (tree ring) (Stuiver et al., 1998) (a) and modelled $\Delta^{14}C$ based on ^{10}Be concentrations (b) and ^{10}Be flux (c) from the GRIP ice core. The GRIP ss08c time scale can be synchronised with the absolute tree ring chronology by matching common features as indicated by the grey vertical lines. All records are detrended by removing the linear trend from the data for the period from 8500 to 7800 yr BP.

allows us to connect the GRIP time scale to the absolutely dated ^{14}C time scale. This comparison indicates that the GRIP time scale agrees with the tree ring chronology around 7900 yr BP. For the 8200 yr BP event it suggests that the GRIP ss08c time scale has to be shifted by approximately 20–30 years towards younger ages which is within the errors of this ice core time scale (Johnsen et al., 1999). The differences between modelled and measured $\Delta^{14}C$ as, for example, around 8350 yr BP complicate the matching when it comes to small $\Delta^{14}C$ wiggles. However, the overall structure of measured and modelled $\Delta^{14}C$ agrees very well. We estimate that this method allows us to synchronise the ice core with the tree ring time scale with uncertainties of less than 20 years during the period of interest.

Fig. 3 shows the comparison of measured and modelled $\Delta^{14}C$ after applying the time correction as indicated by the grey lines in Fig. 2. In most cases, measured and modelled $\Delta^{14}C$ agree within the indicated

$1 - \sigma$ errors. This confirms that the observed variations in both records are mainly due to production changes. Neither changes in the ^{10}Be transport or deposition at Summit nor changes in the carbon cycle seem to play a dominant role. Around 8200 yr BP the ^{10}Be concentration-based $\Delta^{14}C$ agrees better with the tree ring $\Delta^{14}C$ than the $\Delta^{14}C$ based on the ^{10}Be flux. This could indicate that for relatively stable climatic conditions as prevailing during the Holocene the ^{10}Be production is represented better by the ^{10}Be concentration than by the ^{10}Be flux. However, potential changes in the carbon cycle (see following section) could also be responsible for the differences around 8200 yr BP. Panel c in Fig. 3 shows $\delta^{18}O$ measured in the GRIP ice core. The minimum values in $\delta^{18}O$ in the GRIP ice core occur at 8150 yr BP. Fig. 3a indicates that using this method the 8200 yr BP cold event is very well dated in the GRIP ice core with an error smaller than 20 years.

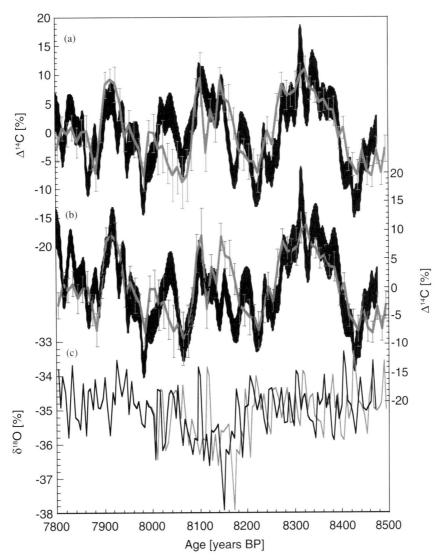

Fig. 3. Measured and modelled Δ^{14}C on the absolute tree ring time scale. Panel a shows the comparison of tree ring and modelled Δ^{14}C which is based on the ^{10}Be concentration in the GRIP ice core. Panel b shows the comparison with modelled Δ^{14}C based on the ^{10}Be flux to Summit. Panel c shows δ^{18}O measured in the GRIP ice core after adjustment of the ice core time scale (ss08c: grey curve, tree ring chronology: black curve).

Even though modelled and measured Δ^{14}C agree to a large extent there are some differences between these curves as for example around 8000 yr BP. To assign these relatively small differences to changes in the carbon cycle or to changes in the ^{10}Be transport or deposition is difficult. In the following chapter the likelihood is discussed that they are due to changes in the carbon cycle.

3. Changes in ocean ventilation around 8200 yr BP?

The 8200 yr BP cold event can be attributed to a freshwater perturbation which modified water and heat transport in the North Atlantic (Klitgaard-Kristensen et al., 1998; Barber et al., 1999). The spatial pattern of

climate change is similar to that of the Younger Dryas event which also points to a role for North Atlantic thermohaline circulation (Alley et al., 1997). This change could be associated with a reduced global deep-water formation leading to a decreased ^{14}C transport to the deep sea and, therefore, to an increased ^{14}C concentration in the atmosphere, biosphere and upper ocean. Here, we will address the question whether such an ocean circulation change associated with the 8200 yr BP cold event causes a detectable Δ^{14}C signal.

Fig. 4 shows the differences between ^{10}Be-based and measured Δ^{14}C derived from the data shown in Fig. 3. Around 8150 yr BP there is a minimum in the difference between ^{10}Be-based and measured Δ^{14}C. This discrepancy is more distinct in panel b where the difference between measured and modelled Δ^{14}C based on the ^{10}Be

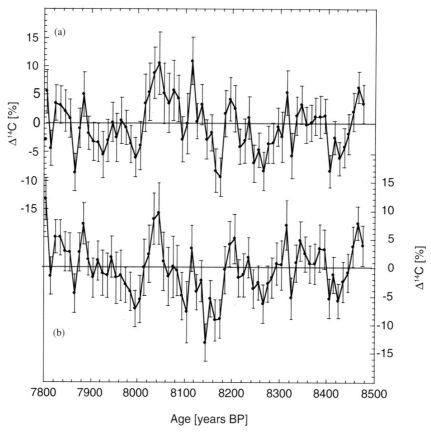

Fig. 4. Differences between tree ring and ^{10}Be-based Δ^{14}C. Panel a shows the discrepancies between measured and modelled Δ^{14}C based on the ^{10}Be concentration. Panel b shows this difference using the modelled Δ^{14}C based on the ^{10}Be flux.

flux is shown. Values lower than 0 mean that the measured atmospheric ^{14}C concentration is higher than indicated by the ^{10}Be-based production. This could be caused, for example, due to decreased deep-water formation. For this reason Fig. 4 is consistent with the scenario of a decreased oceanic heat transport to the North Atlantic and a consequently decreased deep-water formation. However, already Fig. 4 indicates that the evidence for this scenario based on the ^{10}Be and Δ^{14}C data is not very strong. There are other periods with similar differences in Δ^{14}C which show no apparent association with climate changes.

In order to estimate an upper limit of ocean-induced Δ^{14}C changes around 8200 yr BP, we can compare it to the much stronger climate change during the Younger Dryas cold event at the end of the last ice age. During the Younger Dryas a relatively short increase in Δ^{14}C is followed by a subsequent decrease (Hughen et al., 2000). To a large extent these changes can be explained by changes in ^{14}C production in combination with a decreased ocean circulation (Muscheler et al., 2000). The ^{10}Be-based ^{14}C production (Muscheler et al., 2000) as well as sophisticated carbon cycle models (Marchal et al., 2001) point to ventilation-related changes in Δ^{14}C in the order of 40‰ during the Younger Dryas. Using

the information from the Younger Dryas event we can obtain an upper limit of ocean-related Δ^{14}C changes around 8200 yr BP by comparing these two events. Fig. 5a shows the Younger Dryas and the 8200 yr BP climate change as expressed in the Summit δ^{18}O record (Johnsen et al., 1997). Panel b shows the ocean ventilation changes that we have to assume to be able to explain the Younger Dryas Δ^{14}C (Muscheler et al., 2000). For comparison, we assumed that the same change in ocean ventilation occurred around 8200 yr BP. Since the 8200 yr BP event is much shorter we expect a much smaller Δ^{14}C change. Neglecting variations in ^{14}C production we get a Δ^{14}C increase of about 40‰ for a reduction in ocean ventilation to 70% during 1200 years which corresponds to the length of the Younger Dryas (Johnsen et al., 1992). The same change lasting for only 150 years yields a Δ^{14}C increase of only 15‰. The reason for this smaller change is that it takes time until the equilibrium ^{14}C distribution according to the new ocean ventilation scheme is reached. Therefore, a short-term change in ocean ventilation is not able to produce very strong Δ^{14}C changes. In addition, the 30% reduction in ventilation most probably overestimates the global change in ocean ventilation around 8200 yr BP. Therefore, we do not expect to observe a Δ^{14}C

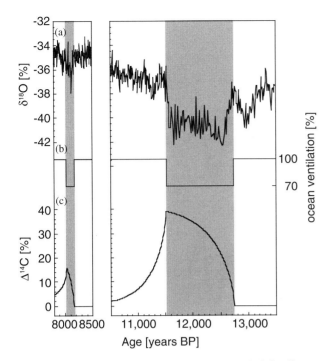

Fig. 5. Comparison of ocean ventilation changes and their effect on $\Delta^{14}C$ during the Younger Dryas and around 8200 yr BP. Panel a shows $\delta^{18}O$ measured in the GRIP ice core (Johnsen et al., 1997). Assuming a decrease in ocean ventilation of 30% (in our model expressed as a reduction in the diffusive exchange within the ocean (Siegenthaler, 1983)) (panel b) causes a $\Delta^{14}C$ increase of about 40‰ during the Younger Dryas and 15‰ around the 8200 yr BP cold event (panel c).

change larger than 15‰ due to ocean ventilation. To detect such an effect is difficult in view of the quoted uncertainties (Fig. 4). Another important point is that ocean ventilation changes and $\Delta^{14}C$ reaction do not exactly mirror each other. Our assumed ventilation changes are square functions but the $\Delta^{14}C$ reactions have different shapes. For example, directly after returning to the original ventilation (100%) at 11,500 and 8000 yr BP the atmospheric ^{14}C excursions are still present and they subsequently decrease in an almost exponential way. Therefore, the indicated $\Delta^{14}C$ difference shown in Fig. 4 cannot be directly transferred to ocean circulation changes.

Fig. 6 shows the ocean ventilation changes which would be required to reconcile tree ring and ^{10}Be-based $\Delta^{14}C$. The ocean ventilation was adjusted to account for the difference between modelled and measured $\Delta^{14}C$ (Muscheler et al., 2004). During each time step within each model run modelled $\Delta^{14}C$ and measured $\Delta^{14}C$ are compared. If ^{10}Be-based $\Delta^{14}C$ is higher compared to the tree ring data the ocean ventilation is increased and vice versa. As a result of the carbon cycle properties shown in Fig. 5 short-term $\Delta^{14}C$ differences cannot be explained by changes in the ocean ventilation. For example, a $\Delta^{14}C$ change of 10‰ within 10 years cannot be obtained with realistic ocean ventilation changes.

Therefore, the short-term variations in the ^{10}Be and $\Delta^{14}C$ data were removed by low-pass filtering the data (cut off frequency = 1/50 yr) before calculating the ocean ventilation. Each ventilation curve in Fig. 6 is based on 100 different Monte-Carlo calculations in order to include the errors in the ^{10}Be and $\Delta^{14}C$ data.

Fig. 6 shows that the difference between modelled and measured $\Delta^{14}C$ indeed points to a decreased ocean ventilation around the 8200 yr BP cold event. The biggest decrease in ocean ventilation is connected with the beginning of the 8200 yr BP cold event. However, the minimum in ocean ventilation occurs slightly before the minimum in $\delta^{18}O$. In addition, there are similar changes not associated with obvious climate deteriorations. Under the assumption that the GRIP ss08c time scale is correct, we would have to attribute the additional differences between ^{10}Be and $\Delta^{14}C$ completely to changes in the ^{10}Be or the radiocarbon system. One might suspect that in such a case we obtain stronger ventilation changes around 8150 yr BP. If we perform the same calculation as before using the GRIP ss08c time scale, we obtain in general stronger changes in oceanic ventilation. However, the minimum in oceanic ventilation around 8150 yr BP is not significantly different from the results shown in Fig. 6. This indicates that other processes have to be considered in this discussion. Changes in ^{10}Be transport or deposition could also explain the differences between ^{10}Be-based and tree ring $\Delta^{14}C$. For example, this is the most likely explanation for the differences around 8000 yr BP. As a consequence, the discrepancies are too small and the current knowledge of potential system effects is too limited to draw final conclusions. New high-resolution ^{10}Be records from other ice cores could help to reduce the uncertainties about climate-induced changes in the ^{10}Be system. With additional constrains for the reconstruction of past ^{14}C production changes, it might be possible to detect or confirm relatively small carbon cycle changes with this method in future.

4. Solar influence on climate around 8200 yr BP?

The solar irradiance is the most important parameter for the Earth's climate. Satellite-based measurements over the last 20 years show changes in solar irradiance over one 11-year sunspot cycle of about 0.1% which is assumed to be too small to cause significant climate changes. However, analysis of sunspot data and other solar proxies points to larger changes in solar activity on longer time scales (Lean et al., 1995; Beer et al., 2000). Furthermore, there is growing evidence that many grand minima (for example, Maunder Minimum) in solar activity can be related to climate changes. In particular, the comparison of radionuclide and climate records provides increasing evidence for an important influence

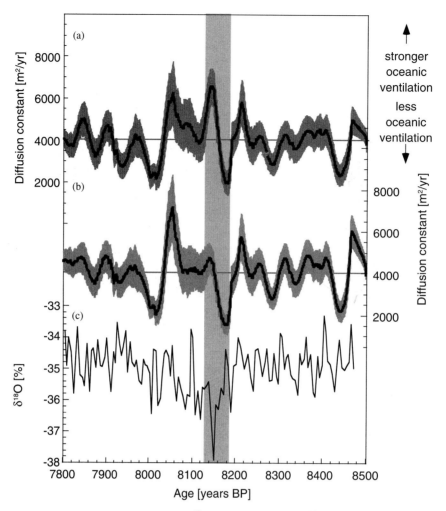

Fig. 6. Ocean ventilation changes as indicated by the comparison of ^{10}Be-based and tree ring Δ^{14}C. Panel a shows the variations in ocean ventilation which reconcile tree ring Δ^{14}C and Δ^{14}C based on the measured ^{10}Be concentration in the GRIP ice core. Panel b shows the inferred ventilation changes based on the ^{10}Be flux to Summit. Panel c shows the GRIP δ^{18}O record.

of solar activity changes on the climate during the Holocene (for example, Denton and Karlén, 1973; Bond et al., 2001).

In this context it is interesting to compare the strongest Holocene climate change recorded in the Greenland ice cores with a potential change in solar forcing. Since climate change (for example, δ^{18}O) and forcing (^{10}Be) is recorded in the same core, it is possible to study this relationship independent of any uncertainties related to the time scale. Fig. 7 shows δ^{18}O, Δ^{14}C, ^{10}Be concentration and flux on the same, the tree ring time scale. Already by looking at the raw data it is obvious that one of the most prominent solar cycles, the 207-year cycle (Damon and Sonett, 1991; Beer et al., 1994; Wagner et al., 2001a), is present in the radionuclide records from 8500 to 7800 yr BP. The ^{10}Be and Δ^{14}C records show maximum values around 8350, 8150 and 7950 yr BP. High ^{10}Be and Δ^{14}C values can be attributed to a lower solar shielding and consequently to

a lower solar activity. Fig. 7 shows that the 8200 yr BP cold event could be associated with decreased solar forcing. However, in this time interval there are again other periods with even stronger solar activity changes which do not exhibit an apparent correlation to changes in δ^{18}O. Therefore, there is no convincing evidence that solar forcing has caused the 8200 yr BP cold event. One could, however, speculate that the decrease in solar forcing around 8200 yr BP was triggering the climate change at 8200 yr BP. During an unstable climatic period it is probably much more likely that small changes in solar forcing can have a strong impact on climate by triggering the climatic changes. This could explain that around 8350 and 7950 yr BP when presumably the climatic preconditions were more stable we do not see a link between radionuclides and δ^{18}O in the GRIP ice core. The concept of stochastic resonance (Benzi et al., 1981) could also explain the non-linear reaction of the climate system to a small change in solar

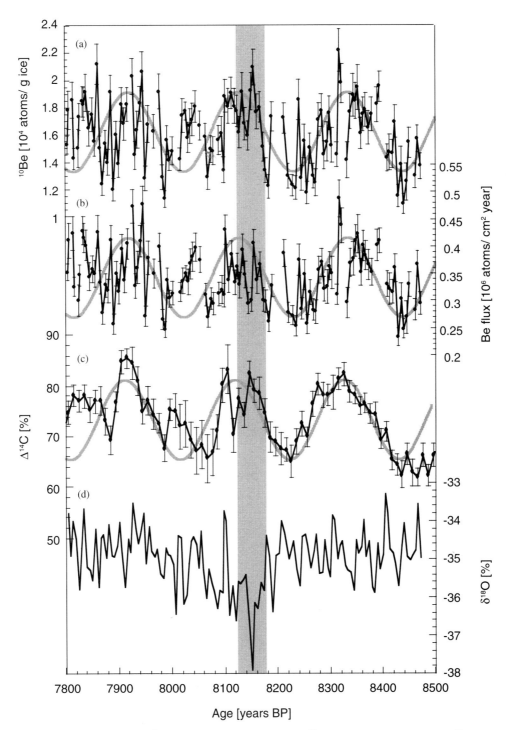

Fig. 7. Comparison of radionuclide records with $\delta^{18}O$ measured in the GRIP ice core. ^{10}Be concentration (panel a) and ^{10}Be flux (panel b) as well as tree ring $\Delta^{14}C$ (panel c) point to a decrease in solar activity at the beginning of the 8200 yr BP cold event. All records are plotted on the tree ring chronology. The sine waves (grey lines in panels a, b and c) illustrate the solar 207 yr cycle which is visible in the radionuclide records.

forcing. In combination with random internal variations a relatively weak external forcing could lift the climate system above a certain threshold which then could lead to a relatively strong change in climate (Benzi et al., 1982; Rahmstorf and Alley, 2002).

5. Conclusions

The comparison of ^{10}Be and $\Delta^{14}C$ yields crucial information on causes and timing of climate change. The example of the period around the 8200 yr BP cold event shows that it is possible to match the ice core time scales with the absolute dated tree ring time scale. This comparison indicates that the minimum $\delta^{18}O$ values in the GRIP ice core occurred at 8150 yr BP. Furthermore, the comparison of ^{10}Be and $\Delta^{14}C$ has the potential to infer carbon cycle changes and their connection to climate change. However, the changes in $\Delta^{14}C$ caused by decreased ocean ventilation are expected to be relatively small around 8200 yr BP. Taking further into account the uncertainties due to potential changes in ^{10}Be transport and deposition the expected effects on $\Delta^{14}C$ are within the uncertainties of the data. The 8200 yr BP cold event starts in a phase of decreasing solar forcing. The variable sun could have been the trigger to start the 8200 yr BP event but is most probably not the main cause for this climate deterioration.

Acknowledgments

We would like to thank Sigfus Johnsen for providing data and two anonymous reviewers for constructive comments. This work was supported by the Swiss National Science Foundation.

References

Alley, R.B., Finkel, R.C., Nishiizumi, K., Anandakrishnan, S., Shuman, C.A., Mershon, G., Zielinski, G.A., Mayewski, P.A., 1995. Changes in continental and sea-salt atmospheric loadings in central Greenland during the most recent deglaciation: model-based estimates. Journal of Glaciology 41, 503–514.

Alley, R.B., Mayewski, P.A., Sowers, T., Stuiver, M., Taylor, K.C., Clark, P.U., 1997. Holocene climatic instability—a prominent, widespread event 8200 yr ago. Geology 25, 483–486.

Barber, D.C., Dyke, A., Hillaire-Marcel, A.E., Jennings, A.E., Andrews, J.T., Kerwin, M.W., Bilodeau, G., McNeely, R., Southon, J., Morehaed, M.D., Gagnon, J.-M., 1999. Forcing of the cold event of 8200 years ago by catastrophic drainage of Laurentide lakes. Nature 400, 344–348.

Beer, J., Siegenthaler, U., Bonani, G., Finkel, R.C., Oeschger, H., Suter, M., Wölfli, W., 1988. Information on past solar activity and geomagnetism from ^{10}Be in the Camp Century ice core. Nature 331, 675–679.

Beer, J., Blinov, A., Bonani, G., Finkel, R.C., Hofmann, H.J., Lehmann, B., Oeschger, H., Sigg, A., Schwander, J., Staffelbach, T., Stauffer, B., Suter, M., Wölfli, W., 1990. Use of ^{10}Be in polar ice to trace the 11-year cycle of solar activity. Nature 347, 164–166.

Beer, J., Joos, C.F., Lukasczyk, C., Mende, W., Siegenthaler, U., Stellmacher, R., Suter, M., 1994. ^{10}Be as an indicator of solar variability and climate. In: Nesme-Ribes, E. (Ed.), ^{10}Be as an Indicator of Solar Variability and Climate. Springer, Berlin, pp. 221–233.

Beer, J., Mende, W., Stellmacher, R., 2000. The role of the Sun in climate forcing. Quaternary Science Reviews 19, 403–415.

Beer, J., Muscheler, R., Wagner, G., Laj, C., Kissel, C., Kubik, P.W., Synal, H.-A., 2002. Cosmogenic nuclides during Isotope Stages 2 and 3. Quaternary Science Reviews 21, 1129–1139.

Benzi, R., Sutera, A., Vuliani, A., 1981. The mechanism of stochastic resonance. Journal of Physics A 14, L453–L457.

Benzi, R., Parisi, G., Sutera, A., Vulpiani, A., 1982. Stochastic resonance in climatic change. Tellus 34, 10–16.

Bond, G., Kromer, B., Beer, J., Muscheler, R., Evans, M.N., Showers, W., Hoffmann, S., Lotti-Bond, R., Hajdas, I., Bonani, G., 2001. Persistant solar influence on North Atlantic climate during the Holocene. Science 294, 2130–2136.

Damon, P.E., Sonett, C.P., 1991. Solar and terrestrial components of the atmospheric C-14 variation spectrum. In: Sonett, C.P., Giampapa, M.S., Matthews, M.S. (Eds.), Solar and Terrestrial Components of the Atmospheric C-14 Variation Spectrum. The University of Arizona, Tucson, pp. 360–388.

Denton, G.H., Karlén, W., 1973. Holocene climatic variations—their pattern and possible cause. Quaternary Research 3, 155–205.

Finkel, R.C., Nishiizumi, K., 1997. Beryllium-10 concentrations in the Greenland Ice Sheet Project 2 ice core from 3–40 ka. Journal of Geophysical Research 102, 26,699–26,706.

Friedrich, M., Kromer, B., Spurk, M., Hofmann, J., Kaiser, K.F., 1999. Paleo-environment and radiocarbon calibration as derived from Lateglacial/Early Holocene tree-ring chronologies. Quaternary International 61, 27–39.

Hughen, K.A., Southon, J.R., Lehmann, S.J., Overpeck, J.T., 2000. Synchronous radiocarbon and climate shifts during the last deglaciation. Science 290, 1951–1954.

Johnsen, S.J., Clausen, H.B., Dansgaard, W., Fuhrer, K., Gundestrup, N., Hammer, C.U., Iversen, P., Jouzel, J., Stauffer, B., Steffensen, J.P., 1992. Irregular glacial interstadials recorded in a new Greenland ice core. Nature 359, 311–313.

Johnsen, S.J., Dahl-Jensen, D., Dansgaard, W., Gundestrup, N., 1995. Greenland palaeotemperatures derived from GRIP bore hole temperature and ice core isotope profiles. Tellus 47B, 624–629.

Johnsen, S.J., Clausen, H.B., Dansgaard, W., Gundestrup, N.S., Hammer, C.U., Andersen, U., Andersen, K.K., Hvidberg, C.S., Dahl-Jensen, D., Steffensen, J.P., Shoji, H., Sveinbjörnsdóttir, Á.E., White, J., Jouzel, J., Fisher, D., 1997. The $\delta^{18}O$ record along the Greenland Ice Core Project deep ice core and the problem of possible Eemian climatic instability. Journal of Geophysical Research 102, 26,397–26,410.

Johnsen, S.J., Clausen, H.B., Jouzel, J., Schwander, J., Sveinbjörnsdóttir, Á.E., White, J., 1999. Stable isotope records from Greenland Deep Ice Cores: the climate signal and the role of diffusion. In: Wettlaufer, J.S., Dash, J.G., Untersterner, N. (Eds.), Ice Physics and the Natural Environment. Springer, Berlin, pp. 89–107.

Klitgaard-Kristensen, D., Sejrup, H.P., Haflidason, H., Johnsen, S.J., Spurk, M., 1998. A regional 8200 cal. yr BP cooling event in northwest Europe, induced by final stages of the Lautentide ice-sheet deglaciation. Journal of Quaternary Science 13, 165–169.

Lal, D., Peters, B., 1967. Cosmic ray produced radioactivity on the Earth. In: Flügge, S. (Ed.), Cosmic Ray Produced Radioactivity on the Earth. Springer, Berlin, 46/2, pp. 551–612.

Lean, J., Beer, J., Bradley, R., 1995. Reconstruction of solar irradiance since 1610: implications for climate change. Geophysical Research Letters 22, 3195–3198.

Marchal, O., Stocker, T.F., Muscheler, R., 2001. Atmospheric radiocarbon during the Younger Dryas: production, ventilation, or both? Earth and Planetary Science Letters 185, 383–395.

Masarik, J., Beer, J., 1999. Simulation of particle fluxes and cosmogenic nuclide production in the Earth's atmosphere. Journal of Geophysical Research 104, 12,099–12,111.

Muscheler, R., Beer, J., Wagner, G., Finkel, R.C., 2000. Changes in deep-water formation during the Younger Dryas cold period inferred from a comparison of ^{10}Be and ^{14}C records. Nature 408, 567–570.

Muscheler, R., Beer, J., Wagner, G., Laj, C., Kissel, C., Raisbeck, G.M., Yiou, F., Kubik, P.W., 2004. Changes in the carbon cycle during the last deglaciation as indicated by the comparison of ^{10}Be and ^{14}C records. Earth and Planetary Science Letters 219, 325–340.

Rahmstorf, S., Alley, R.B., 2002. Stochastic resonance in glacial climate. Eos 83, 129–135.

Raisbeck, G.M., Yiou, F., Fruneau, M., Loiseaux, J.M., Lieuvin, M., Ravel, J.C., 1981. Cosmogenic $^{10}Be/^{7}Be$ as a probe of atmospheric transport processes. Geophysical Research Letters 8, 1015–1018.

Siegenthaler, U., 1983. Uptake of excess CO_2 by an outcrop-diffusion model ocean. Journal of Geophysical Research 88, 3599–3608.

Siegenthaler, U., Beer, J., 1988. Model comparison of ^{14}C and ^{10}Be isotope records. In: Stephenson, F.R., Wolfendale, W. (Eds.), Secular Solar and Geomagnetic Variations in the Last 10,000 Years. Kluwer Academic Publishers, Durham, 1987, pp. 315–328.

Siegenthaler, U., Heimann, M., Oeschger, H., 1980. ^{14}C variations caused by changes in the global carbon cycle. Radiocarbon 22, 177–191.

Snowball, I., Zillén, L., Gaillard, M.-J., 2002. Rapid early-Holocene environmental changes in northern Sweden based on studies of two varved lake-sediment sequences. Holocene 12, 7–16.

Stuiver, M., Braziunas, T.F., 1993. Sun, ocean, climate and atmospheric $^{14}CO_2$, an evaluation of causal and spectral relationships. The Holocene 3, 289–305.

Stuiver, M., Polach, H.A., 1977. Discussion reporting of ^{14}C data. Radiocarbon 19, 355–363.

Stuiver, M., Reimer, P.J., Bard, E., Beck, J.W., Burr, G.S., Hughen, K.A., Kromer, B., McCormac, G., Van der Pflicht, J., Spurk, M., 1998. INTCAL98 radiocarbon age calibration, 24,000-0 cal BP. Radiocarbon 40, 1041–1083.

Wagner, G., Beer, J., Laj, C., Kissel, C., Masarik, J., Muscheler, R., Synal, H.-A., 2000. Chlorine-36 evidence for the Mono Lake event in the Summit GRIP ice core. Earth and Planetary Science Letters 181, 1–6.

Wagner, G., Beer, J., Kubik, P.W., Laj, C., Masarik, J., Mende, W., Muscheler, R., Raisbeck, G.M., Yiou, F., 2001a. Presence of the solar de Vries cycle (205 years) during the last ice age. Geophysical Research Letters 28, 303–306.

Wagner, G., Laj, C., Beer, J., Kissel, C., Muscheler, R., Masarik, J., Synal, H.-A., 2001b. Reconstruction of the paleoaccumulation rate of central Greenland during the last 75 kyr using the cosmogenic radionuclides ^{36}Cl and ^{10}Be and geomagnetic field intensity data. Earth and Planetary Science Letters 193, 515–521.

Yiou, F., Raisbeck, G.M., Baumgartner, S., Beer, J., Hammer, C., Johnsen, S., Jouzel, J., Kubik, P.W., Lestringuez, J., Stiévenard, M., Suter, M., Yiou, P., 1997. Beryllium-10 in the Greenland Ice Core Project ice core at Summit, Greenland. Journal of Geophysical Research 102, 26783–26794.

ELSEVIER

Quaternary Science Reviews 23 (2004) 2113–2126

QSR

Sea surface temperatures and ice rafting in the Holocene North Atlantic: climate influences on northern Europe and Greenland

Matthias Moros[a,*], Kay Emeis[b], Bjørg Risebrobakken[a], Ian Snowball[c], Antoon Kuijpers[d], Jerry McManus[e], Eystein Jansen[a]

[a]Bjerknes Centre for Climate Research, University of Bergen, Allégaten 55, 5007 Bergen, Norway
[b]Baltic Sea Research Institute, Seestrasse 15, 18119 Rostock, Germany
[c]Department of Geology, GeoBiosphere Science Centre, Quaternary Sciences, University of Lund, Sölvegatan 13SE-223 62 Lund, Sweden
[d]Geological Survey of Denmark and Greenland, Øster Voldgade 10, 1350 Copenhagen K, Denmark
[e]Department of Geology and Geophysics, Woods Hole Oceanographic Institution, Clark 121, Woods Hole MA02543-1050, USA

Abstract

The oceanographic conditions in the high-latitude North Atlantic ocean during the Holocene were reconstructed through analyses of sea surface temperature (SST; alkenone unsaturation ratios) and ice rafting (mineralogy and grain size) from two sediment sequences, one recovered from the Reykjanes Ridge at 59°N and the other from the Norwegian Sea at 68°N. Comparison of our records to published ice core and terrestrial proxy-climate data sets suggests that atmospheric temperature changes over Northern Europe and Greenland were coupled to SST variability and ice rafting. The records outline four major climatic phases: (i) an early-Holocene Thermal Maximum that lasted until approximately 6.7 kyr BP, (ii) a distinctly cooler phase associated with increased ice rafting between 6.5 and 3.7 kyr BP, (iii) a transition to generally warmer, but relatively unstable climate conditions between 3.7 and 2 kyr BP and (iv) a second distinct SST decline that took place between 2 and 0.5 kyr BP. In contrast to the dominant control of Northern Hemisphere summer insolation on early-Holocene climate development (via strong seasonality), the trigger for the onset of relatively unstable climatic conditions in the North Atlantic at 3.7 kyr BP is not straightforward. However, it is possible that this change was triggered by late-Holocene winter insolation increase at high northern latitude and/or by inter-hemispheric changes in orbital forcing. The late-Holocene Neoglaciation trend, which is characteristic of numerous terrestrial archives in northern Europe, may not only be attributed to a gradual decrease in orbitally forced summer temperature, but also to increase snow precipitation at high northern latitudes during generally milder winters.
© 2004 Elsevier Ltd. All rights reserved.

1. Introduction

Globally distributed multi-proxy palaeoclimate data sets indicate that Holocene climate was more variable than commonly anticipated. Some investigations point to a cyclicity of about 1.5 kyr for this variability (Campbell et al., 1998; Bond et al., 2001; Viau et al., 2002; Andrews and Giraudeau, 2003), which may represent the imprint of a pervasive, albeit elusive, climate mechanism that also operate throughout the last glacial cycle. Although the existence of a common millennial scale periodicity during the Holocene period has been disputed (e.g. Schulz and Paul, 2002; Risebrobakken et al., 2003), there is no doubt that climate change was a frequent characteristic of the Holocene.

The climate in northern Europe in particular is strongly influenced by ocean/atmosphere processes, i.e. by changes in the northward heat and moisture transport to the North Atlantic. Temperature fluctuations of the high-latitude surface layer of the North Atlantic have been reconstructed using transfer functions derived from diatom (e.g. Koc and Jansen, 1994; Birks and Koc, 2002; Andersen et al., 2004), radiolarian

*Corresponding author. Tel.: +47-555-89829; fax: +47-555-84330.
E-mail address: matthias.moros@bjerknes.uib.no (M. Moros).

0277-3791/$ - see front matter © 2004 Elsevier Ltd. All rights reserved.
doi:10.1016/j.quascirev.2004.08.003

(Dolven et al., 2002) and planktonic foraminiferal assemblages (Andersson et al., 2003; Risebrobakken et al., 2003; Sarnthein et al., 2003). Further sea surface temperature (SST) reconstructions were based on records of alkenone unsaturation ratios (e.g. Calvo et al., 2002), on changes in the isotopic composition of planktonic foraminifers (Duplessy et al., 2001; Risebrobakken et al., 2003; Sarnthein et al., 2003) and coccolith assemblages (Giraudeau et al., 2000; Andrews and Giraudeau, 2003). However, the results from many SST proxies differ significantly, for reasons that are still not fully understood (Marchal et al., 2002). It must also be recognized that the reconstructions of Holocene climatic changes documented in marine and terrestrial archives have been restricted by the fact that most biological proxies reflect ambient growth conditions during the summer months, with the result that winter conditions remain poorly constrained.

Additional information about changing surface water conditions in the North Atlantic can be provided by proxies of ice-rafting (in the form of ice-rafted debris, IRD). However, already published patterns from the sparse Holocene ice-rafting records available (e.g. Bond et al., 2001; Jennings et al., 2002) differ significantly.

Despite the absence of unambiguous data, it is widely agreed that the decrease in summer insolation at northern high latitudes drove a pervasive Holocene cooling trend in the North Atlantic region (Koc and Jansen, 1994; Marchal et al., 2002; Snowball et al., in press). However, many features of Holocene marine (e.g. Giraudeau et al., 2000), terrestrial (e.g. Davis et al., 2003), and ice-core temperature reconstructions (e.g. Dahl-Jensen et al., 1998; Alley et al., 1999) cannot be explained solely by this summer insolation reduction.

Based on new data from the North Atlantic (the Reykjanes Ridge—alkenone-derived SST's; and Norwegian Sea—ice-rafting proxy data), this paper presents evidence for significant long-term SST and ice-rafting fluctuations that occurred coherently over the Reykjanes Ridge at 59°N and the Norwegian Sea, and that were linked to fluctuations observed in published climate records over Scandinavia and parts of Greenland.

1.1. Oceanographic settings

The two coring sites (Fig. 1) underlie the pole-ward moving relatively warm, saline surface waters of the North Atlantic Current (NAC) that enters the Nordic Seas east of Iceland. The surface ocean circulation system in the Nordic Seas consists of three main currents: the warm NAC and two cold water pathways, the East Greenland (EGC) and the East Icelandic Current (EIC). The northeast southwest trending Arctic Front (maximum extent of winter drift ice) separates the warm from the cold, seasonally ice-carrying waters (Fig. 1). The coring site in the Norwegian Sea is located within the recent core of the NAC. Here, the temperature of the NAC is typically varying around 10 °C. The Reykjanes Ridge site is influenced by the north-westward moving Irminger Current, which is also a branch of the warm NAC. Recent summer SST's range from 9 to 12 °C (Wright and Worthington, 1970). The coring site is located near the Sub-Arctic Front in the central North Atlantic, which is not as sharply developed as its equivalent in the Nordic Seas.

2. Material and methods

The Norwegian Sea IMAGES piston core MD95-2011 was taken at 66°58.19N, 07°38.36E in 1048 m water depth. At the Reykjanes Ridge site LO09-14 (59°N 31°W; 1700 m water depth), a large box core (LBC; 0.4 m long; Ø 40 cm), a giant gravity (GGC; 2.7 m; Ø 30 cm), and a gravity core (GC; 5.6 m; Ø 12.5 cm) were taken in order to provide a high-resolution composite record.

2.1. Chronology

Based on correlation and overlapping of distinct features in records of alkenone-derived SST, magnetic susceptibility, calcite concentrations etc., and calibrated AMS^{14}C datings, we established a 6 m Holocene composite core record and timescale for LO09-14. For age control, a total of 34 AMS^{14}C datings were made on *Globigerina bulloides* at the accelerator dating facilities at Kiel University, Aarhus University and Woods Hole Oceanographic Institution (Table 1). The ^{14}C dates were calibrated to calendar years before present by using OxCal Programme v3.8 with Stuiver and Reimer marine calibration table (Bronk Ramsey, 1995, 2001; Stuiver et al., 1998). For the conversion from depth to time domain we used polynomial fits for the GGC and GC (Table 1), and a linear interpolation for the LBC-core between the AMS^{14}C dates. When comparing the results from AMS^{14}C dating of the gravity cores LO09-14GGC (Table 1) we found a loss/squeezing of surface sediment due to the coring process. The LO09-14 composite core thus has a low-resolution interval between about 2.2 cal kyr BP (regression line datum, GGC) and about 1.2 cal kyr BP (LBC) with only four SST data points. The chronology of the IMAGES core MD95-2011 from the Norwegian Sea was taken from Risebrobakken et al. (2003).

2.2. Alkenone analyses

Long-chain alkenones were extracted from aliquots of freeze-dried and homogenized sediment (1–5 g using a Dionex *A*ccelerated *S*olvent *E*xtractor (ASE 200). Each LO09-14 sample was extracted twice for 10 min using

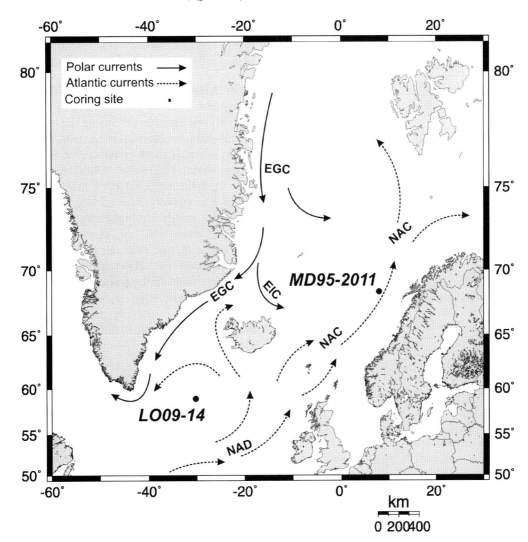

Fig. 1. Core sites, surface water circulation (EGC—cold East Greenland Current; NAC—warm North Atlantic Current) and frontal systems (AF—Arctic Front, SAF—Sub-Arctic Front) in the North Atlantic.

25 ml dichloromethane at a temperature of 75 °C and pressure of 80 bar of nitrogen. The resulting extracts were concentrated to approximately 50 μl with a rotary evaporator and dried with nitrogen gas. After drying the residues were dissolved in 100 μl of hexane. Alkenones were analysed by capillary gas chromatography with a Fisions 8000 series gas chromatograph equipped with an autosampler AS2000, a 30 m fused silica column (DB 5 HT; 0.32 mm i.d. and 0.1 μm film thickness) and a fused silica pre-capillary (5 m long, 0.53 mm internal diameter), on-column injection, and a flame ionization detector (FID). The carrier gas was hydrogen. The oven temperature programme was from 45 °C to 300 °C at 15 °C min^{-1} (isothermal at 300 °C for 15 min), then to 330 °C at 15 °C min^{-1} (isothermal at 330 °C for 10 min). The injector was set at 45 °C and the detector at 330 °C. Quantification of C_{37}, C_{38} and C_{39} was achieved with internal standards (cholestane, n-C_{36}, non-adecanone added before extraction) and detector response to external standards. The analytical precision based on

multiple extraction of sediment samples was better than 0.02 units (or 0.6 °C) for the $U_{37}^{K'}$ index. We calculated the SST from $U_{37}^{K'}$ according to Müller et al. (1998):

$$U_{37}^{K'} = 0.033T + 0.044.$$

2.3. Sedimentological analyses

Fluctuations in ice rafting in the Norwegian Sea (MD95-2011) were reconstructed using the mineralogical quartz-to-plagioclase (Q/Plag) ratio (Moros et al., 2004). The bulk Q/Plag ratio was determined from X-ray diffractometry (XRD) measurements at the Baltic Sea Research Institute. Calcite contents (not presented here) were also estimated from XRD measurements performed on bulk material and the fraction <63 μm. Sand contents of the size fractions >63 μm (wt% >63 μm) and >150 μm (wt% >150 μm) of MD95-2011 were determined by wet sieving of bulk sediment.

Table 1
AMS[14]C dating results on the Reykjanes Ridge LO09-14 LBC (large box core), LO09-14GGC (giant gravity core) and LO09-14GC (gravity core) plus polynomial fits. (Radiocarbon dates and calibrated ages)

Core name/ Depth (cm)	Lab no.	[14]C age ± standard deviation	Calibrated age (Cal. yr BP)
LO09-14LBC			
0	AAR-5049	705±40	310
2	OS-32526	785±35	425
9	OS-32140	1080±25	643
29	OS-32475	1250±65	795
39	OS-32474	1560±30	1112
LO09-14GGC			
0.5	AAR-6671	1190±35	*Not included*
9	OS-32477	2770±45	2499
17	KIA7500	2495±25	2151
30	OS-32478	2990±35	2760
47	AAR-6437	2690±40	2384
69	OS-32524	3030±35	2793
79	OS-32525	2860±35	2627
101	KIA	3165±35	2945
120	OS-32696	3260±45	3081
150	KIA7501	4250±35	4346
163	OS-32697	4680±50	4907
197	OS	5220±50	5577
240	KIA7502	6440±35	6913
276	KIA	7180±40	7636
LO09-14GC			
51	KIA7497	5330±50	5689
103	AAR-4454	6620±80	7125
135	KIA20793	6920±40	7450
159	KIA7498	6605±40	*Not included*
165	KIA20794	7435±40	7860
185	OS-32681	7730±35	8190
216	KIA7499	8005±40	8457
243	OS-32690	8420±40	8926
270	KIA20795	8385±45	8920
276	AAR-5050	8145±45	*Not included*
300	AAR-4455	8790±100	9324
303	OS-32691	9050±40	9653
315	OS-32692	9040±45	9638
333	OS-32693	9220±45	9911
370	OS-32694	9350±40	10029
394	AAR-5051	9310±55	10003
437	OS-32695	9920±55	10705
447	AAR-4456	10410±90	11366

OS = Woods Hole Oceanographic Institute; AAR = Aarhus University; KIA = Kiel.Polynomial fits for the conversion from core depth (cm) to time domain are: (i) *LO09-14GGC*: cal kyr BP = $2474 - 2.9*cm + 0.0222*cm^2 + 0.00084*cm^3 - 2.31\text{E-}006*cm^4$, $R^2 = 0.987$ and (ii) *LO09-14GC*: cal kyr BP = $4221 + 33.95*cm - 0.085*cm^2 + 9.61\text{E-}005*cm^3$, $R^2 = 0.985$.

3. Results

3.1. Fluctuations in Holocene ice rafting in the Norwegian Sea

The XRD and grain size ice-rafting proxy data in MD95-2011 indicate a trisection of the Holocene ice-rafting history when looking at the deviations from average levels of Q/Plag ration and wt% > 63 μm (Fig. 2): (i) an early-Holocene interval of relatively low input by ice-rafting punctuated by high input during a short period at 8.2 kyr BP, (ii) a mid-Holocene increase to highest baseline values between 6.5 and 3.7 kyr BP and (iii) low input levels during the latest Holocene. The MD95-2011 wt% > 63 μm record resembles the XRD curve, whereas there is no correlation to the > 150 μm sieving data. The wt% > 63 μm varies between 2 and 10% and the percentage of material > 150 μm is generally low (lower than 0.5%). Superimposed on the long-term trends are higher-frequency oscillations with the "8.2 event" as the most striking feature in all granulometric and mineralogical data. The *G. quinqueloba* contents of MD95-2011 (Risebrobakken et al., 2003) also indicate the trisection of the Holocene seen in the XRD data (Fig. 2). Maximum *G. quinqueloba* contents are found in the mid-Holocene period, bracketed by low values in the early- and late-Holocene intervals.

3.2. Holocene SST at the Reykjanes Ridge and in the Norwegian Sea

Fig. 3 depicts our new alkenone-derived SST's in LO09-14, and published planktonic oxygen isotope data on *Neogloboquadrina pachyderma* sin. (NPS; Risebrobakken et al., 2003) and diatom-based SST estimates (Birks and Koc, 2002; Andersen et al., 2004) of MD95-2011. The NPS $\delta^{18}O$, corrected for the global ice volume effect, can be interpreted as a temperature signal based on the following arguments: The MD95-2011 coring position is in the centre of the North Atlantic inflow in the Nordic Seas with a salinity of approx. 35.2 PSU. The site is not influenced by coastal water, and changes in salinity will be due to inmixing of colder Arctic water from the West with a salinity of 34.9 PSU. The modern salinity range of Atlantic water is a few tenths of a PSU, and the influence this has on foraminiferal $\delta^{18}O$ is 0.2‰ at most, i.e. equal to less than 1 °C if calculated as a temperature signal. If larger salinity changes were to have occurred in the Holocene, they would be due to inmixing of waters with lower salinity, but colder temperatures. Thus the $\delta^{18}O$ change due to salinity will be reduced or cancelled, due to temperature. The foraminiferal and diatom evidence clearly document dominating Atlantic waters over the location for the entire Holocene, and salinity changes were probably not more than a few tenths of a PSU. Thus the planktonic $\delta^{18}O$ record should to a large extent be regarded as a temperature proxy. This is also supported by the strong resemblance between the $\delta^{18}O$ record and the % polar foraminifers (Andersson et al., 2003; Risebrobakken et al., 2003). In addition, salinity changes might have been occurred rather during the mid-Holocene where an increase in ice rafting and a shift in

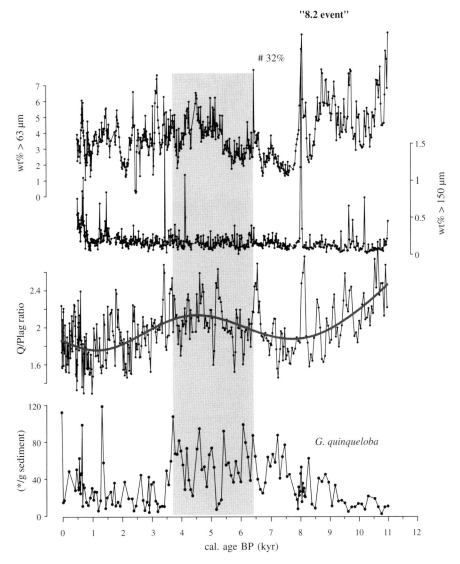

Fig. 2. Holocene fluctuations in ice rafting reconstructed by measuring the quartz-to-plagioclase ratio using XRD in MD95-2011 from the Norwegian Sea. Additionally, MD95-2011 sieving and *G. quinqueloba* content (Risebrobakken et al., 2003) are shown. Note the similar long-term trends in the foraminiferal, XRD and wt% > 63 μm data. The "8.2 event" is a marked feature in all lithological records. The grey area indicates the mid-Holocene phase with increased ice rafting and higher *G. quinqueloba* contents.

the polar front are observed (see below, Fig. 2) than during the late Holocene, where pronounced NPS $\delta^{18}O$ fluctuations are recorded.

The alkenone-based SST variations of LO09-14 have an amplitude of 2.5 °C (between ca 12.5 and 10 °C) throughout the Holocene. Similarly, several SST proxies measured on MD95-2011 gave Holocene surface ocean temperature changes in the order of about 2 °C for alkenone-derived (Calvo et al., 2002), of 2.5 °C for radiolarian-based (Dolven et al., 2002) and of 3 °C for foraminifer-based (Andersson et al., 2003; Risebrobak-ken et al., 2003) SST reconstructions. A more than 5 °C Holocene temperature range is, however, recorded by the diatom-based reconstruction (Fig. 3). The LO09-14 alkenone-derived SST record outlines four phases (Fig. 3): (i) an early-Holocene Thermal Maximum

(HTM) that lasted until 6.7 kyr BP, (ii) a distinctly cooler phase between 6.5 and 3.7 kyr BP, (iii) a phase of highly variable SST between 3.7 and 2 kyr BP and (iv) a distinct SST decline between 2 and 0.5 kyr BP.

In MD95-2011 an early HTM is clearly developed in the diatom-based SST reconstruction, whereas oxygen isotope data do not show this early HTM (Fig. 3). In addition, similar to the diatom-based data an early HTM is recorded in the alkenone-derived SST reconstruction (Calvo et al., 2002), whereas MD95-2011 foraminifer abundance (Risebrobakken et al., 2003) as well as radiolarian-derived SST's (Dolven et al., 2002) do not display the early HTM. Very pronounced in the NPS $\delta^{18}O$ and also the radiolarian-derived SST's (Dolven et al., 2002) is the "8.2 event", but it is not seen in the diatom and alkenone-based SST data (Fig. 3).

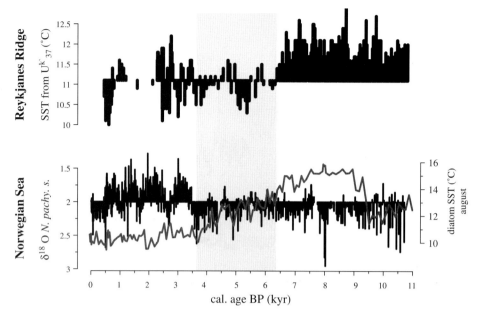

Fig. 3. Holocene temperature fluctuations at the Reykjanes Ridge 59°N and Norwegian Sea diatom-derived SST's (Andersen et al., 2004; Birks and Koc, 2002) plus $\delta^{18}O$ *N. pachyderma* sin. (Risebrobakken et al., 2003) data. Note an early HTM is shown by the diatom and alkenone data but not by the oxygen isotope data. The latter shows an optimum during the late Holocene which is also observed in the Reykjanes Ridge alkenone SST reconstruction. The grey area indicates a phase with generally lower SST's and increased ice rafting during the mid-Holocene.

The mid-Holocene cool interval evident in the LO09-14 alkenone record is in the diatom (Fig. 3) and alkenone-based SST data sets of MD95-2011 similarly characterized by decreasing temperatures, whereas it is not seen in NPS $\delta^{18}O$. For the late-Holocene interval from 3.7 to 2 kyr BP the records display distinct differences in the proxy data: Uniformly relatively low temperatures are indicated by diatom (Fig. 3) and alkenone-derived SST in MD95-2011, whereas the alkenone-based SST in LO09-14 indicate warmer but strongly varying SST's around 11 °C with rapid warm/cold fluctuations of ± 1 °C. At the same time, warmer temperatures are indicated by the NPS oxygen isotopes (Fig. 3). The cooling trend from 2 kyr BP is indicated by NPS $\delta^{18}O$ in the Norwegian Sea, and LO09-14 alkenone SST estimates, whereas diatom-based SST estimates remain uniformly cool.

4. Discussion

4.1. Holocene variations in ice rafting in the northern North Atlantic

The long-term features of our MD95-2011 bulk mineralogical record compare well with the overall trends in Bond et al. (2001) ice-rafting data, which were reconstructed by counting hematite stained grains (Fig. 4): generally lower ice rafting is recorded in the early and late Holocene interrupted by a phase of increased ice rafting during the mid-Holocene time

interval. Similar general ice-rafting trends have been observed also on the Barents Sea shelf (M. Moros, unpublished data) and in the van Mijenfjord of Svalbard (Hald et al., 2004). The long-term trends in Bond et al. (2001) and the MD95-2011 XRD results are also identifiable in the *G. quinqueloba* content data. They indicate an Eastward shift in the Arctic frontal system towards the Norwegian Sea coring site from the early to the mid-Holocene and a movement back to a more distal position during the late Holocene. It is likely that lithic and biogenic parameters, which are independent of each other, reflect coherent changes in the ocean surface and may indicate temperature fluctuations over the coring site: relatively warmer surface ocean temperatures in the early and late Holocene prevented ice from reaching the site, whereas the mid-Holocene interval was cooler and saw ice at the coring locations.

However, when focussing on the centennial to millennial scale events there are pronounced differences between our bulk Q/Plag ratio proxy and Bond et al. (2001) data. For example, the cold "8.2 event", which is believed to be the most pronounced climatic anomaly of the Holocene in the Northern Hemisphere, is clearly documented as an ice-rafting peak in the Q/Plag ratio data. However, it is not evident in the hematite stained grain record. Further support for our XRD results comes from granulometric data where the "8.2 event" is also a striking feature in the wt% > 63 μm as well as > 150 μm data. It is worth noting, that the Q/Plag ratios (plus the *G. quinqueloba* record) generally

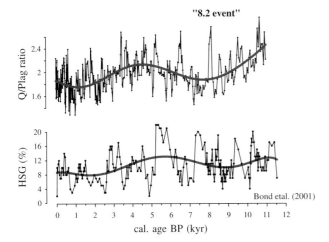

Fig. 4. Holocene fluctuations in ice rafting reconstructed by counting hematite stained grains (HSG, Bond et al., 2001) and by measuring the quartz-to-plagioclase ratio using XRD in MD95-2011 from the Norwegian Sea. Note the similar long-term trends in both records but also the differences on millennial time scale.

correlate with the wt% > 63 μm data, but not with the > 150 μm data. XRD analyses (not shown here) reveal that the wt% > 63 μm and > 150 μm records are not strongly influenced by biogenic calcitic material, hence the sieving data clearly reflect changes in the lithic grain size fractions. The calcite content of the bulk material is generally low (max. 20%) and changes that occur, and the maximum contents found are due to biogenic material in the grain size fraction < 63 μm. Obviously, bulk mineralogical (provenance) data or finer fraction signals may provide a more accurate picture of ice rafting during warmer episodes, like the Holocene, as the main contributor to ice rafting might be sea ice and not iceberg drifts (Dowdeswell et al., 1998; Funder et al., 1998).

4.2. Long-term Holocene SST changes in the Norwegian Sea and at the Reykjanes Ridge, and comparison with other marine and terrestrial records

4.2.1. Interpretation of the SST proxies

As mentioned above, there are significant differences between the SST proxies. The planktonic $\delta^{18}O$ shows a rather flat early- to mid-Holocene and a marked increase in amplitude and decrease in mean values from about 4 kyr BP. This points to both higher near surface temperature variations and a warming of the subsurface waters after 3.7 kyr BP. In contrast, the diatom (Birks and Koc, 2002; Andersson et al., 2003) and alkenone-(Calvo et al., 2002) derived SST data of MD95-2011 and LO09-14 display a temperature maximum in the early Holocene. It is most likely that the individual proxies reflect temperatures recorded in different seasons and/or in different water depths. Thus, the discrepancy between

foraminiferal and diatom SST data can be explained: the foraminifer data at the suggested calcification depth of *N. pachyderma* sin. represent the more stable, long-term mean annual temperature record averaged over a greater depth range, whereas the diatom-based SST reconstructions predominantly reflect the short-term summer season temperature in the shallower euphotic zone (upper 10 m). In addition, the latter values will strongly depend on wind-induced surface water mixing during the plankton bloom, with highest temperatures favoured under (high-pressure) light wind conditions. The same explanation may also apply to the alkenone-derived SST data. In addition, Dolven et al. (2002) use the same arguments by explaining the discrepancy between their radiolarian-based reconstruction and the diatom data of MD95-2011 with differences in the living depth. Radiolarians and foraminifers live deeper (> 50 m) than diatoms and represent therefore a temperature regime where seasonal temperature fluctuations are less marked and the temperatures are influenced by the wintertime ventilation of the thermocline. In experiments with climate models, Liu et al. (2003) show that there is a strong influence of seasonality of the orbital forcing on the development of Holocene temperatures in the surface layer, with colder thermocline temperatures at high latitudes in the early Holocene due to the reduced winter season insolation as compared with the late Holocene, and the opposite trend for the surface due to the high early-Holocene summer insolation. This may also explain the unusual high-temperature shift of more than 5 °C in the diatom-based MD95-2011 SST record throughout the Holocene. Further support for a decoupling of surface and subsurface ocean temperatures during the early-Holocene summers provides the observation that the prominent "8.2 event" is not clearly shown in the real surface ocean temperature proxy records (alkenone, diatoms), but very distinct in the oxygen isotope and radiolarian-based records of MD95-2011. In the Norwegian Sea the alkenone-derived SST data (Calvo et al., 2002; Marchal et al., 2002) generally mimic the diatom SST records throughout the Holocene, whereas at the Reykjanes Ridge the alkenone SST's only follows the Norwegian Sea diatom record during the early to mid-Holocene. In the late Holocene, i.e. between 3.7 and 2 kyr BP, the LO09-14 alkenone temperature record shows features, which are more related to the Norwegian Sea foraminifer-based records. A shift in the coccolith bloom period or generally more instable climate with enhanced cyclone activity (see also below) and associated increased vertical mixing of surface waters may have played an important role in causing such a pattern. In addition, the alkenone data of MD95-2015 (Marchal et al., 2002) south of Iceland show features that are comparable with the Reykjanes Ridge record.

A strong link exists between the North Atlantic long-term SST temperature trends and the intensity of ice rafting in the ice-rafting data which bolster our inferences referring the SST proxy interpretation. The main features of the ice-rafting records match the SST reconstructions, with the ice-rafting maximum coincident with the mid-Holocene cooling episode. Our ice-rafting records differ from IRD data obtained on the East Greenland Shelf (e.g. Jennings et al., 2002) which show an increase in IRD deposition during the late Holocene. However, our data were obtained from areas influenced by the warm NAC and the east Greenland sites underlie the cold EGC. We speculate that the warming after 3.7 kyr BP we infer from e.g. foraminifer-based data might have led to decreased ice rafting in the eastern and higher-latitudinal parts of the northern North Atlantic, but conversely to increasing IRD input off East Greenland via a stronger EGC. The warmer NAC may have prevented significant amounts of drift ice to reach areas in the eastern North Atlantic during the late Holocene.

4.2.2. Comparison with other marine and terrestrial temperature records

There are noticeable similarities between the Norwegian Sea diatom-derived SST data (Birks and Koc, 2002; Andersen et al., 2004) and Greenlands summer temperatures (Alley et al., 1999), but also between the NPS $\delta^{18}O$ in MD95-2011 and Greenlands mean annual temperature during the late Holocene (Fig. 5B and C). In addition, recently published (Davis et al., 2003) Holocene summer and winter temperature reconstructions based on pollen data for Europe give similar results (Fig. 5D). The pollen data show marked long-term features which are largely comparable with our observations. The winter temperature curve of central- and north-western Europe resembles features of our planktonic foraminiferal isotope record with a temperature maximum during the late Holocene and not during the early-Holocene. The summer pollen-derived temperatures follow more the diatom and alkenone SST reconstructions with a temperature maximum in the early to mid-Holocene interval.

There is a relatively close correspondence between the long-term features of the SST reconstruction and the temperature variability over Fennoscandia (Fig. 6C) as derived from a pollen-based transfer function (Seppä and Birks, 2001). The observed two distinct cooling episodes and the renewed warmer temperatures between 3.7 and 2 kyr BP are also recorded by the composition of the terrestrial flora. Differences between these records in the early-Holocene can be attributed to the existing influence of the retreating Scandinavian ice sheet on the adjacent environment, the long-term succession of forest communities and/or the possible existence of non-analogue climatic conditions brought about by the

stronger seasonal insolation. On the other hand, our high early-Holocene SST's are in agreement with finds of similar age tree mega fossils at high-elevation sites in the Scandes mountains, which have been interpreted as evidence of relatively high early-Holocene temperatures (Kullman, 1999). In addition, salinity changes in the Baltic Sea indicate similar hydrographic conditions in the early and late Holocene interrupted by a sudden change between 7 and 3.7 kyr BP (Fig. 6D, Emeis et al., 2003). A distinct cold phase between 6 and 4 kyr BP over Fennoscandia has been noted by a study using quantitative temperature reconstruction based on fossil chironomid larvae assemblages (Korhola et al., 2002) and evidence of climatically induced tree-line lowering during the mid-Holocene exists in Swedish Lapland (Barnekow, 1999).

4.2.2.1. A mid-Holocene temperature minimum.
Evidence for the occurrence of a colder interval that began at ca 6.5 kyr BP has been found on the Scotian shelf in the NW Atlantic (Levac, 2001). Based on coccolith studies, Giraudeau et al. (2000) also found a two-step oceanic cooling trend during the Holocene. However, widespread North Atlantic marine evidence for these changes is lacking and only a long-term cooling has previously been observed (Koc and Jansen, 1994; Marchal et al., 2002) which is most likely due to the proxy methods used and the study area where they have been applied. The cooling episode between 7 and 5.5 kyr BP is noted for increased storminess in the NE United States (Noren et al., 2002). The sea-salt Na and non-sea-salt K contents of the Greenland ice cores which most likely reflect changes in atmospheric circulation (O'Brien et al., 1995) provide evidence for distinct cooling episodes between 7 and 5 kyr BP, and the LIA (Fig. 6). Furthermore, Holocene benthic isotope data may indicate a reduction in deep water intensity around 5 kyr BP which might be linked to a colder period (Oppo et al., 2003). Again, our observed two cooling episodes are also evident in the GISP2 Greenland ice core temperature record (Alley et al., 1999) and the second one clearly in Greenland borehole temperature reconstruction (Dahl-Jensen et al., 1998). The first cooling trend, which started around 7 kyr BP is evident only from the summer temperatures and the second cooling trend only by the mean annual temperature record. Cooling between 6.5 and 5 kyr BP also affected the Tropics as shown by the Kilimanjaro ice core records (Thompson et al., 2002). In addition, outside the North Atlantic sector a marked climatic change between 6.7 and 5 kyr BP has been inferred from marine sediment records off West Africa (deMenocal et al., 2000), in the polar regions, in the tropical Pacific and the tropical Andes (summarized by Steig, 1999), and sudden climatic change, e.g. lake level drops, between

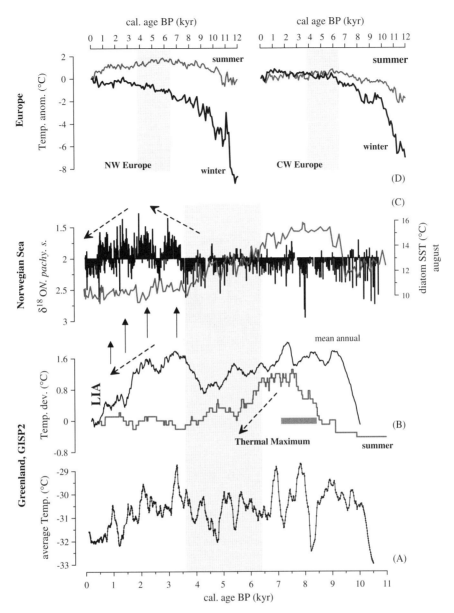

Fig. 5. Holocene (A) average temperatures (Alley, 2000) and (B) mean annual plus summer (grey) temperature over Greenland (Alley et al., 1999), (C) Norwegian Sea diatom-derived SST's (Birks and Koc, 2002; Andersen et al., 2004), grey curve) and δ^{18}O *N. pachyderma* sin. (Risebrobakken et al., 2003) data, and (D) Holocene summer (grey) and winter temperatures over North-Western and Central-Western Europe (Davis et al., 2003) versus a cal. kyr BP timescale. Note whereas the diatom-derived SST data and Greenlands and Europes summer temperatures display a temperature maximum during the early to mid-Holocene, the δ^{18}O *N. pachyderma* sin. and Greenland's and Europe's mean annual and winter temperatures show a maximum during the late Holocene, but apparently not during the early Holocene. The grey area indicates a phase with generally lower SST's and increased ice rafting during the mid-Holocene.

6.5 and 4 kyr BP are observed in the Andes (Rowe et al., 2002; Servant and Servant-Vildary, 2003).

4.2.2.2. Warmer, but unstable climatic conditions from 4 to 2 kyr BP.   Besides the records shown in Figs. 5 and 6, warmer North Atlantic surface temperature from ca 3.5 kyr BP has been observed based on coccolith studies at the inner N. Iceland shelf (Andrews and Giraudeau, 2003) and south of Iceland (Giraudeau et al., 2000). Based on foraminiferal studies warmer SST's in the late Holocene have been noted by Sarnthein et al. (2003), by

Marchal et al. (2002) in core MD95-2015, and by Hald and Aspeli (1997). In addition, warmer late-Holocene temperatures are indicated by European speleothem records (McDermott et al., 1999). As quoted above, warmer mean annual temperatures between 4 and 2 kyr BP were reconstructed from the GISP2 Greenland ice core by Alley et al. (1999) and highest winter temperatures are observed in western Europe based on pollen data (Davis et al., 2003). Moreover, a well-defined warming trend starting around 5 kyr BP is recorded in the Antarctic Byrd ice core (Fig. 7, Johnsen

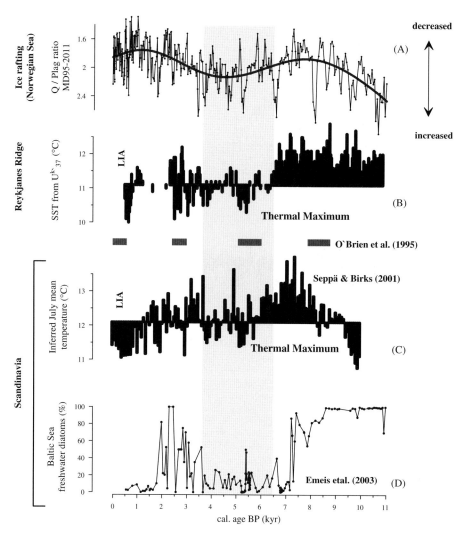

Fig. 6. Holocene (A) ice-rafting proxy data (reverse scaled) from the Norwegian Sea, (B) temperature fluctuations recorded at the Reykjanes Ridge 59°N and (C) over Fennoscandia as derived from pollen-based transfer function (Seppä and Birks, 2001). Additionally, Greenlands sea-salt and non-sea-salt concentration maxima (O'Brien et al., 1995) and (D) changes in the surface salinity in the Baltic Sea (Emeis et al. 2003) are shown. Note the similar long-term trends in the temperature records. The ice rafting and salinity data mirror these trends. The grey area indicates a phase with generally lower SST's and increased ice rafting during the mid-Holocene. A cold phase during the mid-Holocene is also indicated by variations in sea salt concentration (higher concentrations reflect stronger polar circulation). Further integrated is the Little Ice Age (LIA) episode.

et al., 1972). Indications of warmer temperatures over Antarctica are also found in the Vostock and Dome-C ice cores and are supported by findings of an Antarctic thermal maximum during the late Holocene (Björck et al., 1996; Ingolfsson and Hjort, 2002; Yoon et al., 2000).

4.3. Possible mechanisms for long-term Holocene climatic changes in the North Atlantic region

Here, we discuss possible mechanisms for the Holocene climatic changes observed in the North Atlantic. Due to the scarcity of high-resolution Holocene records from the North Atlantic elements of this discussion remains hypothetic at this stage. The long-term reduction of summer insolation during the

Holocene at high northern latitudes has been suggested as the driver of the general Holocene cooling trend that has been reconstructed for the extra-tropical region of the Northern Hemisphere (e.g. Koc and Jansen, 1994; Marchal et al., 2002). This view is supported by our data. The increased seasonality in the Northern Hemisphere during the early Holocene had obviously a strong positive impact on the SSTs during the summer months, but a rather limited influence on the subsurface ocean. Proxies clearly indicative of short-term summer temperature conditions in the shallow euphotic zone (such as diatom census data and alkenone unsaturation ratios) show a distinct early HTM, whereas proxies reflecting a more annual mean/winter temperature and/or a more stable subsurface ocean signal do not reproduce this feature. This observation is in agreement with the

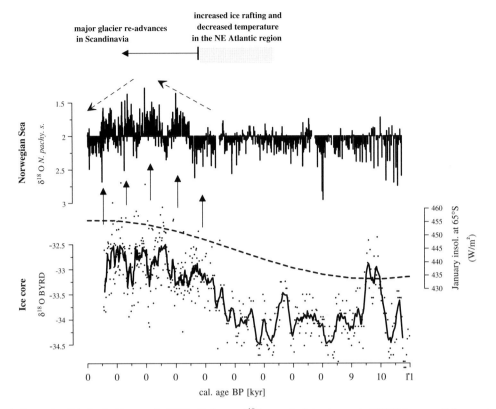

Fig. 7. δ^{18}O *N. pachyderma* sin. (Risebrobakken et al., 2003), Holocene δ^{18}O record of the Antarctic BYRD ice core (Johnsen et al., 1972) versus a cal kyr BP timescale. Additionally, January insolation changes at 65°S (Berger and Loutre, 1991) are shown. Decreasing summer insolation (decreasing seasonality) at high northern latitudes may have lead to the observed temperature minimum between 6.5 and 3.7 kyr BP as well as increasing ice rafting. The Neoglaciation trend recorded in Scandinavia may be linked to a mean annual (winter) temperature increase and associated increased precipitation due to enhanced cyclone activity. Note the similarities between the oxygen isotope and the Byrd ice core records on a long-term scale after ca 6 kyr BP but also the out-of-phase on millennial scale (arrows) which may implicate a link between the hemispheres and an active southern role on northern hemisphere climate after ca 5 kyr BP. However, the mechanism remains unknown.

modelling results of Liu et al. (2003). The reduction in seasonality at northern high latitudes during the course of the Holocene is mainly caused by summer insolation reduction (Liu et al., 2003) and is clearly reflected by summer biased proxy records. The general summer temperature decrease accompanied with less seasonality may have also led to the observed increased ice rafting in the North Atlantic during the mid-Holocene. The summer insolation trend at 65°N may explain the cooling after approximately 7 kyr BP, but it's expected effect is in contrast to the temperature reconstructions from subsurface dwelling microfossils, which point to warmer, more unstable, mean annual temperatures since 3.7 kyr BP. Interestingly, the widely observed Neoglaciation trend (in the form of glacier advances) began approximately at the same time, at around 4.0–3.0 kyr BP (Dahl and Nesje, 1996; Karlen and Kuylenstierna, 1996; Nesje et al., 1991; Matthews et al., 2000), as the North Atlantic mean annual SST (winter) started to rise and some atmospheric temperature reconstructions over Scandinavia and Greenland indicate a warming. We are aware that the evidence for this winter warming is still weak over Scandianavia as

many proxies are summer biased. Glacier expansion is a result of a positive net balance between winter accumulation and summer ablation, with winter precipitation being an important factor especially in maritime glaciers (Dahl and Nesje, 1996). More efficient moisture transport to the Scandinavian mountains during milder winter months, rather than summer temperature changes, can be considered as a potential cause for the glacier re-advances that began there at 4 kyr BP. Proxy climate data from sites in northern (Snowball et al., 1999) and central Sweden (Hammarlund et al., 2002) point to a rapid shift to more humid and unstable climatic conditions over Scandinavia at 3.7 kyr BP.

What caused the apparent mean annual temperature rise that took place at about 4 kyr BP? Decreasing seasonality in the Northern Hemisphere accompanied with northern and southern winter insolation increase may have forced the mean annual temperature to rise in the North Atlantic region and may have led to stronger meridional atmospheric circulation at higher northern latitudes. More frequent and intense low atmospheric pressure systems (storm tracks) would have influenced the northern high latitudes during a period when summer insolation decreased at northern high latitudes and winter temperatures were relatively mild. This scenario suggests a climatic regime that bears resemblance to the modern positive mode of NAO and associated increased moisture transport that occurs over northern Europe and which leads to significant glacier advances.

In general, the reconstructions of the early Holocene indicate that the climate system in the North Atlantic region was modulated by orbitally controlled high summer insolation (strong seasonality). The seasonal temperature difference at northern high latitudes decreased throughout the Holocene, which may have led to a stronger influence of northern winter insolation changes during the late Holocene. We speculate that the decreasing seasonality in the North may have also increased the potential of a stronger impact of southern ocean changes on the North Atlantic climate, which may be indicated by the link between our more mean annual, subsurface related proxies and the southern ocean signal during the late Holocene (Fig. 7).

5. Conclusion

Based on studies of alkenone-derived SST's and lithological parameters from two sediment cores from areas underlying the warm NAC and extensive comparisons to earlier obtained data sets, we postulate that atmospheric temperature changes over Northern Europe and Greenland were linked to long-term Holocene North Atlantic SST variations. Variable degrees of ice-rafting mimic these long-term temperature fluctuations. A distinct mid-Holocene climate change, is registered in the North Atlantic records as a distinct cooling, with a possible low latitude extent. We identify four major climatic features of the Holocene which are (i) an early HTM that lasted until approximately 6.7 kyr BP, (ii) an abrupt lowering of surface layer temperatures associated with an increase in ice rafting between ca 6.5 and 3.7 kyr BP, (iii) a transition to generally warmer, but also relatively unstable climate conditions between 3.7 and 2 kyr BP and (iv) a second distinct surface layer temperatures decline that took place between 2 and 0.5 kyr BP. The early HTM seems to be only clearly evident in marine proxy data which are summer and ocean surface related. Changes that occur in the late Holocene are only reflected by more mean-annual or winter (subsurface) related marine proxies. The latter points to the fact that many climatic reconstructions suffer from the fact that only summer proxies are available.

In contrast to a likely dominant control of Northern Hemisphere summer insolation on early-Holocene climate development via a strong seasonality, the trigger for the onset of relatively unstable climatic conditions in

the North Atlantic at 4 kyr BP is not very clear. Northern winter insolation increases and influence from changes in the tropics or the Southern Hemisphere changes are potential candidates.

Acknowledgements

We thank Svante Björck, Atle Nesje and Jacques Giraudeau for many fruitful discussions, and Gerard Bond, Basil Davis, Richard Alley and Nalan Koc for providing data. The two anonymous reviewers are thanked for their very helpful comments. We are grateful to Christian Blauscha and Claas Meliss for help during the sample preparation. This study was supported (M. Moros) by an EU *Marie-Curie fellowship* (HPMF-CT-2002-01631).

References

Alley, R.B., 2000. The Younger Dryas cold interval as viewed from central Greenland. Quaternary Science Reviews 19, 213–226.

Alley, R.B., Agustsdottir, A.M., Fawcett, P.J., 1999. Ice-core evidence of late-Holocene reduction in North Atlantic Ocean heat transport. Geophysical Monograph 112, 301–312.

Andersson, C., Risebrobakken, B., Jansen, E., Dahl, S.O., 2003. Late Holocene surface ocean conditions of the Norwegian Sea (Võring Plateau). Paleoceanography 18, 1044.

Andersen, C., Koc, N., Jennings, A.E., Andrews, J.T., 2004. Nonuniform response of the major surface currents in the Nordic Seas to insolation forcing: implications for the Holocene climate variability. Paleoceanography 19, PA200310.1029/2002PA000873.

Andrews, J.T., Giraudeau, J., 2003. Multi-proxy records showing significant Holocene environmental variability: the inner N. Iceland shelf (Hunafloi). Quaternary Science Reviews 22, 175–193.

Barnekow, L., 1999. Holocene treeline dynamics and inferred climatic changes in the Abisko area, northern Sweden, based on macrofossil and pollen records. The Holocene 9, 253–265.

Berger, A.L., Loutre, M.F., 1991. Insolation values for the climate of the last 10 million years. Quaternary Science Reviews 10, 297–317.

Birks, C.J.A., Koc, N., 2002. A high-resolution diatom record of late-quaternary sea-surface temperatures and oceanographic conditions from the eastern Norwegian Sea. Boreas 31, 323–344.

Björck, S., Olsson, S., Ellis-Evans, C., Hakonsson, H., Humlum, O., deLirio, J.M., 1996. Late Holocene palaeoclimatic records from lake sediments on James Ross Island, Antarctica. Palaeography, Palaeoclimatology, Palaeoecology 121, 195–220.

Bond, G., Kromer, B., Beer, J., Muscheler, R., Evans, M.N., Showers, W., Hoffmann, S., Bond, R.L., Hajdas, I., Bonani, G., 2001. Persistent solar influence on North Atlantic climate during the Holocene. Science 294, 2130–2136.

Bronk Ramsey, C., 1995. Radiocarbon calibration and analysis of stratigraphy: the OxCal program. Radiocarbon 37, 425–430.

Bronk Ramsey, C., 2001. Development of the radiocarbon program OxCal. Radiocarbon 43, 355–363.

Calvo, E., Grimalt, J., Jansen, E., 2002. High resolution U^k_{37} sea surface temperature reconstruction in the Norwegian Sea during the Holocene. Quaternary Science Reviews 21, 1385–1394.

Campbell, I.D., Campbell, C., Apps, M.J., Rutter, N.W., Bush, A.B.G., 1998. Late Holocene ~1500 yr climatic periodicities and their implications. Geology 26, 471–473.

Dahl, O.S., Nesje, A., 1996. A new approach to calculating Holocene winter precipitation by combining glacier equilibrium-line altitudes and pine-tree limits: a case study from Hardangerjokulen, central southern Norway. The Holocene 6, 381–398.

Dahl-Jensen, D., Mosegaard, K., Gundestrup, N., Clow, G.D., Johnsen, S.J., Hansen, A.W., Balling, N., 1998. Past temperature directly from the greenland ice sheet. Science 282, 268–271.

Davis, B.A.S., Brewer, S., Stevenson, A.C., Guiot, J., Contributors, D., 2003. The temperature of Europe during the Holocene reconstructed from pollen data. Quaternary Science Reviews 22, 1701–1716.

deMenocal, P., Ortiz, J., Guilderson, T., Sarnthein, M., 2000. Coherent high- and low-latitude climate variability during the Holocene warm period. Science 288, 2198–2202.

Dolven, J.K., Cortese, G., Björklund, K.R., 2002. A high-resolution radiolarian-derived paleotemperature record for the late Pleistocene–Holocene in the Norwegian Sea. Paleoceanography 17.

Dowdeswell, J.A., Elverhoi, A., Spielhagen, R.F., 1998. Glacimarine sedimentary processes and facies on the polar North Atlantic margins. Quaternary Science Reviews 17, 243–272.

Duplessy, J.C., Ivanova, E., Murdmaa, I., Paterne, M., Labeyrie, L., 2001. Holocene paleoceanography of the northern Barents Sea and variations of the northward heat transport by the Atlantic Ocean. Boreas 30, 2–16.

Emeis, K.C., Struck, U., Blanz, T., Kohly, A., Voss, M., 2003. Salinity changes in the central Baltic Sea (NW Europe) over the last 10 000 years. The Holocene 13, 413–423.

Funder, S., Hjort, C., Landvik, J.Y., Nam, S.I., Reeh, N., Stein, R., 1998. History of a stable ice margin—East Greenland during the Middle and Upper Pleistocene. Quaternary Science Reviews 17, 77–123.

Giraudeau, J., Cremer, M., Manthe, S., Labeyrie, L., Bond, G., 2000. Coccolith evidence for instabilities in surface circulation south of Iceland during Holocene times. Earth and Planetary Science Letters 179, 257–268.

Hald, M., Aspeli, R., 1997. Rapid climatic shifts of the northern Norwegian Sea during the last glaciation and the Holocene. Boreas 26, 15–28.

Hald, M., Ebbesen, H., Forwick, M., Godtliebsen, F., Khomenko, L., Korsun, S., Ringstad-Olsen, L., Vorren, T.O., 2004. Holocene paleoceanography and glacial history of the West Spitsbergen area, Euro-Arctic margin. Quaternary Science Reviews, this issue (doi: 10.1016/j.quascirev.2004.08.006).

Hammarlund, D., Björck, S., Buchardt, B., Isrealson, C., Thomsen, C.T., 2002. Rapid hydrological changes during the Holocene revealed by stable isotope records of lacustrine carbonates from Lake Igelsjön, southern Sweden. Quaternary Science Reviews 21.

Ingolfsson, O., Hjort, C., 2002. Glacial history of the Antarctic Peninsula since the Last Glacial Maximum—a synthesis. Polar Research 21, 227–234.

Jennings, A.E., Knudsen, K.L., Hald, M., Hansen, C.V., Andrews, J.T., 2002. A mid-Holocene shift in Arctic sea-ice variability on the East Greenland Shelf. The Holocene 12, 49–58.

Johnsen, S.J., Dansgaard, W., Claussen, H.B., Langway Jr., C.C., 1972. Oxygen isotope profiles through the Antarctic and Greenland ice-sheets. Nature 235, 429–434.

Karlen, W., Kuylenstierna, J., 1996. On solar forcing of Holocene climate: evidence from Scandinavia. The Holocene 6, 359–365.

Koc, N., Jansen, E., 1994. Response of the high-latitude Northern hemisphere to orbital climate forcing: evidence from the Nordic Seas. Geology 22, 523–526.

Korhola, A., Vasko, K., Toivonen, H.T.T., Olander, H., 2002. Holocene temperature changes in northern Fennoscandia reconstructed from chironomids using Bayesian modelling. Quaternary Science Reviews 21, 1841–1860.

Kullman, 1999. Early Holocene tree growth at a high elevation site in the northernmost Scandes of Sweden (Lapland): a palaeobiogeographical case study based on megafossil evidence. Geografiska Annaler 81, 63–74.

Levac, E., 2001. High resolution Holocene palynological record from the Scotian Shelf. Marine Micropaleontology 43, 179–197.

Liu, Z., Brady, E., Lynch-Stieglitz, J., 2003. Global ocean response to orbital forcing in the Holocene. Paleoceanography 18, 10.1029/2002PA000819.

Marchal, O., et al., 2002. Apparent long-term cooling of the sea surface in the northeast Atlantic and the Mediterranean during the Holocene. Quaternary Science Reviews 21, 455–483.

Matthews, J.A., Dahl, S.O., Nesje, A., Berrisford, M.S., Andersson, C., 2000. Holocene glacier variations in central Jotunheimen, southern Norway based on distal glaciolacustrine sediment cores. Quaternary Science Reviews 19, 1625–1647.

McDermott, F., Frisia, S., Huang, Y., Longinelli, A., Spiro, B., Heaton, T.H.E., Hawkesworth, C.J., Borsato, A., Keppens, E., Fairchild, I.J., van der Borg, K., Verheyden, S., Selmo, E., 1999. Holocene climate variability in Europe: evidence from $\delta^{18}O$, textural and extension-rate variations in three speleothems. Quaternary Science Reviews 18, 1021–1038.

Moros, M., McManus, J., Rasmussen, T., Kuijpers, A., Snowball, I., Dokken, T., Nielsen, T., Jansen, E., 2004. Quartz content and the quartz-to-plagioclase ratio determined by X-ray diffraction: proxies for ice rafting in the northern North Atlantic? Earth and Planetary Science Letters 218, 389–401.

Müller, P.J., Kirst, G., Ruhland, G., von Storch, I., Rosell-Mélé, A., 1998. Calibration of the alkenone paleotemperature index UK'37 based on core tops from the eastern South Atlantic and the global ocean (60°N–60°S). Geochimica Cosmochimica Acta, 1757–1772.

Nesje, A., Kvamme, M., Rye, N., Reidar, L., 1991. Holocene glacial and climate history of the Jostedalsbreen region, western Norway: evidence from lake sediments and terrestrial deposits. Quaternary Science Reviews 10, 87–114.

Noren, A.J., Bierman, P.R., Steig, E.J., Lini, A., Southon, J., 2002. Millennial-scale storminess variability in the northeastern United States during the Holocene epoch. Nature 419, 821–824.

O'Brien, S.R., Mayewski, P.A., Meeker, L.D., Meese, D.A., Twickler, M.S., Whitlow, S.I., 1995. Complexity of Holocene climate as reconstructed from a Greenland ice core. Science 270, 1962–1964.

Oppo, D.W., McManus, J.F., Cullen, J.R., 2003. Palaeo-oceanography: deepwater variability in the Holocene epoch. Nature 422, 277.

Risebrobakken, B., Jansen, E., Andersson, C., Mjelde, E., Hevroy, K., 2003. A high-resolution study of Holocene paleoclimatic and paleoceanographic changes in the Nordic Seas. Paleoceanography 18, 1017–1031.

Rowe, H.D., Dunbar, R.B., Mucciarone, D.A., Seltzer, G.O., Baker, P.A., Fritz, S.C., 2002. Insolation, moisture balance and climate change on the South American Altiplano since the last glacial maximum. Climate Change 52, 175–199.

Sarnthein, M., van Kreveld, S., Erlenkeuser, H., Grootes, P.M., Kucera, M., Pflaumann, U., Schulz, M., 2003. Centennial-to-millennial-scale periodicities of Holocene climate and sediment injections off the western Barents shelf, 75°N. Boreas 32, 447–461.

Schulz, M., Paul, A., 2002. Holocene climate variability on centennial-to-millennial time scales: 1. Climate records from the North-Atlantic realm. In: Wever, G.E.A. (Ed.), Climate Development and History of the North Atlantic Realm. Springer, New York, pp. 41–54.

Seppä, H., Birks, H.J.B., 2001. July mean temperature and annual precipitation trends during the Holocene in the Fennoscandian tree-line area: pollen-based climate reconstructions. The Holocene 11, 527–539.

Servant, M., Servant-Vildary, S., 2003. Holocene precipitation and atmospheric changes inferred from river paleowetlands in the Bolivian Andes. Palaeogeography, Palaeoclimatology, Palaeoecology 194, 187–206.

Snowball, I., Korholla, A., Briffa, K.R., Koc, N. Holocene climate dynamics in Fennoscandia and the North Atlantic. In: Batterbee, R., Gasse, F. (Eds.), IGBP/PAGES PEP III. Kluwer Academic Publisher, Dordrecht.

Snowball, I., Sandgren, P., Pettersen, G., 1999. The mineral magnetic properties of an annually laminated Holocene lake-sediment sequence in northern Sweden. The Holocene 9, 353–362.

Steig, E.J., 1999. Mid-Holocene climate change. Science 286, 1485–1487.

Stuiver, M., Reimer, P.J., Bard, E., Beck, J.W., Burr, G.S., Hughen, K.A., Kromer, B., McCormac, G., Van der Plicht, J., Spurk, M., 1998. INTCAL98 Radiocarbon age calibration, 24,000-0 cal BP. Radiocarbon 3, 1041–1083.

Thompson, L.G., Mosley-Thompson, E., Davis, M.E., Henderson, K.A., Brecher, H.H., Zagorodnov, V.S., Mashiotta, T.A., Lin, P.N., Mikhalenko, V.N., Hardy, D.R., Beer, J., 2002. Kilimanjaro Ice Core records: evidence of Holocene Climate change in Tropical Africa. Science 298, 589–593.

Viau, E.A., Gajewski, K., Fines, P., Atkinson, D.E., Sawada, M.C., 2002. Widespread evidence of 1500 yr climate variability in North America during the past 14 000 yr. Geology 30, 455–458.

Wright, W.R., Worthington, L.V., 1970. The water masses of the North Atlantic Ocean: a volumetric census of temperature and salinity. Serial atlas of the Marine Environment, Folio 19, American Geographical Society.

Yoon, H.I., Park, B.-K., Kim, Y., Kim, D., 2000. Glaciomarine sedimentation and its paleoceanographic implications along the fjord margins in the South Shetland Islands, Antarctica during the last 6000 years. Palaeogeography, Palaeoclimatology, Palaeoecology 157, 189–211.

Quaternary Science Reviews 23 (2004) 2127–2139

Timing and mechanisms of surface and intermediate water circulation changes in the Nordic Seas over the last 10,000 cal years: a view from the North Iceland shelf

J. Giraudeau[a,*], A.E. Jennings[b], J.T. Andrews[b]

[a]*Département de Géologie et Océanographie, UMR CNRS 5805, Université Bordeaux 1, Avenue des Facultés, 33405 Talence cedex, France*
[b]*INSTAAR and Department of Geological Sciences, University of Colorado, Boulder, CO 80309-0450, USA*

Abstract

The North Iceland shelf bears essential components of the present surface and intermediate circulation of the northern North Atlantic. Instrumental and historical data give evidence of the sensitivity of this domain to broad, regional-scale oceanic and atmospheric anomalies. Our investigation of the paleohydrological variability off Northern Iceland throughout the last 10 000 cal yr suggests that atmospheric forcing alone, through combined changes in strength of the wind stress curl and sea-level atmospheric pressure pattern over the Nordic Seas, is sufficient to explain the recorded changes in origins and dynamics of surface and intermediate water masses. Our biotic proxies, coccoliths and benthic foraminifera, were extracted from a giant piston core (MD99-2269) collected in a shelf trough where sediment accumulated at an excess rate of 2 m/kyr. The mid-Holocene from 6.5 to 3.5 cal kyr BP was a time of peaked carbonate production and subsequent sedimentation, and strong water-column stratification with a thick layer of cold-fresh Arctic surface water overlapping an enhanced flow of Irminger/Atlantic Intermediate water. Applying conditions triggering present-time carbonate plankton blooms in the studied area, we infer that a lowered cyclonic activity associated with decreased winter storms and reduced production of Arctic Intermediate Water in the Iceland Sea were conductive of the recorded mid-Holocene water column structure. The opposite situation (warm Atlantic surface water, low vertically-integrated inflow of Irminger water, abutment of Arctic Intermediate water in deep shelf troughs) characterized the early Holocene as well as a shorter late Holocene period centred at 2 cal kyr BP. The Little Ice Age (ca. 0.2–0.6 cal kyr BP) and a short event at around 3 cal kyr BP stand as times of extreme advection of polar waters and extended sea–ice development. A comparison of the recorded long-term Holocene evolution of water column structure off Northern Iceland with climate and hydrological changes in the north-eastern Atlantic suggests that the strength of Atlantic inflow into the Nordic Seas was subjected to a balance between the Irminger and the Norwegian branches. This balance is thought to be mostly related to changes in the intensity and location of westerly winds and associated atmospheric pressure gradients in the North Atlantic.

1. Introduction

The North Iceland shelf has recently been the subject of intensive marine geological investigations which showed that this restricted area encapsulated, throughout the last deglaciation and Holocene periods, oceanographic and atmospheric variability that occurred over a much broader area (Andrews et al., 2000; Jennings et al., 2002; Andrews and Giraudeau, 2003; Andersen et al., 2004). Historical, as well as instrumental records have highlighted the sensitivity of this domain to recent oceanic and atmospheric anomalies such as the Great Salinity Anomaly (GSA) in the late 1960s, or the Little Ice Age (LIA) cold spell dated off northern Iceland at 750–100 cal yr BP (Knudsen and Eiriksson, 2002). Both events shared common features off northern Iceland including surface water cooling and freshening associated with increased influence of arctic and polar waters

*Corresponding author. Tel.: +33-540-008-860; fax: +33-556-840-848.

E-mail address: j.giraudeau@epoc.u-bordeaux1.fr (J. Giraudeau).

0277-3791/$ - see front matter © 2004 Elsevier Ltd. All rights reserved.
doi:10.1016/j.quascirev.2004.08.011

(Dickson et al., 1988; Olafsson, 1999; Eiriksson et al., 2000). Those surface water changes have a profound effect on both primary productivity and sea-ice extent as evidenced by hydrographic survey conducted in these waters over the last 50 yr (Thordardottir, 1984). In addition, it has been shown that surface water changes on the North Iceland shelf are tightly associated with altered overturning of deep and intermediate waters in the Iceland and Greenland seas (Malmberg and Jonsson, 1997), as well as with variations in the flux of Atlantic waters entering the Norwegian Sea (Blindheim et al., 2000). Atmospheric forcing, through variability in strength of the wind stress curl over the Nordic Seas (Jonsson, 1992), as well as changes in the state of the atmospheric pressure system over the North Atlantic (Dawson et al., 2003; Blindheim et al., 2000), are seen as the main drivers of changes in the water-mass structure of the northern North Atlantic.

The North Iceland shelf bears essential components of the present surface and deep circulation of the northern North Atlantic (Fig. 1; Hopkins, 1991; Stefansson, 1962). It is located close to the Arctic Front which separates Arctic/Polar water masses carried by the south-eastward flowing East Iceland Current (EIC) from the North Iceland Irminger Current (NIIC), a branch of the Irminger Current which rounds the western side of Iceland and feeds the North Iceland shelf with warm Atlantic waters. The NIIC reaches the inner part of the North Iceland shelf while the deeper

realms are occupied by Arctic Intermediate waters (AIW), down to approximately 500 m, which is formed by convection in the Iceland and Greenland seas (Rytter et al., 2002; Malmberg and Jonsson, 1997). Starting from the mid-shelf area, Atlantic Water carried in the NIIC is submerged beneath colder, fresher Arctic Water carried by the EIC (Fig. 1). This water-column stratification is well displayed in Hunafloaall, where the wedge of Atlantic Water between 100 and 350–400 m water depth is cooled to 3.5 °C and freshened to 34.9‰. This submerged vestige of the NIIC is termed Irminger Intermediate Water (IIW). Beneath the IIW, temperatures decline with depth from 3.5 °C to slightly lower than 0 °C in the deepest parts of the northern shelf troughs. Salinities are 34.8‰, less than those in the IIW. The bottom water mass represents upper Arctic Intermediate Water (Swift, 1986), although the deepest parts of the northern troughs may contain Norwegian Sea Deep Water (Rytter et al., 2002).

We investigate the water-column dynamics off NW Iceland throughout the last 10 000 cal yr using the combined records of coccoliths, proxies for surface water conditions, and benthic foraminifera as tracer of intermediate and bottom water masses in the nearby Nordic Seas. Our higher time resolution proxy in the present study, coccoliths, has been successfully tested in recent works dealing with the Holocene short and long-term evolution of the North Atlantic Drift south of Iceland (Giraudeau et al., 2000), and of its NIIC branch

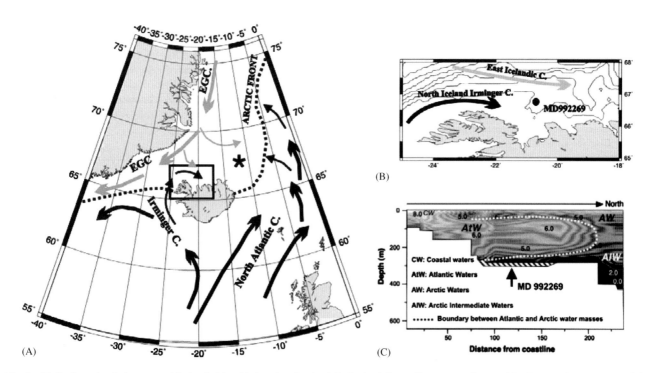

Fig. 1. (A) Surface circulation around Iceland (A), with location (star) of the Iceland Sea sediment trap discussed in the text. (B) Location of the studied core. (C) Water masses across the NW Iceland shelf as depicted from the water column distribution of potential temperature in the vicinity of the core location (after Andrews et al., 2003b).

over the inner North Iceland shelf (Andrews and Giraudeau, 2003). It complements another surface water proxy, diatoms, which have been recently investigated in MD99-2269 (Andersen et al., 2004). Recent works on the modern and fossil distribution of benthic foraminifera around Iceland highlighted the close correspondence of species distribution with bottom water-masses and their characteristics (Rytter et al., 2002; Jennings et al., 2002; Jennings et al., 2004).

2. Material, core chronology and methods

Giant piston core MD99-2269 (66°37′N–20°51′W) was collected as part of the IMAGES V cruise of RV *Marion Dufresne*. The core was retrieved at 365 m water depth from a 30 m thick sediment unit on the floor of Hunafloaall, a north–south orientated depression off N/NW Iceland. Previous works have shown that this core contains a continuous Holocene sedimentary series which accumulated at a rate close to 2 m/kyr (Andrews et al., 2003a). The construction of the shelf sediment unit and the recorded excess rate of sedimentation are mainly explained by the combination of bottom currents focussing and an ample continuous supply of volcanic material from the nearby submarine Kolbeinsy Ridge, a fraction which exceeds both the detrital and biogenic components over the North Iceland shelf (Oehmig and Wallrabe-Adams, 1993).

The stratigraphic framework has recently been revised by Andrews et al. (2003b). It is essentially based on 11 AMS14C dates measured on molluscs, which were ultimately converted to sidereal years with the CALIB 4.3 program (Stuiver et al., 1998) in applying a constant 400 yr reservoir age ($\Delta R = 0$). Additional age control was provided by the occurrence of the Saksunarvatn tephra (10.18 cal ka; Grönvold et al., 1995) toward the base of the core. These 12 dates (Fig. 2) define an age/depth model given by the linear equation: age (cal yrs BP) = $-15.8 + 4.9 \times$ depth (cm) ($R^2 = 0.997$). This best-fit equation suggests that sediment accumulated at the core location at the constant rate of ~2 m/ky (Andrews et al., 2002). The hypothesis of a constant reservoir age for the North Iceland shelf throughout the last 10 000 yr has recently been challenged by Eiriksson et al. (2000) and Knudsen and Eiriksson (2002) who, on the ground of coupled AMS14C and tephra-based chronologies, suggested a 150 yr increase in reservoir age between 3 and ca. 1 cal kyr BP from a shelf-trough farther east on the shelf. However, ongoing research using tephra horizons (Hekla tephras) on core MD99-2269 did not result in significant changes in reservoir age at the core location, probably because Hunafloaall is closer to the influence of Atlantic/Irminger waters than are the more arctic-influenced sedimentary archives studied by Eiriksson and collaborators (Kristjansdottir,

2002). Additional radiocarbon dates combined with the identification of tephras will undoubtedly lead to modifications in the details of the chronology. Indeed, although a linear sedimentation rate has been used in this and previous papers (Andrews et al., 2003a), new but unpublished AMS14C dates suggest departures from this model, especially for the period > 8 cal ka.

Sample preparation for the study of coccoliths and benthic foraminifera involved standard techniques summarized in Andrews and Giraudeau (2003) and Jennings et al. (2002), respectively. Coccolith observations and census counts were conducted using a light microscope at 1250 × magnification, following a series of dilution and filtration treatments of a pre-weighed amount of dry bulk sediment as described by Andruleit (1996). Census counts were expressed as coccolith concentrations (specimens/g of dry bulk sediment), and subsequently transformed into accumulation rates (specimens/cm^2/time) using the estimated mass accumulation rate (Andrews et al., 2003b). A calculation of coccolith carbonate contribution to the bulk sediment was done following the method described by Beaufort and Heussner (1999) and data set given in Young and Ziveri (2000), which is based on estimates of the mean carbonate mass of the various species found in core M99-2269. Sample resolution for the coccolith analyses varied from 5 cm for samples representative of the last 3000 yr and of the 7.5–8.5 kyr BP interval, to 10 cm for the rest of the core. This translates into a 20–50 yr time resolution.

Samples for foraminiferal analysis were taken at 50 cm spacing (about 250 yr) throughout the length of the core. The samples were wet-sieved at 63 and 106 µm. The > 106 µm fractions were divided with a microsplitter until the split contained between 200 and 300 benthic foraminifers. Foraminifera in the entire split were identified to genus and species levels and tallied into percentages.

Data on total carbonate content, measured at 5 cm interval using a Coulometer in the Sedimentology Laboratory at INSTAAR (Boulder, CO), were taken from Andrews et al. (2003b).

3. Results

3.1. Coccolith species distribution and coccolith carbonate

Coccolith species diversity is typically low as expected for this arctic/subarctic setting (Baumann et al., 2000; Andrews and Giraudeau, 2003). Dominance is equally shared between *Coccolithus pelagicus*, the cold-end member of the coccolithophore community in North Atlantic waters, and the ubiquitous *Emiliania huxleyi* (Fig. 2) which is presently responsible for extensive

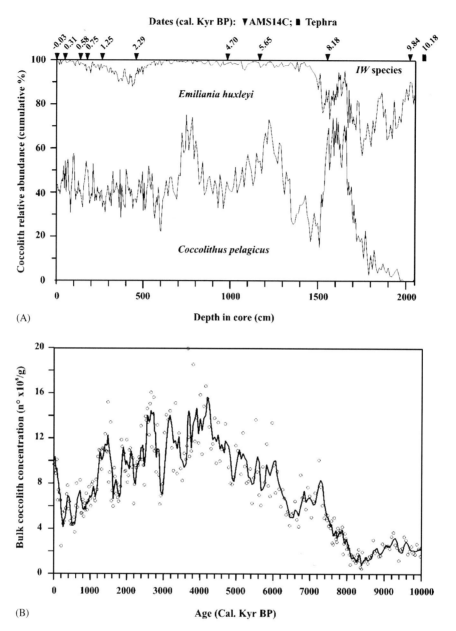

Fig. 2. (A) Coccolith species relative abundance versus depth in core; AMS14C dates and the location of the Saksunarvatn tephra layer (all expressed in cal kyr BP) are given on the top of the plot. (B) Bulk coccolith concentration versus age; the black curve is a smooth record (3 pt running average) of original data (open dots).

blooms in transitional/subarctic waters as well as in shallow settings along Norway and NE America (Brown and Yoder, 1994). The subordinate species *Gephyrocapsa mullerae*, *Calcidiscus leptoporus* and *Syracosphaera* sp., together account for an average 7.5% of the total assemblage throughout the studied interval (Fig. 2). They were grouped as "Irminger water (IW) species" considering their present biogeography (Samtleben et al., 1995) and following the conclusion of Andrews and Giraudeau (2003) that this species group can be used off northern Iceland as a tracer of surface Irminger/Atlantic water inflow.

Total coccolith concentrations follow a simple bell-shape trend with low values in the early Holocene, a regular increase toward maximum concentrations centred at 4–4.5 cal kyr BP., and a subsequent decrease with rapid high amplitude changes toward the late Holocene (Fig. 2). The range of bulk coccolith sedimentation (mean = 7.4×10^8 specimens/g) falls within Holocene values given for the Iceland Sea (Andruleit and Baumann, 1998) but is nearly double the Holocene values of bulk coccolith carbonate sedimentation estimated at a nearby shallower setting due south of the studied core (Andrews and Giraudeau, 2003). This difference suggests that the deeper, offshore setting of core MD99-2269 may be more favourable to cocco-lithophore production and subsequent sedimentation than nearshore, shallow environments off northern Iceland. Our estimates of bulk coccolith accumulation rates (Fig. 2; mean $= 120 \times 10^6$ specimens/cm^2/yr) are on average 10 times higher than values given for surface sediments of the Nordic Seas (Andruleit, 2000), an indication of the degree of sediment focussing respon-sible for the construction of the shelf sediment body where the studied core was retrieved.

The linear sedimentation model implies that sediment focussing did not disturb the information on surface water changes given by the coccolith concentration, i.e. that there was no apparent changes in both the degree of focussing, and the amount of dilution by non-coccolith (mainly terrigenous) components throughout the last 10 000 yr. The record of bulk carbonate content at the core site supports this idea, as Andrews et al. (2003b) showed that the carbonate record of MD99-2269 has a true regional significance, being similar, both in trend and absolute values to nearby core sites barely affected by strong sediment focussing. We will therefore argue in the following section that the record of coccolith concentration is a reliable tracer of surface water dynamics.

The record of total carbonate weight % indicates that net carbonate accumulation peaked between ca. 3.5 and 6.5 cal kyr BP (Fig. 3). Our independently calculated values of coccolith carbonate content indicate that changes in total carbonate content can be attributed largely to changes in this biogenic fraction which contributes on average to more than 50% of the total CaCO$_3$ wt%. Large peaks in the numbers of foramini-fera per gram of bulk sediment (not shown) also occur in this interval, supporting carbonate peaks at 6 and 4.2 cal kyr BP. The overwhelming contributor to the coccolith carbonate mass weight is *C. pelagicus*, whose large, highly calcified coccoliths account for more than 95% of the total coccolith mass fraction throughout the last 8 cal kyr BP. In particular, sedimentation of *C. pelagicus*, alone, explains between 70% and 90% of the peaks in bulk carbonate accumulation at 6 and 3.8 cal kyr BP (Fig. 3). The hydrological conditions which lead to the mid-Holocene (3.5–6.5 cal kyr BP.) high carbonate content over the North Icelandic shelf sediments must therefore be discussed in view of the physical and chemical status of the photic layer which promotes an enhanced primary production of *C. pelagicus*.

The downcore record of IW species (Fig. 4) defines two distinct periods of enhanced advection of warm surface Irminger waters off northern Iceland, as proposed earlier by Andrews and Giraudeau (2003). The high resolution stratigraphic framework of core MD99-2269 helps to refine the timing of these decoupled

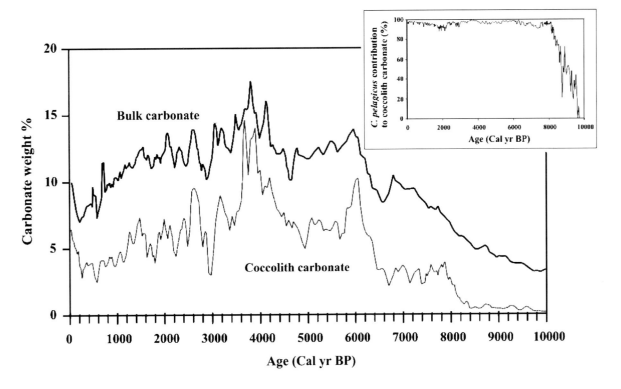

Fig. 3. Relative contribution of carbonates (bulk and coccolith fractions) to the bulk sediment in core MD99-2269. Both records have been smoothed (3 pt running average). Inset: contribution of *Coccolithus pelagicus* to the coccolith carbonate fraction.

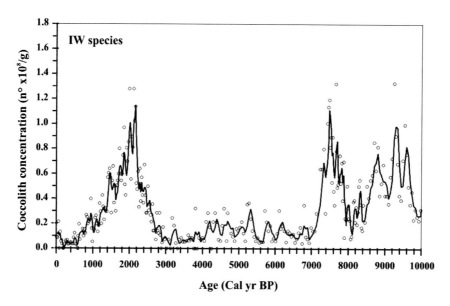

Fig. 4. Coccolith concentrations of Irminger Water (IW) species; the black curve is a smooth record (3pt running average) of original data (open dots).

events—the early Holocene interval ending abruptly at 7.2 cal kyr BP., and the more recent period of peak advection of Irminger waters occurring between 2.6 and 0.8 cal kyr BP, an interval including both the European Roman and Medieval Warm Periods (Briffa et al., 1990). Minimum concentrations of IW species in the studied core are centred at 3–3.5 kyr BP. on one hand, and 0.2–0.6 kyr BP. (i.e. Little Ice Age) on the other hand. The mid-Holocene interval of enhanced bulk carbonate and associated peak sedimentation of C. pelagicus is characterized by medium to low concentrations of IW species.

3.2. Bottom water conditions: the benthic foraminiferal record

Previous studies on modern benthic foraminifera off northern Iceland and East Greenland (Jennings and Helgadottir, 1994; Rytter et al., 2002; Jennings et al., 2004) show that modern foraminiferal assemblage compositions and foraminiferal concentrations are related to both hydrographical (salinity, temperature and water depths) and biological (food availability) factors, and that these parameters are well resolved in terms of the progression of bottom water-masses from the coastal and inner trough areas influenced by Atlantic waters carried in the Irminger Current to the hydrographically stratified areas farther offshore in deeper areas such as the shelf troughs.

The benthic foraminiferal record in the last 10 000 cal yr of MD99-2269 is dominated by two inversely varying arctic species, C. neoteretis and Cassidulina reniforme, whose abundances together average 60% (Fig. 5). Cassidulina reniforme shows very high relative abundances during the early Holocene (from ca.10 to 8 cal kyr BP.). After 9 cal kyr BP its abundances steadily decline to a minimum between 4.5 and 3 cal kyr BP. After 3 cal kyr BP its abundances begin to rise again, but never to such high values as in the early Holocene (Fig. 5). In no modern Iceland shelf samples does C. reniforme occur in such high abundances as it does in the early Holocene interval of MD99-2269 (Jennings et al., 2004; Rytter et al., 2002). Both C. reniforme intervals are also characterized by subsidiary contributions of other arctic species including Nonionellina labradorica and Elphidium excavatum forma clavata. These assemblages are consistent with cold bottom waters with salinities greater than 30‰ generally in areas of seasonal sea–ice cover (Polyak et al., 2002; Jennings and Helgadottir, 1994).

Countering the decline of C. reniforme, C. neoteretis rises steadily from moderate values of about 15% at 10 cal kyr BP to values close to 60% between 4.5 and 3.5 cal kyr BP. The interval of maximum contribution of C. neoteretis coincides with the peak in bulk carbonate and coccolith carbonate contents during the mid-Holocene, from 6.5 to 3.5 cal kyr BP (Fig. 5). Peaks in bulk foraminifera concentration (not shown) coincide in general with peaks in C. neoteretis and bulk carbonate percentages. Subsidiary species co-occurring with C. neoteretis during its peak abundance include Pullenia bulloides and M. barleeanus, both infaunal species reflecting high marine productivity and burial of organic matter (e.g. Corliss, 1985, 1991; Wollenberg and Mackensen, 1998a, 1998b). The cold arctic species that attended C. reniforme, are absent or occur in very low percentages. Cassidulina neoteretis has been described as characteristic of normal marine water masses with stable salinity and temperature (Seidenkrantz, 1995), as well as

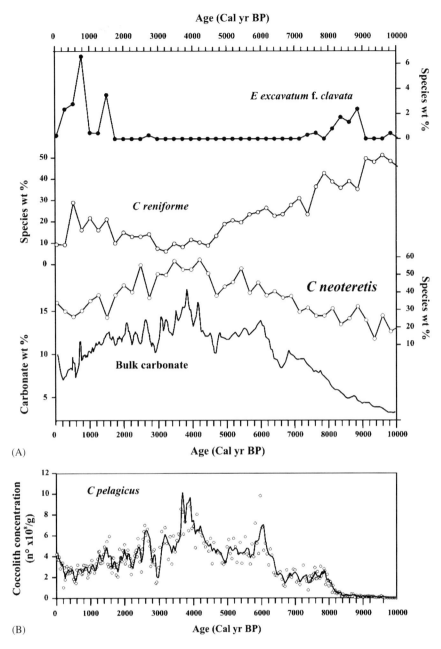

Fig. 5. (A) Relative abundances of the benthic foraminiferal species discussed in the text. The record of bulk carbonate wt % is added for comparison with the *Cassidulina neoteretis* record. (B) Coccolith concentrations of *Coccolithus pelagicus*; the black curve is a smoothed record (3pt running average) of original data (open dots).

a tracer of relatively warm bottom waters of Atlantic/ Irminger origins within the East Greenland Current on the Greenland shelf (Jennings and Weiner, 1996) or intermediate Atlantic waters (IIW) off northern Iceland (Eiriksson et al., 2000). Its present distribution over the North Iceland shelf, specifically in hydrographically stratified shelf trough environments where the inflowing NIIC reaches the bottom of the water column, clearly confirms its affinity for modified Atlantic waters (Rytter et al., 2002; Jennings et al., 2004). The natural interpretation for the high percentages of *C. neoteretis* from at least 6 to 3 kyr would be an increasing influence

of IIW on the north Iceland shelf during the mid-Holocene, and associated decreased influence of cold Arctic bottom waters.

4. Discussion and conclusions

The combined coccolith and benthic foraminifera Holocene records are indicative of a clear antagonism in physical–chemical status and sources of water masses between the surface and bottom layers of the water column. In the following discussion, we interpret the

Holocene evolution of water-column structure over the North Iceland shelf as a direct consequences of coupled changes in outflow of Arctic bottom and surface waters from the Nordic Seas and inflow of Atlantic waters around western Iceland.

4.1. Hydrological conditions triggering peak carbonate accumulation in the Iceland Sea

As stated earlier, the very high carbonate sedimentation identified during the ca. 6.5–3.5 cal kyr BP Holocene period is correlative with the peak accumulation of the coccolith species *C. pelagicus*. As reviewed by Baumann et al. (2000), *C. pelagicus* is the cold-end member of the coccolithophore community presently thriving in the Nordic Seas. It is one of the few species able to sustain temperatures close to 0 °C, which explains its overwhelming dominance, albeit with low standing stocks, in the polar community of the East Greenland Current (Samtleben and Schröder, 1992). Further east, toward the Norwegian Sea, as well as south of Iceland, the dominance is progressively shared with *E. huxleyi*, a species which is responsible for the summer development of massive blooms in the subarctic North Atlantic (Brown and Yoder, 1994). Beside its classical temperature constraint and northernmost high latitude distribution, *C. pelagicus* has been identified as an abundant taxon in recently upwelled, nutrient rich waters off Southern Africa (Giraudeau et al., 1993) and Portugal (Cachao and Moita, 2000), as well as in surface sediments below or close to the highly productive Polar Front Zone in the southern ocean (Roth, 1994). This species is therefore considered in paleostudies as a proxy of both cold and highly productive/nutrient-rich waters.

A time series particle flux experiment conducted in the Iceland Sea since 1986 (Fig. 1; Olafsson et al., 2000; joint Iceland Marine Research Institute and Woods Hole Oceanographic Inst. research program) is shedding light into the hydrological processes which presently result in excess sedimentation of *C. pelagicus* in the Nordic Seas. This sediment trap mooring, located to the north-east of the studied core location, captured in the year 1999 a large flux of biogenic carbonate, 86% of which being composed of monospecific coccoliths. This excess carbonate flux (up to 10 times higher than the annual mean) was related to a summer bloom of *C. pelagicus* with an areal extent of ca. 30 000 km^2, i.e. the size of Iceland. According to hydrological data collected in the area for the last 3 decades, conditions which led to this monospecific coccolith bloom are associated with nutrient supply, utilization and depletion, coupled with a recent infiltration of a low salinity surface lens of Arctic water into the Iceland Sea (see Dorinda Ostermann, WHOI Project highlights, www.whoi.edu/science/GG). Beside the classical succession of silicate depletion by siliceous plankton, and bloom progression by non siliceous plankton on the remaining pool of nitrates and phosphates, the presence of a 100 m thick low salinity water inducing deeper mixed layer is thought to be the key element triggering the excess production of *C. pelagicus* at the trap location (Dorinda Ostermann, pers. comm.; Ostermann et al., 2000).

4.2. Implications for the mid-Holocene period of excess carbonate accumulation, and for the long-term Holocene evolution of surface and bottom hydrology

We apply these present-time observations to the mid-Holocene period of excess carbonate accumulation at the core location to imply that this 3.5–6.5 cal kyr interval was characterized by the increased influence of cool, low salinity arctic/polar surface waters. The low contribution of Irminger water species to the coccolith assemblage during this time interval (Fig. 4) indeed suggests that surface waters over the North Iceland shelf were not of Atlantic origin. Such excess carbonate production necessitates a nutrient pool which is hardly consistent when considering the sole nutrient-poor, fresh surface layer of arctic origin.

As shown by Thordardottir (1984), on the basis of hydrological data collected over the last 30 yr, Atlantic waters constitute the main source of nutrients to the North of Iceland. The benthic foraminiferal record from core MD99-2269 suggests that bottom waters over the north Iceland shelf carried a clear Atlantic/Irminger signature during the mid-Holocene period of peak carbonate sedimentation (Fig. 5). We therefore argue that (1) a deepened mixed layer induced by the input of cool, fresh arctic surface waters, coupled with an intermediate/bottom water source of nutrient is capable of inducing excess production and subsequent sedimentation of *C. pelagicus* coccolith carbonates; (2) the flow of nutrient-rich Irminger Intermediate waters around NW Iceland was strongly enhanced during this mid-Holocene period to the prejudice of outflow and abutment of AIW over the North Iceland shelf.

Both assumptions can be reconciled and explained based on recent observations and a conceptual framework of hydrographic and atmospheric variations in the Iceland Sea as reviewed by Malmberg and Jonsson (1997). According to these authors, and following earlier ideas elaborated by Aagaard (1972) and Jonsson (1992), the strength of the wind stress curl over the Iceland Sea is crucial in the mechanisms of intermediate convection in this area. In this, a positive wind stress curl over the Iceland Sea gives rise to an enhanced cyclonic circulation in the gyre and upward Ekman pumping. Such a mechanism pumps salty water into the area and effectively reduces stability of the water column in the gyre making intermediate water production more likely to take place.

In addition, instrumental records (Malmberg and Kristmansson, 1992) show that a strengthened wind stress curl, and therefore more efficient convection in the Iceland Sea, is positively correlated to the amount of surface Irminger waters in the area. Finally, on a more regional basis, this enhanced cyclonic activity in the Iceland Sea is associated with increased winter storms in the Nordic Seas (Dawson et al., 2003).

Conversely, the late 1960s severe ice conditions in North Iceland waters were that of low wind stress curl as well as reduced water column mixing and intermediate water production in the Iceland Sea, but very strong convection in the Greenland Sea (Malmberg and Jonsson, 1997). Instrumental data also revealed a southward shift in winter storm activity, as well as reduced advection of Irminger surface waters, and concomitant increased influence of polar/arctic surface waters north of Iceland (Meincke, 2002; Malmberg and Kristmannsson, 1992). All these conditions were associated with a strengthening and eastward extension of the Greenland High (Dickson et al., 1988).

We use the hydrological/atmospherical conditions conducive to these opposite situations as analogues to explain the recorded long term Holocene evolution of surface and bottom hydrology, as well as changing carbonate production over the North Iceland shelf. The early Holocene up to ca. 7 cal kyr BP as well as a shorter late Holocene period centred at 2 cal kyr BP are seen as times of enhanced atmospheric circulation over the Nordic Seas, enhanced production of AIW in the Iceland Sea, and surface water warming off northern Iceland linked to reduced influence of the arctic/polar water-bearing EIC. Our micropaleontological data additionally suggest that such conditions promote an overall rather limited inflow of Irminger water around NW Iceland, this diminished influence of IIW at the bottom of shelf troughs being filled by AIW. These conditions are thought to be primarily constrained by a positive NAO-type atmospheric pattern of enlarged meridional pressure differences over the North Atlantic.

Conversely, the mid-Holocene (ca. 3.5–6.5 cal kyr BP) interval of enhanced carbonate accumulation is viewed as a time of reduced atmospheric and gyre circulation in the Iceland Sea, both conditions affecting intermediate water production (lowered) and cooling/freshening of the surficial waters (increased) in this area. This surface water cooling refers to the "Neoglacial" cooling starting at ca. 6 cal kyr BP as evidenced in the Denmark Strait and south of Iceland on the ground of Ice Rafted Detritus (Bond et al., 2001; Jennings et al., 2002) and coccolith proxies (Giraudeau et al., 2000). Intermediate waters over the North Iceland shelf during this interval had a strong Atlantic signature, in relation with an enhanced flow of Irminger waters which affected the deeper part of shelf troughs. The peculiar situations at ca. 3 cal kyr BP and 0.2–0.6 cal kyr BP (Little Ice Age)

which are identified by our micropaleontological proxies (Figs. 4 and 5) as periods of both strongly reduced Irminger surface water influence and low coccolith carbonate sedimentation, stand as times of extreme advection of polar, ice bearing waters, and extended sea-ice development (additional evidences are given by Knudsen and Eiriksson, 2002; see also Lamb, 1979; Jennings et al., 2002).

The diatom record constructed from MD99-2269 (Andersen et al., 2004) is generally consistent with the coccolith dataset presented herein. Diatom assemblages are indicative of a general surface water thermal optimum off northern Iceland from 10 to 6.5 cal kyr BP, but fluctuating sea-surface conditions with evidences for seasonal sea–ice conditions which is consistent with the early Holocene benthic foraminiferal assemblages containing N. labradorica (see previous chapter). Andersen et al. (2004) identified the 6.5–3 cal kyr BP period as the "Holocene Transition Period" with a gradual cooling of surface waters of about 4 °C suggesting increased influence of arctic surface waters carried by the EIC. As suggested by our data set, the afore-mentioned authors argued that this surface water temperature changes are mostly driven by changes in the strength and location of the Icelandic Low pressure cell and associated westerly winds.

4.3. Extending our findings to the Holocene evolution of the northern North Atlantic

Assessing that the Holocene history of the water mass structure over the North Iceland shelf is primarily driven by atmospheric processes acting at the scale of the northern North Atlantic implies that the long term hydrological evolution of other areas of the Nordic Seas, in particular the eastern margin, might have been affected by the same processes.

In this regard it is important to note that the previously discussed summer AD 1999 episode of peak coccolith carbonate production in the Iceland Sea (Olafsson et al., 2000), was part of a larger pattern of increased carbonate production in other high latitude regional seas. This year of excess production of carbonate plankton in the Iceland Sea, shortly followed the year AD 1996 which saw an anomalous drop of the NAO index, a unique return over the last 40 years to the negative values typical of the late 1960's. Based on hydrographic transects across 48°N, Meincke (2002) indicated that the northward oceanic heat transport across this latitude dropped by 50% the year immediately following AD 1996, thereby suggesting that atmospheric anomalies of this kind have the potential of inducing global cooling in the Nordic Seas. Following this AD 1997 drop in northward heat transport, the Bering, Norwegian and Barents Seas each experienced major coccolithophore blooms (Ostermann, pers.

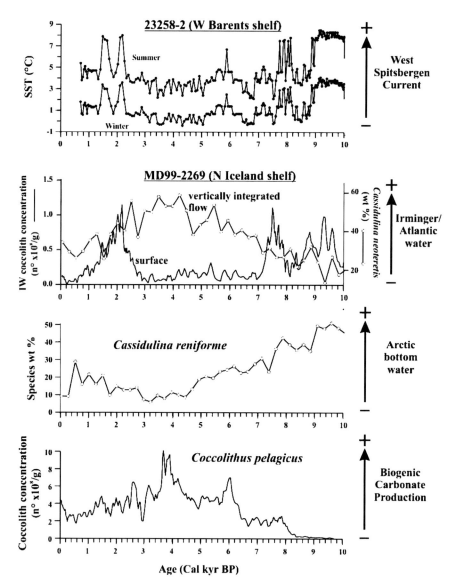

Fig. 6. Synthetic plot showing (bottom) selected coccolith and benthic foraminiferal records in core MD99-2269 as proxies of water column dynamics and carbonate productivity, compared (top) with SSTs estimates over the W Barents shelf (Sarnthein et al., 2003).

comm.) during the summer of AD 1998. The Iceland Sea peak carbonate production event in summer AD 1999 might stand as the ultimate manifestation (not repeated in the Nordic seas in AD 2000 or AD 2001) of a 3-year suite of hydro-biological changes induced by a drastic modification in atmospheric circulation pattern.

Instrumental data and observations of present hydrological/climatological anomalies in the Nordic Seas therefore suggest an in-phase response of the southwestern and northeastern domains to changes in the North Atlantic atmospheric pressure patterns. We hypothesize that this relationship is valid when considering the Holocene long-term evolution of hydrology in the Nordic Seas. The recently published record of Holocene SST's changes off the Western Barents shelf (Sarnthein et al., 2003) indeed correlates very well with the main phase of surface water temperature changes off

Northern Iceland (positive relationship) as well as with the overall inflow of Iminger water round NW Iceland (negative relationship) (Fig. 6). This planktic foraminiferal-based estimation is indicative of a much enhanced West Spitsbergen Current during the early Holocene up to 7.7 cal kyr BP as well as around 2 cal kyr BP, both periods being characterized off Northern Iceland by surface water warming and by limited inflow of intermediate Irminger water (Fig. 6). Additional evidence for the regional-scale correlation of changes in atmospheric pressure pattern comes from the estimates of winter precipitation in Southern Norway over the last 10 cal kyr as summarized by Nesje (2002), which shows two main episodes of dry conditions at ca. 8.2 cal kyr BP as well as within the mid-Holocene between 6 and 4 cal kyr BP. Such dry (and cold) conditions over central southern Norway are thought to reflect reduced westerlies and cyclone activity, i.e. a predominantly "negative NAO index weather mode" (Nesje, 2002).

The close correlation between Atlantic water circulation changes and precipitation changes in the northeastern North Atlantic on the one hand, and the pattern of hydrological changes in the Iceland Sea on the other hand, is a clear manifestation that these far-off areas are reacting to the same changes in regional-scale atmospheric patterns. Data presented in this paper suggest that the strength of Atlantic inflow into the Nordic seas was subjected during the Holocene to a balance between the Irminger and the Norwegian (and West Spitsbergen) branches, and that it is mostly related to patterns of atmospheric circulation, via the intensity and location of westerly winds. Our conclusions slightly differ from Andersen et al. (2004) who, based on diatom assemblages only, argue for a common dynamics of both branches of the Atlantic inflow into the Nordic Seas. This disagreement might come from the summer surface ocean temperature signature of diatom records which is more likely to be explained in the light of the Holocene trend in summer insolation at high northern latitudes, than in view of winter-related atmospheric processes (NAO). We argue that the main reason for the observed discrepancy lays in the use, in the present work, of a coupled surface and benthic proxy record, which suggests that the strength of the vertically integrated Irminger water flow (ca. the top 500 m of water column over the North Iceland shelf) was rather negatively correlated with the temperature signature of the surface layer where coccolithophore and diatom populations thrive, over the last 10 000 cal yr.

Whereas it is tempting to consider that these Holocene ocean circulation changes were induced by a "NAO-like" modulation of wind-regime, it is important to notice that contrary to the Norwegian Sea, instrumental records do not seem to show any connection between the NAO index and the Atlantic inflow off Northern Iceland (Olafsson, 1999). Beside the standard problem of using instrumentally verified, short-terms (multi-year in the case of NAO) events as analogs for much longer and persistent events as detected in the sedimentological records, local atmospheric forcing other than NAO might influence ocean circulation around Iceland.

Acknowledgements

Core MD99-2269 was recovered in 1999 as part of the GINNA/IMAGES V, Leg 3 cruise of the RV Marion Dufresne. We gratefully acknowledge Yvon Balut (IPEV) and J.-L. Turon (DGO, chief scientist) for the outstanding coring operations. The cruise was supported by the French Polar Institute (IPEV) and USA participation was funded through grant NSF-OCE98-09001 and by NSF-OPP-0004233. This paper is also a contribution to an Earth System History (ESH) grant from NSF ATM-0317832. We thank D. Ostermann (WHOI) for fruitful discussions on extant blooms of coccolithophores in the Iceland Sea. A. Jennings and J. Andrews thank Nancy Weiner for her work on the foraminifera. Thanks are extended to E. Jansen and P. de Menocal, coordinators of the IMAGES Holocene Working Group Workshop (Hafslo, August 2003) for a superbly organized meeting. This is DGO-UMR 5805 EPOC contribution no. 1520.

References

Aagaard, K., 1972. On the drift of the Greenland pack ice. In: Karlsson, T. (Ed.), Sea Ice. Proceedings of an International Conference, National Research Council, pp. 17–22.

Andersen, C., Koç, N., Jennings, A., Andrews, J.T., 2004. Non uniform response of the major surface currents in the Nordic Seas to insolation forcing: implications for the Holocene climate variability. Paleoceanography 19 PA 2003.

Andrews, J.T., Giraudeau, J., 2003. Multi-proxy records showing significant Holocene environmental variability: the inner N. Iceland shelf (Hunafloi). Quaternary Science Reviews 22, 175–193.

Andrews, J.T., Hardardottir, J., Helgadottir, G., Jennings, A.E., Geirsdottir, A., Sveinbjornsdottir, A.E., Schoolfield, S., Kristjansdottir, G.B., Smith, L.M., Thors, K., Syvitski, J.P.M., 2000. The N and W Iceland Shelf: insights into Last Glacial Maximum ice extent and deglaciation based on acoustic stratigraphy and basal radiocarbon AMS dates. Quaternary Science Reviews 19, 619–631.

Andrews, J.T., Geirsdottir, A., Hardardottir, J., Principato, S., Gronvold, K., Kristjansdottir, G.B., Helgadottir, G., Drexler, J., Sveinbjornsdottir, A., 2002. Distribution, sediment magnetism, and geochemistry of the Saksunarvatn (10.18 ± cal ka) tephra in marine, lake, and terrestrial sediments, NW Iceland. Journal of Quaternary Science 17, 731–745.

Andrews, J.T., Hardardottir, J., Stoner, J.S., Mann, M.E., Kristjansdottir, G.B., Koç, N., 2003a. Decadal to millennial-scale periodicities in North Iceland shelf sediments over the last 12 000 cal yr: long-term North Atlantic oceanographic variability and solar forcing. Earth and Planetary Science Letters 210, 453–465.

Andrews, J.T., Hardardottir, J., Kristjansdottir, G.B., Grönvold, K., Stoner, J.S., 2003b. A high-resolution Holocene sediment record from Hunafloaall, N Iceland margin: century- to millennial-scale variability since the Vedde tephra. The Holocene 13, 625–638.

Andruleit, H.A., 1996. A filtration technique for quantitative studies of coccoliths. Micropaleontology 42, 403–406.

Andruleit, H.A., 2000. Dissolution-affected coccolithophore fluxes in the central Greenland Sea (1994/1995). Deep-Sea Research II 47, 1719–1742.

Andruleit, H.A., Baumann, K.-H., 1998. History of the last deglaciation and Holocene in the Nordic Seas as revealed by coccolithophore assemblages. Marine Micropaleontology 35, 179–201.

Baumann, K.-H., Andruleit, H.A., Samtleben, C., 2000. Coccolithophores in the Nordic Seas: comparison of living communities with surface sediment assemblages. Deep-Sea Research II 47, 1743–1772.

Beaufort, L., Heussner, S., 1999. Coccolithophorids on the continental slope of the Bay of Biscay—production, transport and contribution to mass fluxes. Deep-Sea Research II 46, 2147–2174.

Blindheim, J., Borovkov, V., Hansen, B., Malmberg, S.-A., Turrell, W.R., Osterhus, S., 2000. Upper layer cooling and freshening in the Norwegian Sea in relation to atmospheric forcing. Deep-Sea Research I 47, 655–680.

Bond, G., Kromer, B., Beer, J., Muscheler, R., Evans, M.N., Showers, W., Hoffman, S., Lotti-Bond, R., Hajdas, I., Bonani, G., 2001. Persistent solar influence on North Atlantic climate during the Holocene. Science 294, 2130–2135.

Briffa, K.R., Bartholin, T.S., Eckstein, D., Jones, P.D., Karlén, W., Schweingruber, F.H., Zetterberg, P., 1990. A 1400-year tree-ring record of summer temperatures in Fennoscandia. Nature 346, 434–439.

Brown, C.W., Yoder, J.A., 1994. Coccolithophorid blooms in the global ocean. Journal of Geophysical Research 99, 7467–7482.

Cachao, M., Moita, M.T., 2000. Coccolithus pelagicus, a productivity proxy related to moderate fronts off Western Iberia. Marine Micropaleontology 39, 131–155.

Corliss, B.H., 1985. Microhabitats of benthic foraminifera within deep-sea sediments. Nature 314, 435–438.

Corliss, B.H., 1991. Morphology and microhabitat preferences of benthic foraminifera from the northwest Atlantic Ocean. Marine Micropaleontology 17, 195–236.

Dawson, A.G., Elliott, L., Mayewski, P., Lockett, P., Noone, S., Hickey, K., Holt, T., Wadhams, P., Foster, I., 2003. Late-Holocene North Atlantic climate "seesaws", storminess changes and Greenland ice sheet (GISP2) palaeoclimates. The Holocene 13, 383–394.

Dickson, R.R., Meincke, J., Malmberg, S., Lee, A., 1988. The "Great Salinity Anomaly" in the northern North Atlantic 1968–1982. Progress in Oceanography 20, 103–151.

Eiriksson, J., Knudsen, K.L., Haflidason, H., Heinemeier, J., 2000. Chronology of late Holocene climatic events in the northern North Atlantic based on AMS 14C dates and tephra markers from the volcano Hekla, Iceland. Journal of Quaternary Science 15, 573–580.

Giraudeau, J., Monteiro, P.S., Nikodemus, K., 1993. Distribution and malformation of living coccolithophores in the northern Benguela upwelling system off Namibia. Marine Micropaleontology 22, 93–110.

Giraudeau, J., Cremer, M., Manthé, S., Labeyrie, L., Bond, G., 2000. Coccolith evidence for instabilities in surface circulation south of Iceland during Holocene times. Earth and Planetary Science Letters 179, 257–268.

Grönvold, K., Oskarsson, N., Johnsen, S.J., Clausen, H.B., Hammer, C.U., Bond, G., Bard, E., 1995. Ash layers from Iceland in the Greenland GRIP ice core correlated with oceanic and land sediments. Earth and Planetary Science Letters 135, 149–155.

Hopkins, T.S., 1991. The GIN Sea—a synthesis of its physical oceanography and literature review 1972–1985. Earth Science Reviews 30, 175–318.

Jennings, A.E., Helgadottir, G., 1994. Foraminiferal assemblage from the fjords and shelf of Eastern Greenland. Journal of Foraminiferal Research 24, 123–144.

Jennings, A.E., Weiner, N.J., 1996. Environmental change in eastern Greenland during the last 1300 years: evidence from foraminifera and lithofacies in Nansen Fjord, 68°N. The Holocene 6, 179–191.

Jennings, A.E., Knudsen, K.L., Hald, M., Hansen, C.V., Andrews, J.T., 2002. A mid-Holocene shift in Arctic Sea ice variability on the East Greenland shelf. The Holocene 12, 49–58.

Jennings, A.E., Weiner, N.J., Helgadottir, G., Andrews, J.T., 2004. Modern foraminiferal faunas of the SW to N Iceland shelf: oceanographic and environmental controls. Journal of Foraminiferal Research 34, 180–207.

Jonsson, S., 1992. Sources of fresh water in the Iceland sea and the mechanisms governing its interannual variability. ICES Marine Science Symposia 196, 62–67.

Knudsen, K.L., Eiriksson, J., 2002. Application of tephrochronology to the timing and correlation of palaeoceanographic events recorded in Holocene and Late Glacial shelf sediments off North Iceland. Marine Geology 191, 165–188.

Kristjansdottir, G.B., 2002. Holocene Hekla tephras: a stratigraphic tool for estimating changes in reservoir age of seawater, core MD99-2269, NW Iceland Shelf. 32nd Arctic Workshop, University of Colorado, Boulder, abstract vol., 103–104.

Lamb, H.H., 1979. Climatic variations and changes in the wind and ocean circulation: The Little Ice Age in the Northeast Atlantic. Quaternary Research 11, 1–20.

Malmberg, S.-A., Jonsson, S., 1997. Timing of deep convection in the Greenland and Iceland Seas. ICES Journal of Marine Science 54, 300–309.

Malmberg, S.-A., Kristmannsson, S.S., 1992. Hydrographic conditions in Icelandic waters, 1980–1989. ICES Marine Science Symposia 195, 76–92.

Meincke, J., 2002. Climate dynamics of the North Atlantic and NW-Europe: an observation-based overview. In: Wefer, G., Berger, W., Behre, K.-E., Jansen, E. (Eds.), Climate Development and History of the North Atlantic Realm. Springer, Berlin, Heidelberg, pp. 25–40.

Nesje, A., 2002. Late Glacial and Holocene glacier fluctuations and climatic variations in Southern Norway. In: Wefer, G., Berger, W., Behre, K.-E., Jansen, E. (Eds.), Climate Development and History of the North Atlantic Realm. Springer, Berlin, Heidelberg, pp. 233–258.

Oehmig, R., Wallrabe-Adams, H.J., 1993. Hydrodynamics properties and grain-size characteristics of volcaniclastic deposits on the Mid-Atlantic Ridge North of Iceland. Journal of Sedimentary Petrology 63, 104–152.

Olafsson, J., 1999. Connections between oceanic conditions off N-Iceland, Lake Myvatn temperature, regional wind direction variability and the North Atlantic Oscillation. Rit Fiskideildar 16, 41–57.

Olafsson, J., Ostermann, D., Curry, W.B., Honjo, S., Manganini, S.J., 2000. Time series particle fluxes from the Iceland Sea: 1986 to present. Eos Trans. AGU, 81 (48), Fall Meet. Suppl., Abstract OS22B-08.

Ostermann, D., Curry, W.B., Honjo, S., Olafsson, J., Manganini, S.J., 2000. Time series foraminiferal fluxes from the Iceland Sea: 1987 to present. AGU Ocean Sciences Meeting 2000, Abstract OS32E-11.

Polyak, L., Korsun, S., Febo, L., Stanovoy, V., Khusid, T., Mald, M., Paulsen, B.E., Lubinski, D.A., 2002. Benthic foraminiferal assemblages from the southern Kara Sea, a river-influenced arctic marine environment. Journal of Foraminiferal Research 32, 252–273.

Roth, P.H., 1994. Distribution of coccoliths in oceanic sediments. In: Winter, A., Siesser, W.G. (Eds.), Coccolithophores. Cambridge University Press, Cambridge, pp. 199–210.

Rytter, F., Knudsen, K.L., Seidenkrantz, M.-S., Eiriksson, J., 2002. Modern distribution of benthic foraminifera on the North Iceland shelf and slope. Journal of Foraminiferal Research 32, 217–244.

Samtleben, C., Schröder, A., 2002. Living coccolithophore communities in the Norwegian-Greenland sea and their record in sediments. Marine Micropaleontology 19, 333–354.

Samtleben, C., Schäfer, P., Andruleit, H., Baumann, A., Baumann, K.-H., Kohly, A., Matthiessen, J., Schröder-Ritzrau, A., 1995. SYNPAL Working Group. Plankton in the Norwegian-Greenland Sea: from living communities to sediment assemblages—an actualistic approach. Geologische Rundschau 84, 108–136.

Sarnthein, M., Van Kreveld, S., Erlenkeuser, H., Grootes, P.M., Kucera, M., Pflaumann, U., Schulz, M., 2003. Centennial-to-millenial-scale periodicities of Holocene climate and sediment injections off the western Barents shelf, 75°N. Boreas 32, 447–461.

Seidenkrantz, M.-S., 1995. *Cassidulina teretis* Tappan and *Cassidulina neoteretis* new species (Foraminifera): stratigraphic markers for deep sea and outer shelf areas. Journal of Micropalaeontology 14, 145–157.

Stefansson, U., 1962. North Icelandic Waters. Rit Fiskildeidar III Bind, vol. 3, 269pp.

Stuiver, M., Reimer, P.J., Bard, E., Beck, J.W., Hughen, K.A., Kromer, B., McCormack, F.G., van des Plicht, J., Spurk, M., 1998. INTCAL98 radiocarbon age calibration 24,000–0 cal BP. Radiocarbon 40, 1041–1083.

Swift, J.H., 1986. The Arctic waters. In: Hurdle, B.G. (Ed.). The Nordic Seas. Springer, New York, 777pp.

Thordardottir, T. 1984. Primary production north of Iceland in relation to water masses in May–June 1970–1989. ICES CM 1984/ L:20, 17p.

Wollenburg, J.E., Mackensen, A., 1998a. Living benthic foraminifers from the central Arctic Ocean: faunal composition, standing stock and diversity. Marine Micropaleontology 34, 153–185.

Wollenburg, J.E., Mackensen, A., 1998b. On the vertical distribution of living (rose bengal stained) benthic foraminifers in the Arctic Ocean. Journal of Foraminiferal Research 28, 268–285.

Young, J.R., Ziveri, P., 2000. Calculation of coccolith volume and its use in calibration of carbonate flux estimates. Deep-Sea Research II 47, 1679–1700.

Quaternary Science Reviews 23 (2004) 2141–2154

North Pacific and North Atlantic sea-surface temperature variability during the Holocene

Jung-Hyun Kim[a,*], Norel Rimbu[a], Stephan J. Lorenz[b], Gerrit Lohmann[a,c],
Seung-Il Nam[d], Stefan Schouten[e], Carsten Rühlemann[f], Ralph R. Schneider[g]

[a]*FB5 Geowissenschaften, Universität Bremen, Klagenfurterstraße, D-28359 Bremen, Germany*
[b]*Max-Planck-Institut für Meteorologie, Modelle und Daten, Bundesstrasse 53, D-20146 Hamburg, Germany*
[c]*Alfred-Wegener-Institut für Polar- und Meeresforschung (AWI), Bussestraße 24, D-27570 Bremerhaven, Germany*
[d]*Petroleum and Marine Resources Division, Korea Institute of Geoscience and Mineral Resources, 305-350 Taejon, Korea*
[e]*Royal Netherlands Institute for Sea Research (NIOZ), P.O. Box 59, 1790 AB, Den Burg, Texel, The Netherlands*
[f]*Bundesanstalt für Geowissenschaften und Rohstoffe, Referat B 3.23—Meeresgeologie, Stilleweg 2, D-30655 Hannover, Germany*
[g]*Département de Géologie et Océanographie, UMR5805-EPOC, CNRS/Université de Bordeaux1, 33405 Talence Cedex, France*

Abstract

Holocene climate variability is investigated in the North Pacific and North Atlantic realms, using alkenone-derived sea-surface temperature (SST) records as well as a millennial scale simulation with a coupled atmosphere-ocean general circulation model (AOGCM). The alkenone SST data indicate a temperature increase over almost the entire North Pacific from 7 cal kyr BP to the present. A dipole pattern with a continuous cooling in the northeastern Atlantic and a warming in the eastern Mediterranean Sea and the northern Red Sea is detected in the North Atlantic realm. Similarly, SST variations are opposite in sign between the northeastern Pacific and the northeastern Atlantic. A 2300 year long AOGCM climate simulation reveals a similar SST seesaw between the northeastern Pacific and the northeastern Atlantic on centennial time scales. Our analysis of the alkenone SST data and the model results suggests fundamental inter-oceanic teleconnections during the Holocene.
© 2004 Elsevier Ltd. All rights reserved.

1. Introduction

The oxygen isotope composition of polar ice sheets (e.g., Grootes et al., 1993) suggests a relatively stable Holocene climate when compared to the last glacial and viewed in a long-term perspective. In the context of short-term variability however, significant climate changes have also been identified for the Holocene although their amplitudes are remarkably small in comparison with those during the last glacial (e.g., Bond et al., 2001; Moy et al., 2002; Schulz and Paul, 2002; Oppo et al., 2003).

Recently, reconstructions of sea-surface temperatures (SSTs) for the North Atlantic also showed significant

long-term, regionally coherent climate changes throughout the Holocene (Marchal et al., 2002; Rimbu et al., 2003). For the northeastern Atlantic, Marchal et al. (2002) identified a cooling trend over the last 10,000 years. Rimbu et al. (2003) suggested that this cooling trend is part of a regional SST pattern that resembles a modern SST-pattern, which is related to the Arctic Oscillation/North Atlantic Oscillation (AO/NAO) (e.g., Hurrell, 1995; Thompson and Wallace, 1998; Thompson et al., 2000). Analyzing global SST records, Lorenz and Lohmann (2004) show that the AO/NAO-like SST pattern in the North Atlantic realm is part of a global SST pattern showing a slight warming over almost the entire tropics and a cooling in the extra-tropics from 7 cal kyr BP to the present. Based on long-term transient climate simulations, Lorenz and Lohmann (2004) attributed these trends to solar insolation forcing associated with the Earth's precessional variation during

---

*Corresponding author. Tel.: +49-421-218-8929; fax: +49-421-218-8916.

E-mail address: jungkim@allgeo.uni-bremen.de (J.-H. Kim).

0277-3791/$ - see front matter © 2004 Elsevier Ltd. All rights reserved.
doi:10.1016/j.quascirev.2004.08.010

the Holocene. The Holocene climate changes caused by the external forcing seem to be connected by climate modes (Rimbu et al., 2003, 2004; Lorenz et al., 2004). However, it is not well known how these climate modes induce inter-oceanic teleconnected climate variability during the Holocene.

The aim of this study is to investigate the spatial and temporal SST variations in the North Pacific and North Atlantic realms for the middle to late Holocene and the teleconnections between the two oceans. For this purpose, we used a set of North Pacific and North Atlantic alkenone-derived SST records covering the last 7000 years. The information in this data set is combined with the data analysis of a coupled atmosphere–ocean general circulation model (AOGCM) simulation. This will provide useful information on the dominant modes of climate variability in the North Pacific and the North Atlantic as well as on a common climate mode that the two oceans share.

The paper is organized as follows. Data and methods are described in Section 2. The main results follow in Section 3 with the SST trends and related spatial pattern in the Northern Hemisphere and with the Pacific–Atlantic teleconnections inferred from a control integration with a coupled AOGCM. Possible underlying mechanisms and summary and main conclusions are presented in Sections 4 and 5, respectively.

2. Data and methods

The inspection of Holocene climate trends is based on the statistical analysis of alkenone-derived SST records from the North Atlantic and North Pacific realms and of data from a control integration of a coupled atmosphere–ocean–sea ice model.

In this study, four new SST records and 32 published records compiled from the literature (Table 1) covering North Pacific and North Atlantic regions were analysed. In order to avoid potential biases due to using different SST proxies, the paleotemperature estimates of all sediment cores were based solely on the alkenone method, converting the abundance ratios of long-chain unsaturated alkenones with two to three double bounds into annual mean SST (e.g., Brassell et al., 1986; Prahl et al., 1988; Müller et al., 1998). The alkenone SST proxy has been internationally calibrated and standardized amongst 24 laboratories worldwide (Rosell-Melé et al., 2001). The error for alkenone temperature estimates is in the order of 0.5 and 1.5 °C, for the calibration with culture samples (Prahl and Wakeham, 1987) and surface sediments (Müller et al., 1998), respectively. Analytical precision for each record considered here, however, was much better than the calibration errors reported by Prahl and Wakeham (1987) and Müller et al. (1998) (Table 1). Although different alkenone unsaturation

indices (U_{37}^K or $U_{37}^{K\prime}$) and calibrations were applied for each alkenone SST record (Table 1), it does not affect the comparison of relative SST changes. The core chronologies have been described in previous studies (Table 1). All alkenone SST records compiled in this study will be archived in the WDC-MARE/PANGAEA data base (http://www.pangaea.de/Projects/GHOST, Kim and Schneider, 2004).

In order to identify coherent temporal and spatial variations in the Holocene SST data set, different statistical methods were used. At first, the linear regression analysis for each individual record was applied in order to calculate the linear trend. Then, the spatial distribution of the SST variability was investigated separately for the North Pacific and North Atlantic regions, using an Empirical Orthogonal Function (EOF) analysis (von Storch and Zwiers, 1999). The SST time series were reduced into spatially coherent orthogonal eigenvectors. These eigenvectors, together with their corresponding time coefficients (Principal Component, PC) and eigenvalues (a measure of the variance described by each eigenvector) are referred to in our study as modes of SST variability. The EOF method also served as a data-filtering procedure to smooth the noise and uncertainties in the age models of the individual SST records.

The temporal patterns of the time coefficients of the dominant modes of alkenone SST variability were investigated using the Singular Spectrum Analysis (SSA) (Ghil et al., 2002). The SSA is designed to decompose a short and noisy time series into three statistically independent components: trends, oscillatory patterns, and noise. The trends do not need to be linear and the oscillations can be amplitude and phase modulated (Ghil et al., 2002). In our analysis all time components with a time scale longer than 3 kyr (i.e. the size of the window) are referred to as trends while the time components with time scales smaller than 3 kyr are referred to as centennial to millennial scale variations.

The EOF and SSA methods used here require that the SST values from different records are available for identical time resolution. Here, mean SST values in 100-year time resolution were derived using linear interpolation. The SST anomalies against the SST mean over the considered period were calculated for each record and normalized with the corresponding standard deviation. The normalized time series were combined to obtain two separate space–time SST data sets, one for the North Pacific realm and the other for the North Atlantic realm.

In addition to the alkenone SST data, we employed the analysis from a 3000 year control simulation of the pre-industrial era with the ECHO-G coupled atmosphere–ocean model (Lorenz and Lohmann, 2004). The ECHO-G consists of the ECHAM-4 atmospheric general circulation model, which is coupled through the OASIS program with the HOPE ocean circulation

Table 1
Core locations and information on the chronology, alkenone index, alkenone calibration, and analytical reproducibility for each SST record discussed in this paper

No.	Core name	Lat.	Long.	Water depth (m)	Dating	14C dating no. used for age model	Mean sample interval (min.-max.) (cm)	Meantime resolution (min.-max.) (yr)	Alkenone index	Alkenone calibration	Reproducibility (°C)	Reference
New data from this study												
2	SSDP-102	34.953	128.881	40	AMS14C	3	~13 (1–35)	~61 (5–190)	UK'37	Prahl et al. (1988)	0.1 (±1σ)	This study
19	GeoB 5901-2	36.380	−7.071	574	AMS14C	13	~1(1–2)	~80 (35–317)	UK'37	Prahl et al. (1988)	0.3 (±1σ)	This study
35	74KL	14.321	57.347	3212	AMS14C	3	~3 (2.5–5)	~281(208–722)	UK'37	Prahl et al. (1988)	0.2 (±1σ)	This study: Sirocko et al. (1993)
36	TY93-905	11.067	51.950	1567	AMS14C	4	5	~216 (140–273)	UK'37	Prahl et al. (1988)	0.3 (±1σ)	This study
Data from literature												
North Pacific Realm												
1	GGC15	48.168	151.337	1980	AMS 14C	2	5	~1058 (920–1200)	UK'37	Müller et al. (1998)	0.025 UK'37 (±1σ)	Keigwin (1998) and Ternois et al. (2000)
3	ST.14	32.668	138.455	3252	AMS 14C	2	~6 (0.8–19,2)	~660(116–910)	UK'37	Prahl et al. (1988)	—	Sawada and Handa (1998)
4	ST.19	31.095	138.665	3336	AMS 14C	1	~10 (7.2–28.8)	~883 (369–1090)	UK'37	Prahl et al. (1988)	—	Sawada and Handa (1998)
5	17940-2	20.117	117.383	1727	AMS 14C	10	~8 (4–13)	~153 (70–275)	UK'37	Müller et al. (1998)	0.15 (±1σ)	Pelejero et al. (1999a) and Wang et al. (1999)
6	SCS90-36	17.995	111.494	2050	AMS 14C	1	~3 (1–5)	~800 (255–1274)	UK'37	Prahl et al. (1988)	—	Huang et al. (1997)
7	18252-3	9.233	109.383	1273	AMS 14C	0	~14 (10–30)	~587 (427–1281)	UK'37	Pelejero and Grimalt (1997)	0 2< (±1σ)	Kienast and McKay (2001)
8	17964	6.158	112.213	1556	AMS 14C	2	~19 (9–20)	~712 (353–784)	UK'37	Müller et al. (1998)	0.15 (±1σ)	Pelejero et al. (1999a, b) and Wang et al. (19990)
9	JT96-0909pc	48.912	−126.890	920	AMS 14C	0	~2 (0.5–4)	~406 (90–723)	UK'37	Müller et al. (1998)	0.3 (±1σ)	Kienast and McKay (2001)
10	ODP 1019C	41.682	−124.930	980	AMS 14C, isotope stratigraphy[a]	3	~5(4–11)	~132 (85–289)	UK'37	Prahl et al. (1988)	015 (±1σ)	Barron et al. (2003)
11	ODP 1017E	34.535	−121.107	955	AMS 14C	0	~8 (2.9–17.2)	~363 (129–793)	UK'37	Prahl et al. (1988)	0.005 UK'37 (±1σ)	Kennett et al. (2000) and Ostertag-Henning and Stax (2000)
12	LAPAZ21P	22.990	−109.47	624	SPECMAP chronology[b]	—	~5 (4–5)	~769 (633–792)	UK'37	Prahl et al. (1988)	0.15 (±1σ)	Herbert et al. (2001)
North Atlantic and Indian Ocean Realm												
13	M23258-2	74.995	13.970	1768	AMS14C	5	~3(2–6)	~196(119–349)	UK37	Rosell-Melé et al. (1995)	—	Marchal et al. (2002) and Sarnthein et al. (2003)
14	MD952011	66.967	7.633	1048	AMS14C	9	~4 (1–15)	~58 (4–245)	UK37	Prahl and Wakeham (1987)	0.15 (±1σ)	Calvo et al. (2002)
15	MD952015	58.762	−25.958	2630	AMS14C	?	~5 (5–10)	~101 (75–201)	UK'37	Müller et al. (1998)	—	Marchal et al. (2002)
16	IOW225514	57.838	8.704	420	AMS 14C	6	~5 (1–5)	~68 (8–271)	UK'37	Müller et al. (1998)	—	Emeis et al. (2003)
17	IOW225517	57.667	7.091	293	AMS14C	6	~5 (1–10)	~94 (18–179)	UK'37	Müller et al. (1998)	—	Emeis et al. (2003)
IS	SU81-18	37.767	−10.183	3135	AMS14C	5	~10 (9–10)	~736 (568–837)	UK'37	Prahl et al. (1988)	0.3	Bard et al. (2000)
20	M39-008	36.382	−7.077	576	AMS 14C	4	5	~545 (178–1265)	UK'37	Müller et al. (1998)	0.15 (±1σ)	Cacho et al. (2001)
21	MD952043	36.143	−2.622	1841	AMS14C	4	~5 (2–10)	~133 (66–223)	UK'37	Müller et al. (1998)	0.15 (±1σ)	Cacho et al. (1999)

Table 1 (continued)

No.	Core name	Lat.	Long.	Water depth (m)	14C dating no. used for age model	Dating	Mean sample interval (min.-max.) (cm)	Meantime resolution (min.-max.) (yr)	Alkenone index	Alkenone calibration	Reproducibility (°C)	Reference
22	BS79-38	38.412	13.577	1489	—	Isotope stratigraphy [c]	~7 (5-10)	~329 (216-520)	UK'37	Müller et al. (1998)	0.15 (±1σ)	Cacho et al. (2001)
23	M40-4-SL78/78MUC8	37.036	13.19	470	1	AMS 14C, Isotope stratigraphy [d]	~3 (1-12)	~124 (33-498)	UK'37	Müller et al. (1998)	—	Emeis and Dawson (2003) and Emeis (unpublished)
24	AD91-17	40.870	18.636	844	3	AMS 14C, isotope stratigraphy [c]	~4 (2-6)	~158 (45-442)	UK'37	Müller et al. (1998)	—	Giunta et al. (2001)
25	RL11	36.746	17.718	3376	2	AMS14C	~1 (1-3)	~261 (105-804)	UK'37	Müller et al. (1998)	—	Emeis et al. (2000)
26	M44-KL71	40.842	27.763	566	0	AMS14C	~4 (2-6)	~324 (157-619)	UK'37	Müller et al. (1998)	—	Sperling et al. (2003)
27	ODP 658C	20.750	-18.583	2263	5	AMSI4C	~2(2-11)	~100 (30-370)	UK'37	Prahl et al. (1988)	—	Zhao et al. (1995) and deMenocal et al. (2000)
28	BOFS 31K	19.000	-20.167	3300	0	Isotope stratigraphy [f]	2	~909 (900-910)	UK'37	Prahl et al. (1988)	0.017 UK'37	Zhao et al. (1995) and Chapman et al. (1996)
29	ODP 1002C	10.712	-65.17	893		SPECMAP chronology [g]	~23 (18-40)	~503 (400-890)	UK'37	Prahl et al. (1988)	0.3	Herbert and Schuffert (2000) and Peterson et al. (2000)
30	M35003-4	12.083	-61.250	1299	3	AMSI4C	5	~291 (256-297)	UK'37	Müller et al. (1998)	0.3 (±1σ)	Rühlemann et al. (1999)
31	ODP 967D	34.071	32.726	2551	—	SPECMAP chronology [h]	~5 (1-15)	~309 (89-793)	UK'37	Müller et al. (1998)	—	Emeis et al. (2000)
32	GeoB 5844-2	27.714	34.682	963	5	AMS 14C	2	~307 (215-364)	UK'37	Prahl et al. (1988)	0.6 (±1σ)	Arz et al. (2003)
33	SO90-39KG/56KA	24.834	65.917	695	19	Varve chronology, AMS and conventional 14C	~2(1-15)	~20 (0.2-86)	UK'37	Sonzogni et al. (1997)	0.3 (±1σ)	von Rad et al. (1999) and Doose-Rolinski et al. (2001)
34	SO93-126KL	19.973	90.034	1253	—	SPECMAP chronology [g]	~5 (3-6)	~654 (364-1097)	UK'37	Sonzogni et al. (1997)		Kudrass et al. (2001)

UK'37 = C37:2/(C37:2 + C37:3) according to Prahl and Wakeham (1987); UK37 = (C37:2-C37:4)/(C37:2 + C37:3 + C37:4) according to Brassell et al. (1986); Prahl and Wakeham (1987): T (°C = (UK37 + 0.11)/0.04; Müller et al. (1998): T(°C) = (UK'37-0.044)/0.033; Prahl et al. (1988): T (°C) = (UK'37-0.039)/0.034; Sonzogni et al. (1997): T(°C) = (UK'37-0.317)/0.023; Pelejero and Grimalt (1997): T (°C) = (UK'37-0.092)/0.031; Rosell-Melé et al. (1995): T(°C) = (uk37-0.093)/9.03.

[a]Correlation with GISP2 oxygen isotope record (Grootes and Stuiver, 1997).

[b]The age model was obtained by correlation of the benthic δ18O record with the composite records of Shackleton (2000).

[c]The age model was obtained by correlation of the planktonic δ18O records with those reported from Capotondi el al. (1999).

[d]Comparison with ODP 963 data from the same location (Sprovieri et al., pers.comm.)

[e]The age model of the core AD91-17 is based on AMS 14C, planktonic isotopic record from Capotondi et al. (1999) and the correlation with MD90-917 planktonic isotopic record (Siani et al., 2001).

[f]The chronology was obtained by attributing ages to the key stratigrahic events, which have been identified in other cores from the North Atlantic with detailed radiocarbon chronologies (Bard et al., 1989).

[g]The age model was obtained by correlation of the planktonic δ18O records with the standard SPECMAP composite record from Imbrie et al. (1984).

[h]The age model was obtained by correlation of the planktonic δ18O records with lhe standard SPECMAP composite record from Martinson et al. (1987).

model including a dynamic sea-ice model (Legutke and Voss, 1999). It was forced with present boundary conditions such as solar radiation, sea and continental ice, vegetation, and distribution of land. The atmospheric greenhouse gas concentrations were prescribed at pre-industrial values (280 ppm CO_2, 700 ppb CH_4, and 265 ppb N_2O) (Lorenz and Lohmann, 2004). The last 2300 years of this experiment with constant pre-industrial conditions provide the basis for the statistical analysis of spatial and temporal patterns on centennial time scales. The surface temperature data consist of water temperatures over open water whereas temperatures of the overlying sea ice or snow layer over sea ice covered areas are taken into account. Prior to the

statistical analysis, we calculated the 50-year averages. The averaged data were linearly detrended.

3. Results

3.1. SST trends and related spatial patterns in the North Pacific realm

In general, except for two sites, the alkenone-derived SST records from the North Pacific realm show a warming or no pronounced Holocene temperature trend (temperature plateau) from 7 cal kyr BP to the present (Fig. 1a). The magnitudes of cooling and warming

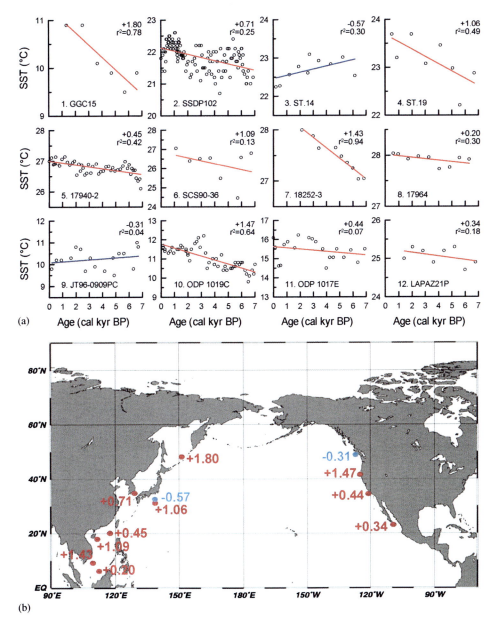

Fig. 1. (a) Data points (open circles) and linear trends of alkenone SST reconstructions from the North Pacific realm. The magnitudes of SST change over the last 7000 years (°C/7 kyr) are indicated in the upper right corner and the core names are given in the lower left corner of each panel. The numbers refer to Table 1. r is the correlation coefficient. (b) The spatial distribution of the magnitude of SST change (°C/7kyr) as shown in (a).

varied between −0.57 and −0.31 °C/7 kyr and +0.20 and +1.80 °C/7 kyr, respectively. The SST records sometimes show very different trends, even within small distance and across oceanic fronts in the eastern and western boundary current systems (Fig. 1b).

The first EOF describes about 42% of the field variance (Fig. 2a). The sign of leading EOF of SST variability in the North Pacific realm indicates a spatial pattern similar to that of linear trend coefficients except for one site in the northeastern Pacific (Fig. 1a). The cooling trend in the northeastern Pacific is not captured by the first EOF. The associated principal component (PC1) shows an increasing trend with an abrupt transition between 4 and 3 cal kyr BP (Fig. 2b). This indicates a clear large-scale warming pattern in the northeastern Pacific over the last 7000 years. In the northwestern Pacific, the pattern also indicates a warming trend except for one site off central Japan, which shows a cooling from 7 cal kyr BP to the present.

In order to identify the temporal patterns of North Pacific SST variability over the last 7000 years, we decomposed the PC1 associated to the first EOF of SST variability from this region (Fig. 2c and d), using the SSA. The reconstructed signal from the first two SSA components (i.e. trends) of the PC1, which describes 97% of the variance, captures the long-term transition from relatively cold conditions to relatively warm conditions over the last 7000 years (Fig. 2c). The reconstructed signal from the next six SSA components, which describes 3% of the variance, shows enhanced millennial scale variability (Fig. 2d).

3.2. SST trends and related spatial patterns in the North Atlantic realm

The alkenone-derived SST records from the North Atlantic realm show different linear trends over the last 7000 years (Fig. 3a). The magnitude of cooling or warming varied between −4.41 and −0.04 °C/7 kyr or between +0.13 and +2.44 °C/7 kyr, respectively. The zonal distributions of SST trends show east–west differences for the same latitudes (Fig. 3b). The northeastern Atlantic and the western Mediterranean Sea cooled while the eastern Mediterranean Sea, the northern Red Sea, the northern Arabian Sea, and the Gulf of Bengal experienced a warming (Fig. 3b).

The first EOF describes 43% of the field variance (Fig. 4a). The sign of leading EOF of SST variability in the North Atlantic realm indicates a spatial pattern similar to that of linear trend coefficients (Fig. 3b). The associated PC1 (Fig. 4b) emphasizes a strong linear trend, indicating a continuous cooling in the northeastern Atlantic and the western Mediterranean Sea contrary to a warming in the eastern Mediterranean Sea, the northern Red Sea, the northern Arabian Sea, and the Gulf of Bengal over the last 7000 years (Fig. 4a).

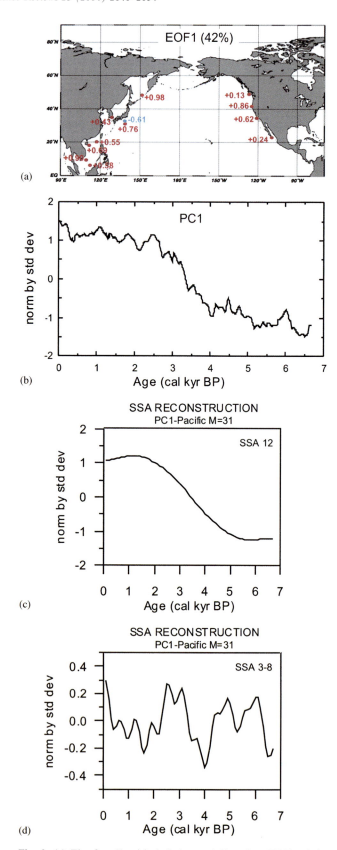

Fig. 2. (a) The first Empirical Orthogonal Function (EOF) of the North Pacific alkenone SST reconstructions and (b) its associated time coefficients (PC1) for the last 7000 years. (c) Reconstruction of the time coefficients of the first EOF of the North Pacific SST variability from the first two Singular Spectrum Analysis (SSA) components (i.e., trends) and (d) from the next six SSA components. $M = 31$ indicates the length of the window in the SSA which is 3.1 kyr.

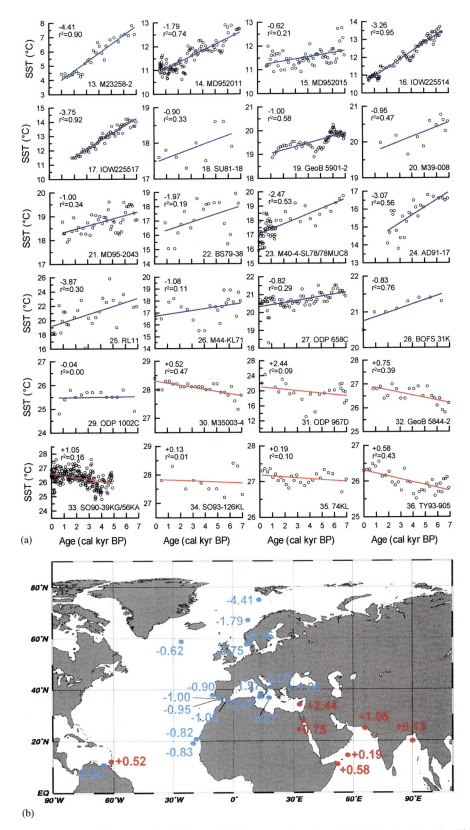

Fig. 3. (a) Data points (open circles) and linear trends of alkenone SST reconstructions from the North Atlantic realm. The magnitudes of SST change over the last 7000 years (°C/7 kyr) are indicated in the upper right corner and the core names are given in the lower left corner of each panel. The numbers refer to Table 1. *r* is the correlation coefficient. (b) The spatial distribution of the magnitude of SST change (°C/7 kyr) as shown in (a).

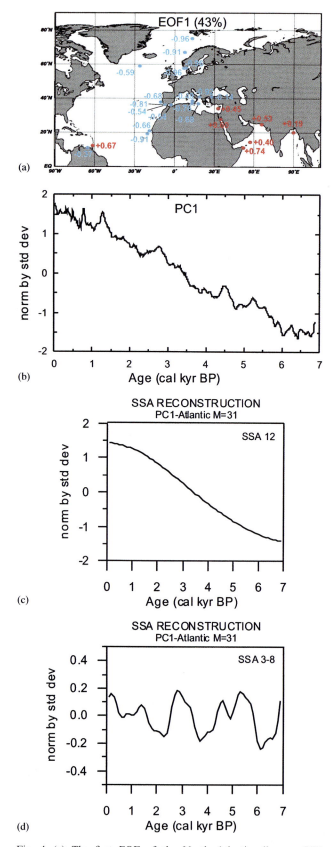

(a)

(b)

(c)

(d)

Fig. 4. (a) The first EOF of the North Atlantic alkenone SST reconstructions and (b) its associated time coefficients (PC1) for the last 7000 years. (c) Reconstruction of the time coefficients of the first EOF of the North Atlantic SST variability from the first two SSA components (i.e., trends) and (d) from the next six SSA components. $M = 31$ indicates the length of the window in the SSA which is 3.1 kyr.

Interestingly, the PC1 for the North Atlantic realm does not indicate an abrupt shift at 4–3 cal kyr BP, which is the case of the PC1 for the North Pacific realm.

In order to identify the temporal patterns of North Atlantic SST variability over the last 7000 years like in the North Pacific, we decomposed the PC1 associated to the first EOF of SST variability from this region (Figs. 4c and d), using the SSA. The trend signal from the first two SSA components of the PC1, which describes 96% of the variance, captures the long-term transition over the last 7000 years (Fig. 4c). The reconstructed signal from the next six SSA components, which shows 4% of the variance, show enhanced millennial scale variability (Fig. 4d).

3.3. Coupled atmosphere–ocean general circulation model experiment

To better assess the North Pacific and North Atlantic SST spatial patterns derived from the analysis of the alkenone SST data, the centennial variability in the 2300 year long control integration of the ECHO-G model (Lorenz and Lohmann, 2004) was investigated. Noteworthy is that in this simulation the variability is generated only by internal processes while the alkenone SST variability is assumed to follow both internal and external forcing mechanisms.

Based on the spatial pattern of the North Pacific SST as derived from alkenone SST data (Fig. 2a), a northeastern Pacific surface temperature (ST) index was defined by ST anomalies from the 2300 year long simulation. The index was obtained by averaging the simulated STs over the area 140–110 °W and 20–60 °N and subtracting the mean value. In Fig. 5a, the time series of this index (red box in Fig. 5b) is represented. Over this region, the alkenone SST data indicate a clear warming trend from 7 cal kyr BP to the present (Fig. 2a). The correlation map between the northeastern Pacific ST index (Fig. 5a) and model STs over the entire Northern Hemisphere indicates that positive temperature anomalies in the northeastern Pacific are accompanied by negative temperature anomalies in the northeastern Atlantic (Fig. 5b).

Motivated by the distribution of the 7000 year cooling trend in alkenone SSTs from this region (Fig. 4a), a northeastern Atlantic ST index was defined by ST anomalies from the 2300 year long ST data (Fig. 6a). The index was obtained by averaging the simulated STs over the area 30 °W–10 °E and 30–70 °N (red box in Fig. 6b) and subtracting the mean value. The correlation map between the northeastern Atlantic ST index and simulated Northern Hemisphere STs shows that a warming in the northeastern Atlantic accompanies a cooling in the northeastern Pacific (Fig. 6b).

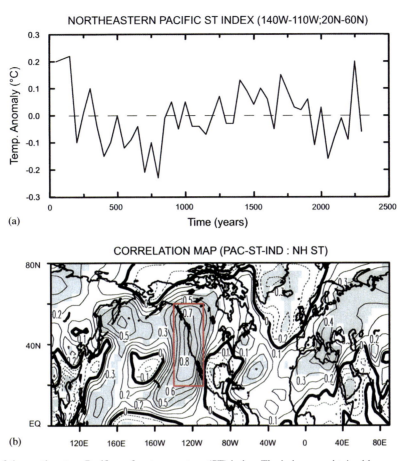

Fig. 5. (a) The time series of the northeastern Pacific surface temperature (ST) index. The index was obtained by averaging the simulated STs over the area (140–110°W:20–60°N) and subtracting the mean value. The 50-yr mean ST data were used for the analysis. (b) The correlation map of the ST index with Northern Hemisphere ST from the pre-industrial simulation of the ECHO-G coupled model. Shading indicates the areas with statistical significance of the correlations at the 95% confidence level based on a *t*-test. The red box shows the ST index area.

4. Possible underlying mechanisms

The correlation coefficient between the northeastern Pacific and northeastern Atlantic ST indices defined above is −0.25 (significant at 95% level). In order to identify atmospheric circulation patterns that may connect the northeastern Pacific and northeastern Atlantic SST variability at centennial time scales and that may also be related to the observed spatial distribution of positive and negative long-term SST trends as shown by the alkenone SST data, a third ST index was defined. It is the difference between the northeastern Pacific and northeastern Atlantic ST indices described above, where the mean value was subtracted (Fig. 7a). A high value of the ST index indicates a simultaneous warming in the northeastern Pacific and a cooling in the northeastern Atlantic.

Based on this ST index, a composite map of the Northern Hemisphere sea level pressure (SLP) was constructed (Fig. 7b). The SLP composite map consists of an average over those years when the values of the ST index are higher (lower) than 1 standard deviation. The

SLP composite map, which is the difference between the averaged SLP maps, indicates an SLP pattern which bears similarities of the Pacific North American (PNA) and NAO atmospheric circulation patterns. The PNA pattern represents a large-scale atmospheric teleconnection between the North Pacific and North America (Wallace and Gutzler, 1981). It is characterized by an atmospheric flow in which the west coast of North America is out of phase with the northeastern Pacific and southeastern United States. During the positive PNA phase, low SLP is centred over the North Pacific and high SLP over western North America. The NAO is the dominant mode of atmospheric behaviour in the North Atlantic. It is a large-scale seesaw in atmospheric mass between the polar low-pressure centre (Icelandic Low) and the subtropical high-pressure centre (Azores High). During the positive NAO, the Icelandic Low is deeper than normal and the Azores High is stronger than usual. The SLP composite map (Fig. 7b) shows the positive phase of the PNA in the North Pacific realm and the negative phase of the NAO in the North Atlantic realm. This suggests that the opposing SST

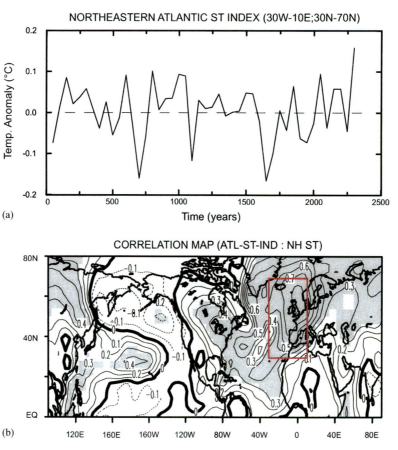

Fig. 6. (a) The time series of the northeastern Atlantic surface temperature (ST) index. The index was obtained by averaging the simulated STs over the area (30 °W–10 °E:30–70 °N) and subtracting the mean value. The 50-yr mean ST data were used for the analysis. (b) The correlation map of ST index with Northern Hemisphere ST from the 2300-year long simulation of ECHO-G coupled model. Shading indicates the areas with statistical significance of the correlations at the 95% confidence level based on a t-test. The red box shows the ST index area.

pattern between the northeastern Pacific and the north-eastern Atlantic as seen in the alkenone SST data was probably caused by an interaction of the positive PNA and the negative NAO phases.

A composite map of the Northern Hemisphere ST (Fig. 7c) was also constructed based on the ST index. Like the SLP composite map, the ST composite map consists of an average over those years when the values of the ST index are higher (lower) than 1 standard deviation. The ST composite map, which is the difference between the averaged ST maps, shows a temperature distribution pattern which bears some resemblance of the Pacific-Decadal Oscillation (PDO) for the instrumental era (Mantua et al., 1997; Mantua and Hare, 2002). The PDO represents decadal changes in SST patterns in the North Pacific. A positive PDO phase equates to anomalously warm surface water in the equatorial Pacific and along the west coast of North America, but anomalously cool surface water in the west Pacific. The PDO atmospheric circulation anomalies extend through the depth of the troposphere and are well expressed as persistence in the PNA pattern

described by Wallace and Gutzler (1981). The ST composite map (Fig. 7c) shows the positive PDO phase consistent with the positive PNA phase (Fig. 7b). The SST pattern in the North Atlantic realm (Fig. 7c) resembles the NAO-related SST pattern (e.g., Hurrell, 1995), showing an opposing pattern between the north-eastern Atlantic and the eastern Mediterranean Sea and the northern Red Sea. The SST pattern in the ST composite map fits relatively well to the alkenone-derived SST pattern in the northeastern Pacific and northeastern Atlantic realms (Figs. 2a and 4a).

In summary, the inverse SST pattern between the northeastern Pacific and the northeastern Atlantic in the model simulation is connected to an atmospheric circulation field that comprises the elements of the PNA pattern and the NAO in opposite phases. The similar SST seesaw observed in both the alkenone SST data and the simulated data suggests that the PNA and NAO-like atmospheric circulation pattern are also responsible for the inverse long-term SST trends between the northeastern Pacific and the northeastern Atlantic during the Holocene.

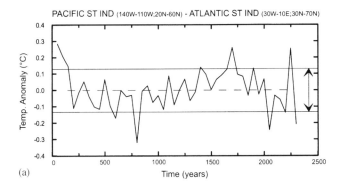

(a)

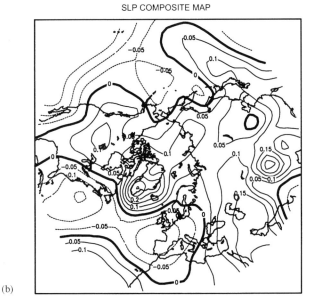

(b)

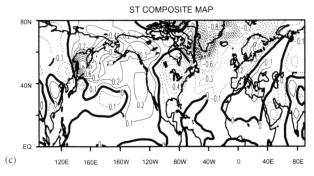

(c)

Fig. 7. (a) The time series of the surface temperature (ST) index. The index was defined as the difference between the northeastern Pacific (Fig. 5a) and northeastern Atlantic (Fig. 6a) ST indices where the mean value was subtracted. The arrow bar indicates the range of the ± 1 standard deviation. (b) Composite map of SLP based on the ST index (a) (see text for details); unit is hPa. (c) Composite map of surface temperature (ST) based on the ST index (a) (see text for details); unit is °C. The 50-yr mean ST and SLP data were used for the analysis.

5. Summary and conclusion

A set of alkenone SST records is analysed for the Holocene period. In general, the alkenone SST records show a basin scale warming in the North Pacific and a cooling in the North Atlantic over the last 7000 years. In the North Atlantic realm, the alkenone SST spatial pattern fits the modern NAO-related SST pattern: SST variations in the northeastern Atlantic are opposite in trend with those from the eastern Mediterranean Sea and the northern Red Sea. This trend is consistent with the northeastern Atlantic SST trend pattern for the Holocene period (Marchal et al., 2002), which has been attributed to a continuous weakening of the NAO-like atmospheric circulation during the Holocene (Rimbu et al., 2003). The opposing SST trends between the northeastern Atlantic and the northern Red Sea were also consistently accompanied by a warming in the tropical regions over the last 7000 years. Indeed, such heterogeneity of SST trends is coherent with coupled atmosphere–ocean circulation model simulations driven solely by the Earth's orbital parameters (Lorenz and Lohmann, 2004) and a continuous weakening of the Icelandic Low and altered winds in the Nordic Seas during the Holocene (Lohmann et al., 2004).

The most clearly defined alkenone SST pattern is an opposing trend with a long-term warming in the northeastern Pacific and a cooling in the northeastern Atlantic over the last 7000 years. The 2300 year long control simulation of the ECHO-G coupled atmosphere–ocean model also suggests that positive temperature anomalies in the northeastern Pacific were accompanied by negative anomalies in the northeastern Atlantic and vice versa. The comparable northeastern Pacific–Atlantic SST seesaw pattern between the alkenone SST data and the model results implies that the long-term SST variations were influenced by similar processes that caused centennial climate variability in the model simulation. A similar Pacific–Atlantic seesaw has also been reported by Kiefer et al. (2001) during Dansgaard–Oeschger events and by Weijer and Dijkstra (2003) in an eigenmode analysis of a global ocean circulation model. Further studies are necessary to better understand how inter-oceanic teleconnections are related to large-scale atmospheric teleconnenntions and ocean circulation changes.

Acknowledgements

We would sincerely like to thank the GHOST contributors for providing their alkenone data, F. Sirocko for the 74KL sediment samples, and R. Kreutz and D. Grotheer for technical assistance in the Bremen University laboratory. Thanks go also to reviewers for their thorough comments. Financial support was provided by grants from the German Ministry of Research and Education (BMBF) through the GHOST project of DEKLIM (RCOM contribution No. 0151).

References

Arz, H.W., Lamy, F., Pätzold, J., Müller, P.J., Prins, M., 2003. Mediterranean moisture soure for an Early-Holocene humid period in the Northern Red Sea. Science 300, 118–121.

Bard, E., Fairbanks, R.G., Arnold, M., Maurice, P., Duprat, J., Moyes, J., Duplessy, J.-C., 1989. Sea level estimates during the last deglaciation based on $\delta^{18}O$ and accelerator mass spectrometry ^{14}C ages measured on Globigerina bulloides. Quaternary Research 31, 381–391.

Bard, E., Rostek, F., Turon, J.-L., Gendreau, S., 2000. Hydrological impact of Heinrich Events in the subtropical northeast Atlantic. Science 289, 1321–1324.

Barron, J.A., Heusser, L., Herbert, T., Lyle, M., 2003. High resolution climatic evolution of coastal Northern California during the past 16,000 Years. Paleoceanography 18, 1020, March 2003, doi:10.1029/2002PA000768.

Bond, G., Kromer, B., Beer, J., Muscheler, R., Evans, M.N., Showers, W., Hoffmann, S., Lotti-Bond, R., Hajdas, I., Bonani, G., 2001. Persistent solar influence on North Atlantic climate during the Holocene. Science 294, 2130–2133.

Brassell, S.C., Eglinton, G., Marlowe, I.T., Pflaumann, U., Sarnthein, M., 1986. Molecular stratigraphy: a new tool for climatic assessment. Nature 320, 129–133.

Cacho, I., Grimalt, J.O., Pelejero, C., Canals, M., Sierro, F.J., Flores, J.A., Shackleton, N.J., 1999. Dansgaard-Oeschger and Heinrich event imprints in Alboran Sea paleotemperatures. Paleoceanography 14, 698–705.

Cacho, I., Grimalt, J.O., Canals, M., Sbaffi, L., Schackleton, N.J., Shonfeld, J., Zahn, R., 2001. Variability of the Western Mediterranean sea surface temperatures during the last 25 000 years and its connection with the Northern Hemisphere climatic changes. Paleoceanography 16, 40–52.

Calvo, E., Grimalt, J., Jansen, E., 2002. High resolution UK37 sea surface temperature reconstruction in the Norwegian Sea during the Holocene. Quaternary Science Reviews 21, 1385–1394.

Capotondi, L., Borsetti, A.M., Morigi, C., 1999. Foraminiferal ecozones: a high resolution proxy for the late Quaternary biochronology in the central Mediterranean Sea. Marine Geology 153, 253–274.

Chapman, M.R., Shackleton, N.J., Zhao, M., Eglington, G., 1996. Faunal and alkenone reconstructions of subtropical North Atlantic surface hydrography and paleotemperature over the last 28 kyr. Paleoceanography 11, 343–357.

deMenocal, P., Ortiz., J., Guilderson, T., Adkins, J., Sarnthein, M., Baker, L., Yaarusinsky, M., 2000. Abrupt onset and termination of the African Humid Period: rapid climate responses to gradual insolation forcing. Quaternary Science Reviews 19, 347–361.

Doose-Rolinski, H., Rogalla, U., Scheeder, G., Lücke, A., von Rad, U., 2001. High resolution temperature and evaporation changes during the late Holocene in the northeastern Arabian Sea. Paleoceanography 16, 358–367.

Emeis, K.-C., Struck, U., Schulz, H.-M., Rosenberg, R., Bernasconi, S., Erlenkeuser, H., Sakamoto, T., Martinez-Ruiz, F., 2000. Temperature and salinity variations of Mediterranean Sea surface waters over the last 16,000 years from records of planktonic stable oxygen isotopes and alkenone unsaturation ratios. Palaeogeography, Palaeoclimatology, Palaeoecology 158, 259–280.

Emeis, K.-C., Struck, U., Blanz, T., Kohly, A., Voß, M., 2003. Salinity changes in the central Baltic Sea (NW Europe) over the last 10000 years. The Holocene 13, 411–421.

Emeis, K.-C., Dawson, A., 2003. Holocene paleoclimate records over Europe and the North-Atlantic. The Holocene 13, 305–309.

Ghil, M., Allen, R.M., Dettinger, M.D., Ide, K., Kondrashov, D., Mann, M.E., Robertson, A., Saunders, A., Tian, Y., Varadi, F.,

Yoiou, O., 2002. Advanced spectral methods for climatic time series. Reviews of Geophysics 40 (1), 3.1–3.41.

Giunta, S., Emeis, K.-C., Negri, A., 2001. Sea-surface temperature reconstruction of the last 16,000 years in the Eastern Mediterranean Sea. Rivista Italiana di Paleontologia e Stratigrafia 107, 463–476.

Grootes, P.M., Stuiver, M., 1997. Oxygen 18/16 variability in Greenland snow and ice with 10^3–10^5-year time resolution. Journal Geophysical Research 102, 26455–26470.

Grootes, P.M., Stuiver, M., White, J.M.C., Johnsen, S., Jouzel, J., 1993. Comparison of oxygen isotope records from the GISP2 and GRIP Greenland ice cores. Nature 366, 552–554.

Herbert, T.D., Schuffert, J.D., 2000. 16. Alkenone unsaturation estimates of sea-surface temperatures at site 1002 over a full glacial cycle. Proceedings of the Ocean Drilling Program, Scientific Results 165, 239–247.

Herbert, T.D., Schuffert, J.D., Andreasen, D., Heusser, L., Lyle, M., Mix, A., Ravelo, A.C., Stott, L.D., Herguera, J.C., 2001. Collapse of the California Current during glacial maxima linked to climate change on land. Science 293, 71–76.

Huang, C.-Y., Wu, S.-F., Zhao, M., Chen, M.-T., Wang, C.-H., Tu, X., Yuan, P.B., 1997. Surface ocean and monsoon climate variability in the south China Sea since last glaciation. Marine Micropaleontology 32, 71–94.

Hurrell, J.W., 1995. Decadal trends in the North Atlantic Oscillation: regional temperatures and precipitation. Science 269, 676–679.

Imbrie, J., Hays, J.D., Martinson, D.G., McIntyre, A., Mix, A.C., Morley, J.J., Pisias, N.G., Prell, W.L., Shackleton, N.J., 1984. The orbital theory of Pleistocene climate: support from a revised chronology of the marine $\delta^{18}O$ record. In: Berger, A.L., Imbrie, J., Hays, J.D., Kulka, J., Saltzman, J. (Eds.), Milankovitch and Climate (Part. 1), NATO Advanced Science Institutes Series C: Mathematical and Physical Sciences, vol. 126. Reidel, Hingham, MA, pp. 269–305.

Kiefer, T., Sarnthein, M., Erlenkeuser, H., Grootes, P., Roberts, A., 2001. North Pacific response to millennial-scale changes in ocean circulation over the last 60 ky. Paleoceanography 16, 179–189.

Keigwin, L.D., 1998. Glacial-age hydrography for the far northwest Pacific Ocean. Paleoceanography 13, 323–339.

Kennett, J.P., Roark, E.B., Cannariato, K.G., Ingram, B.L., Tada, R., 2000. 21. Late Quaternary paleoclimatic and radiocarbon chronology, Hole 1017E, southern California Margin. Proceedings of the Ocean Drilling Program, Scientific Results 167, 249–254.

Kienast, S.S., McKay, J.L., 2001. Sea surface temperature in the subarctic Northeast Pacific reflect millennial-scale Climate Oscillations during the last 16 kyrs. Geophysical Research Letters 28, 1563–1566.

Kim, J.-H., Schneider, R.R., 2004. GHOST global database for alkenone-derived Holocene sea-surface temperature records. http://www.pangaea.de/Projects/GHOST/Holocene.

Kudrass, H.R., Hofmann, A., Doose, H., Emeis, K., Erlenkeuser, H., 2001. Modulation and amplification of climatic changes in the Northern Hemisphere by the Indian summer monsoon during the past 80 ky. Geology 29, 63–66.

Legutke, S., Voss, R., 1999. The Hamburg atmosphere-ocean coupled circulation model ECHO-G. DKRZ Report No. 18, Hamburg, Germany.

Lohmann, G., Lorenz, S., Prange, M., 2004. Northern high-latitude climate changes during the Holocene as simulated by circulation models. AGU Monographs, Bjerknes book about the Nordic Seas, in press.

Lorenz, S.J., Lohmann, G., 2004. Acceleration technique for Milankovitch type forcing in a coupled atmosphere-ocean circulation model: method and application for the Holocene. Climate Dynamics, doi:10.1007/S00382-004-0469-y.

Lorenz, S.J., Kim, J.-H., Rimbu, N., Schneider, R.R., Lohmann, G., 2004. Orbitally driven insolation forcing on Holocene climate trends reassessed. Paleoceanography, submitted for publication.

Mantua, N., Hare, S., 2002. The Pacific Decadal Oscillation. Journal of Oceanography 58, 35–44.

Mantua, N.J., Hare, S.R., Zhang, Y., Wallace, J.M., Francis, R.C., 1997. A Pacific interdecadal climate oscillation with impacts on salmon production. Bulletin of the American Meteorological Society 78, 1069–1079.

Marchal, O., Cacho, I., Stocker, T.F., Grimalt, J.O., Calvo, E., Martrat, B., Shackleton, N., Vautravers, M., Cortijo, E., van Kreveld, S., Andersson, C., Koç, N., Chapman, M., Sbaffi, L., Duplessy, J.-C., Sarnthein, M., Turon, J.-L., Duprat, J., Jansen, E., 2002. Apparent long-term cooling of the sea surface in the Northeast Atlantic and Mediterranean during the Holocene. Quaternary Science Reviews 21, 455–483.

Martinson, D.G., Pisias, N.G., Hays, J.D., Imbrie, J., Moore, T.C., Shackleton, N.J., 1987. Age dating and the orbital theory of the ice ages: development of a high-resolution 0 to 300,000 year chronostratigraphy. Quaternary Research 27, 1–29.

Moy, C.M., Seltzer, G.O., Rodbell, D.T., Anderson, D.M., 2002. Variability of El Niño/Southern Oscillation activity at millennial timescales during the Holocene epoch. Nature 420, 162–165.

Müller, P.J., Kirst, G., Ruhland, G., von Storch, I., Rosell-Melé, A., 1998. Calibration of the alkenone palaeotemperature index $U_{37}^{K'}$ based on core-tops from the eastern South Atlantic and the global ocean (60°N–60°S). Geochimica et Cosmochimica Acta 62, 1757–1772.

Oppo, D.W., McManus, J.F., Cullen, J.L., 2003. Deepwater variability in the Holocene epoch. Nature 422, 277–278.

Ostertag-Henning, C., Stax, R., 2000. 26. Data report: Carbonate records from sites 1012, 1013, 1017, and 1019 and alkenone-based sea-surface temperatures from site 1017. Proceedings of the Ocean Drilling Program, Scientific Results 167, 297–302.

Pelejero, C., Grimalt, J.O., 1997. The correlation between the UK37 index and sea surface temperatures in the warm boundary: the South China Sea. Geochimica et Cosmochimica Acta 61, 4789–4797.

Pelejero, C., Grimalt, J.O., Heilig, S., Kienast, M., Wang, L., 1999a. High-resolution UK37 temperature reconstructions in the South China Sea over the past 220 kyr. Paleoceanography 14, 224–231.

Pelejero, C., Kienast, M., Wang, L., Grimalt, J.O., 1999b. The flooding of Sundaland during the last deglaciation: imprints in hemipelagic sediments from the southern South China Sea. Earth Planetary Science Letters 171, 661–671.

Peterson, L.C., Haug, G.H., Murray, R.W., Yarincik, K.M., King, J.W., Bralower, T.J., Kameo, K., Rutherford, S.D., Pearce, R.B., 2000. 4. Late Quaternary stratigraphy and sedimentation at site 1002, Cariaco basin (Venezuela). Proceedings of the Ocean Drilling Program, Scientific Results 165, 85–99.

Prahl, F.G., Wakeham, S.G., 1987. Calibration of unsaturation patterns in long-chain ketone compositions for paleotemperature assessment. Nature 330, 367–369.

Prahl, F.G., Muehlhausen, L.A., Zahnle, D.L., 1988. Further evaluation of long-chain alkenones as indicators of paleoceanographic conditions. Geochimica et Cosmochimica Acta 52, 2303–2310.

Rimbu, R., Lohmann, G., Kim, J.-H., Arz, H.W., Schneider, R.R., 2003. Arctic/North Atlantic Oscillation signature in Holocene sea surface temperature trends as obtained from alkenone data. Geophysical Research Letters 30, 1280.

Rimbu, N., Lohmann, G., Lorenz, S.J., Kim, J.-H., Schneider, R.R., 2004. Holocene climate variability as derived from alkenone sea surface temperature and coupled ocean-atmosphere model experiments. Climate Dynamics, doi:10.1007/S00382-004-0435-8.

Rosell-Melé, A., Eglinton, G., Pflaumann, U., Sarnthein, M., 1995. Atlantic core-top calibration of the UK37 index as a sea-surface palaeotemperature indicator. Geochimica et Cosmochimica Acta 59, 3099–3107.

Rosell-Melé, A., Bard, E., Emeis, K.-C., Grimalt, J.O., Müller, P., Schneider, R., Bouloubassi, I., Epstein, B., Fahl, K., Fluegge, A., Freeman, K., Goñi, M., Güntner, U., Hartz, D., Hellebust, S., Herbert, T., Ikehara, M., Ishiwatari, R., Kawamura, K., Kenig, F., de Leeuw, J., Lehman, S., Mejanelle, L., Ohkouchi, N., Pancost, R.D., Pelejero, C., Prahl, F., Quinn, J., Rontani, J.-F., Rostek, F., Rullkötter, J., Sachs, J., Blanz, T., Sawada, K., Schulz-Bull, D., Sikes, E., Sonzogni, C., Ternois, Y., Versteegh, G., Volkman, J.K., Wakeham, S., 2001. Precision of the current methods to measure the alkenone proxy UK'37 and absolute alkenone abundance in sediments: results of an interlaboratory comparision study. Geochemistry, Geophysics, Geosystems G 2, 2000GC000141.

Rühlemann, C., Mulitza, S., Müller, P.J., Wefer, G., Zahn, R., 1999. Warming of the tropical Atlantic Ocean and slowdown of thermohaline circulation during the last deglaciation. Nature 402, 511–514.

Sarnthein, M., van Kreveld, S., Erlenkeuser, H., Grootes, P.M., Kucera, M., Pflaumann, U., Schulz, M., 2003. Centennial-to-millennial-scale periodicities of Holocene climate and sediment injections off the western Barents shelf, 75°N. Boreas 32, 447–461.

Sawada, K., Handa, N., 1998. Variability of the path of the Kuroshio ocean current over the past 25,000 years. Nature 392, 592–595.

Schulz, M., Paul, A., 2002. Holocene climate variability on centennial-to-millennial time scales: 1. Climate records from the North-Atlantic Realm. In: Wefer, G., Berger, W., Behre, K.-E., Jansen, E. (Eds.), Climate development and history of the North Atlantic Realm. Springer, Berlin, Heidelberg, pp. 41–54.

Shackleton, N.J., 2000. The 100,000-year ice-age cycle identified and found to lag temperature, carbon dioxide and orbital eccentricity. Science 289, 1897–1902.

Siani, G., Paterne, M., Michel, E., Sulpizio, R., Sbrana, A., Arnold, M., Haddad, G., 2001. Mediterranean sea-surface radiocarbon reservoir age changes since the last glacial maximum. Science 294, 1917–1920.

Sirocko, F., Sarnthein, M., Erlenkeuser, H., Lange, H., Arnold, M., Duplessy, J.C., 1993. Century-scale events in monsoonal climate over the past 24,000 years. Nature 364, 322–324.

Sonzogni, C., Bard, E., Rostek, F., Lafont, R., Rosell-Melé, A., Eglinton, G., 1997. Core-top calibration of the alkenone index vs sea surface temperture in the Indian Ocean. Deep Sea Research II 44, 1445–1460.

Sperling, M., Schmiedl, G., Hemleben, C., Emeis, K.-C., Erlenkeuser, H., Grootes, P., 2003. Black Sea impact on the formation of eastern Mediterranean sapropel S1: evidence from the Marmara Sea. Palaeogeography, Palaeoclimatology, Palaeoecology 190, 9–21.

Ternois, Y., Kawamura, K., Ohkouchi, N., Keigwin, L., 2000. Alkenone sea surface temperature in the Okhotsk Sea for the last 15 kyr. Geochemical Journal 34, 283–293.

Thompson, D.W.J., Wallace, J.W., 1998. The Arctic Oscillation signature in the wintertime geopotential height and temperature fields. Geophysical Research Letters 25, 1297–1300.

Thompson, D.W.J., Wallace, J.M., Hegerl, G.C., 2000. Annular modes in the extratropical circulation, Part II: trends. Journal of Climate 13, 1018–1036.

von Rad, U., Schaaf, M., Michels, K.H., Schulz, H., Berger, W., Sirocko, F., 1999. A 5000-yr record of climate change in varved sediments from the oxygen minimum zone off Pakistan, Northeastern Arabian Sea. Quaternary Research 51, 39–53.

von Storch, H., Zwiers, F.W., 1999. Statistical Analysis in Climate Research. Cambridge University Press, Cambridge, UK 735pp.

Wallace, J.M., Gutzler, D.S., 1981. Teleconnections in the geopotential height field during the Northern Hemisphere winter. Monthly Weather Reviews 109, 784–812.

Wang, L., Sarnthein, M., Erlenkeuser, H., Heilig, S., Ivanova, E., Kienast, M., Pflaumann, U., Pelejero, C., Grootes, P., 1999. East Asian monsoon during the late Quaternary: high-resolution sediment records from the South China Sea. Marine Geology 156, 245–284.

Weijer, W., Dijkstra, H.A., 2003. Multiple oscillatory modes of the global ocean. Journal of Physical Oceanography 33, 2197–2213.

Zhao, M., Beveridge, N.A.S., Shackleton, N.J., Sarnthein, M., Eglinton, G., 1995. Molecular stratigraphy of cores off northwest Africa: sea surface temperature history over the last 80 ka. Paleoceanography 10, 661–675.

Quaternary Science Reviews 23 (2004) 2155–2166

A highly unstable Holocene climate in the subpolar North Atlantic: evidence from diatoms

C. Andersen[a,b], N. Koç[a,*], M. Moros[c]

[a]Norwegian Polar Institute, N-9296 Tromsø, Norway
[b]Department of Earth Science, University of Bergen, Allégt. 41, N-5007 Bergen, Norway
[c]Bjerknes Centre for Climate Research, Allégt. 55, 5007 Bergen, Norway

Abstract

A composite record (LO09-14) of three sediment cores from the subpolar North Atlantic (Reykjanes Ridge) was investigated in order to assess surface ocean variability during the last 11 kyr. The core site is today partly under the influence of the Irminger Current (IC), a branch of the North Atlantic Drift continuing northwestward around Iceland. However, it is also proximal to the Sub-Arctic Front (SAF) that may cause extra dynamic hydrographic conditions. We used statistical methods applied to the fossil assemblages of diatoms to reconstruct quantitative sea surface temperatures (SSTs). Our investigations give evidence for different regional signatures of Holocene surface oceanographic changes in the North Atlantic. Core LO09-14 reveal relatively low and highly variable SSTs during the early Holocene, indicating a weak IC and increased advection of subpolar water over the site. A mid-Holocene thermal optimum with a strong IC occurs from 7.5 to 5 kyr and is followed by cooler and more stable late Holocene surface conditions. Several intervals throughout the Holocene are dominated by the diatom species *Rhizosolenia borealis*, which we suggest indicates proximity to a strongly defined convergence front, most likely the SAF. Several coolings, reflecting southeastward advection of cold and ice-bearing waters, occur at 10.4, 9.8, 8.3, 7.9, 6.4, 4.7, 4.3 and 2.8 kyr. The cooling events recorded in the LO09-14 SSTs correlate well with both other surface records from the area and the NADW reductions observed at ODP Site 980 indicating a surface-deepwater linkage through the Holocene.
© 2004 Elsevier Ltd. All rights reserved.

1. Introduction

The North Atlantic circulation, through its heat transport and deepwater formation, is an important contributor in the climate system. Energetic water masses like the North Atlantic Drift (NAD) and its end-members transport heat from low to high latitudes, provide northwestern Europe and Iceland with mild climate, and control the freshwater budget in the area. Through numerous paleoclimatic investigations these important oceanic mechanisms have been shown to experience pronounced short-term and long-term changes during the last deglacial period (Lehman and

Keigwin, 1992; Koç Karpuz and Jansen, 1992; Bauch and Weinelt, 1997). Our present interglacial period has long been regarded relatively stable based on ice-core records (e.g., Dansgaard et al., 1993). This view of the Holocene period in the North Atlantic and adjacent areas is, however, being revised after evidence of millennial-scale climate fluctuations in atmospheric (O'Brien et al., 1995; Alley et al., 1997), marine (Bond et al., 1997; Bianchi and McCave, 1999; Giraudeau et al., 2000; Klitgaard-Kristensen et al., 2001; Jiang et al., 2002; Andersen et al., 2004) and terrestrial records (Denton and Karlén, 1973; Nesje et al., 2000). Even though the mechanisms behind millennial-scale climate fluctuations is still very much under debate, Bond et al. (1997, 2001) argued that the abrupt climate shifts during the deglacial and the Holocene cooling events occurred at intervals of about 1500 years and that they were a

*Corresponding author. Tel.: +47-77-750-654; fax: +47-77-750-501.

E-mail address: nalan.koc@npolar.no (N. Koç).

0277-3791/$ - see front matter © 2004 Elsevier Ltd. All rights reserved.
doi:10.1016/j.quascirev.2004.08.004

result of climate cyclicity independent from glacial influence.

In this study we present a reconstruction of Holocene sea surface temperatures (SSTs) from a composite of sediment cores located on the Reykjanes Ridge in the subpolar North Atlantic in order to quantify surface stability in this climatically sensitive region. Diatoms, which are siliceous photosynthetic algae, are utilized as surface monitors of paleo-oceanic conditions as they have been shown to reflect well the watermasses and temperatures of the area (Koç Karpuz and Schrader, 1990; Koç et al., 1993). Previous studies based on diatoms and alkenones from the Nordic Seas describe the general Holocene surface ocean climatic development as a period of cooling in step with the decreasing insolation at the northern hemisphere since 11,000 years B.P. (11 kyr) (Koç et al., 1993; Koç and Jansen, 1994; Birks and Koç, 2002; Marchal et al., 2002; Andersen et al., 2004). A Holocene climate optimum lasting from 9.5 kyr until ~6.5 kyr is followed by a Holocene transition period from 6.5–3.0 kyr during which time the climate was deteriorating. During the Late Holocene Cooling from 3.0 kyr to present SSTs stabilized around a mean with a variability of 1–2 °C (Andersen et al., 2004). The presence of millennial-scale coolings observed in different circum North Atlantic regions (Denton and Karlén, 1973; O'Brien et al., 1995; Bond et al., 1997; Giraudeau et al., 2000) do however indicate of a complex system with regions of different sensitivity.

The subpolar North Atlantic is an oceanographically dynamic area. The surface regime is strongly imprinted by the subpolar gyre, characterized by meridional (northward) and zonal (eastward) circulation (Fratantoni, 2001). The core site is situated on the western flank of the Reykjanes Ridge south of Iceland and is today at the margin of the warm Irminger Current (IC) (Fig. 1). The IC is the northwestern branch of the NAD and separates into two branches west of Iceland. While a small branch continues northward and flows around Iceland (Stefánsson, 1962; Hopkins, 1991) most of it turns southwest where it is incorporated into the West Greenland Current (WGC) (Hurdle, 1986; Fratantoni, 2001). The core site is located close to the SAF, which separates subpolar/subarctic water from warm Atlantic water. The front has a meander-like shape and is most strongly defined along the 53°N parallel. Further north the configuration of the front is less well constrained. Summer SSTs are normally between 11 and 12 °C (Dietrich, 1969; Johannessen, 1986), but can vary due to the varying amounts of influence from subpolar water originating from the EGC or from the conjunction area between the WGC and the Labrador Current (LC) (Fig. 1). The site is thereby in a position to record the past strength and variability of the IC branch of the NAD and the SAF. The record provides both further evidence for, but also quantifies the magnitude of a series of Holocene cooling events that can be correlated to repeated southward intrusion of cold water into the North Atlantic reported in earlier studies (Bond et al., 1997, 2001; Giraudeau et al., 2000).

2. Material and methods

2.1. Material

The investigated material consists of one large box-core (LO09-14 LBC; 0.4 m long), one giant gravity-core (LO09-14 GGC; 2.7 m long) and one gravity-core (LO09-14 GC; 5.6 m long), which were recovered from the same site on the Reykjanes Ridge (58°56.3 N, 30°24.5 W) (Fig. 1). The age chronology is based on a total of 34 AMS ^{14}C dates of the foraminifera *Globigerina bulloides* (Table 1). The ^{14}C dates were calibrated to calendar years before present by using OxCal Program v3.8 with Stuiver and Reimer marine calibration table (Bronk Ramsey, 1995, 2001; Stuiver et al., 1998). A marine reservoir age correction of 400 years is used in the calibration. The establishment of an age-depth model was done by linear interpolation between dated levels in the LBC and by polynomial fits for the GGC and the GC (Fig. 2a–c). Due to soft and water-rich sediments at the top of GGC sediments were squeezed out during sampling, hence there is a gap in the composite record between the LBC (at 1.25 kyr) and the GGC (at 2.2 kyr). Stacking GGC and GC make up a composite record covering the period from ~2.2 to 11 kyr. We sampled the LBC every 2 cm, which produced a 15–60 yr resolution. The GGC and the GC cover the time period from 2.2 and 11.2 kyr and sampling intervals of 5–10 cm resulted in a time resolution varying from 50 and 350 years.

2.2. Methods

The diatom samples were prepared according to the methods described by Koç et al. (1993) and references therein. Laboratory processes involve acid treatment to remove carbonate and organic matter, neutralization of remaining acid, clay separation and preparation of quantitative slides. A Leica Orthoplan microscope with 100/1.32 magnification was used for identification and counting of the diatoms. Counting procedures described in Schrader and Gersonde (1978) were followed.

The downcore diatom data were analysed and described in terms of the previously expanded work of Koç Karpuz and Schrader (1990) (Andersen et al., 2004) (Fig. 3). Factor analysis of diatoms from the surface sediments of the Nordic Seas and the North Atlantic produced eight significantly different assemblages (factors). Mapping of the eight factors showed their close affinity to hydrographic regime of the area. Factor 1 is

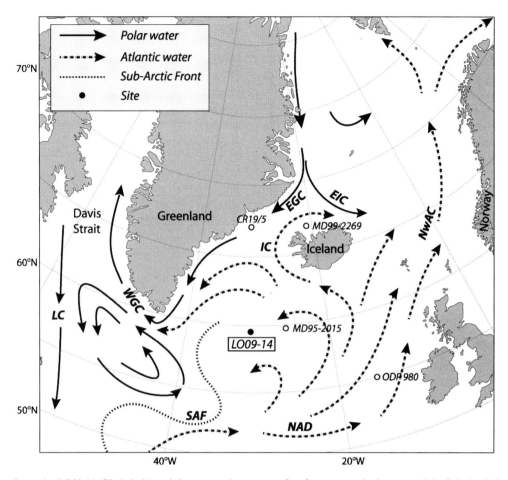

Fig. 1. Location of core site LO09-14 (filled circle) and the present-day pattern of surface currents in the area and the Sub-Arctic Front (SAF). Open circles show the sites of other records discussed in the text. NAD = North Atlantic Drift; IC = Irminger Current; NwAC = Norwegian Atlantic Current; EGC = East Greenland Current; EIC = East Icelandic Current; WGC = West Greenland Current; LC = Labrador Current. The figure is modified from Ruddiman and Glover (1975).

the Greenland Arctic Waters assemblage, which consists primarily of *Thalassiosira anguste-lineata* and *Thalassiosira trifulta*. Highest loadings of this factor occur in sediments underlying the Arctic waters of the Greenland Sea. Factor 2 is the North Atlantic Current assemblage and strongly reflects the distribution of a single species, *Thalassiosira oestrupii*. Other contributors to this assemblage are *Thalassionema nitzschioides*, *Nitzschia bicapitata*, *Rhizosolenia bergonii*, *Roperia tesselata*, and *Nitzschia marina*. Highest loadings of this factor occur under the warm and saline North Atlantic Current Waters. Factor 3 is the Subarctic Waters assemblage consisting primarily of *Rhizosolenia hebetata* f. *semispina* and to a lesser degree *Rhizosolenia borealis* and *Thalassiothrix longissima*. Highest loadings of this factor occur under the Arctic waters of the Iceland Sea and the Subarctic waters of the western North Atlantic. Factor 4 is the Norwegian-Atlantic Current assemblage consisting of *Thalassionema nitzschioides* as the main contributor, and *Proboscia alata* and *Thalassiosira angulata* as secondary contributors. Factor 5 is the Sea-ice assemblage consisting primarily of *Nitzschia grunowii*.

Other important species in this assemblage are *Nitzschia cylindra*, *Thalassiosira hyalina*, *Thalassiosira gravida* spores, *Thalassiosira nordenskioeldii* and *Bacterosira fragilis*. The spatial distribution of this assemblage mirrors the limit of the sea-ice edge in winter. Factor 6 is the Arctic Water assemblage consisting primarily of *Thalassiosira gravida* spores. Other important contributors to this assemblage are *Thalassiosira gravida* vegetative cells, *Actinocyclus curvatulus*, *Rhizosolenia hebetata* f. *semispina* and *Rhizosolenia hebetata* f. *hebetata*. In the Nordic Seas highest loadings of this factor occur under the East Icelandic Current and the Jan Mayen Polar Current. In the North Atlantic highest loadings of this factor is found as a belt following the Sub-Arctic front. Factor 7 is the East- and West-Greenland Current Assemblage consisting mainly of *Thalassiosira gravida* vegetative cells. In the North Atlantic highest loadings of this factor is found under the Subarctic Waters of the Labrador Sea. Factor 8 is the Transitional Waters assemblage consisting mainly of *Rhizosolenia borealis*. Because of their close affinity to modern hydrographic regimes these factors enable us to

Table 1
Radiocarbon dates and calibrated ages

Core name/Depth (cm)	Lab no.	^{14}C age ± standard deviation	Calibrated Age (Cal yr B.P.)
LO09-14LBC			
0	AAR-5049	705 ± 40	310
2	OS-32526	785 ± 35	425
9	OS-32140	1080 ± 25	643
29	OS-32475	1250 ± 65	795
39	OS-32474	1560 ± 30	1112
LO09-14GGC			
0.5	AAR-6671	1190 ± 35	not included
9	OS-32477	2770 ± 45	2499
17	KIA7500	2495 ± 25	2151
30	OS-32478	2990 ± 35	2760
47	AAR-6437	2690 ± 40	2384
69	OS-32524	3030 ± 35	2793
79	OS-32525	2860 ± 35	2627
101	KIA	3165 ± 35	2945
120	OS-32696	3260 ± 45	3081
150	KIA7501	4250 ± 35	4346
163	OS-32697	4680 ± 50	4907
197	OS	5220 ± 50	5577
240	KIA7502	6440 ± 35	6913
276	KIA	7180 ± 40	7636
LO09-14GC			
51	KIA7497	5330 ± 50	5689
103	AAR-4454	6620 ± 80	7125
135	KIA20793	6920 ± 40	7450
159	KIA7498	6605 ± 40	not included
165	KIA20794	7435 ± 40	7860
185	OS-32681	7730 ± 35	8190
216	KIA7499	8005 ± 40	8457
243	OS-32690	8420 ± 40	8926
270	KIA20795	8385 ± 45	8920
276	AAR-5050	8145 ± 45	not included
300	AAR-4455	8790 ± 100	9324
303	OS-32691	9050 ± 40	9653
315	OS-32692	9040 ± 45	9638
333	OS-32693	9220 ± 45	9911
370	OS-32694	9350 ± 40	10029
394	AAR-5051	9310 ± 55	10003
437	OS-32695	9920 ± 55	10705
447	AAR-4456	10,410 ± 90	11366

OS = Woods Hole Oceanographic Institute; AAR = Aarhus University; KIA = Kiel.
All measurements are made on the foraminifer species *Globigerina bulloides*.
Polynomial fits for the conversion from core depth (cm) to time domain are: (i) *LO09-14GGC*: cal kyr BP = 2474 − 2.9 cm + 0.0222 cm^2 + 0.00084 cm^2 − 2.31E-006 cm^4, R^2 = 0.987 and (ii) *LO09-14GC*: cal kyr BP = 4221 + 33.95 cm − 0.085 cm^2 + 9.61E-005 cm^3, R^2 = 0.985.

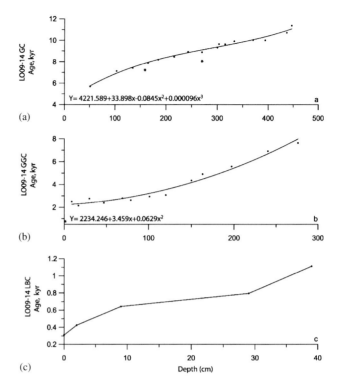

Fig. 2. The age-depth models for LO09-14 GC (a), GGC (b) and LBC (c) based on the dates presented in Table 1. Asterisk mark the excluded dates.

inferred SST of 0.89, and a maximum bias of 0.92 °C (John Birks, personal communication).

3. Results

3.1. Diatom assemblages as proxy for paleo-water masses

Warm Atlantic water carried northward by the IC is the major component of present day surface hydrography over the Reykjanes Ridge (Hopkins, 1991; Fratantoni, 2001). However, proximity to the SAF increase the susceptibility of the core site to subpolar water masses originating from the EGC. In this region the warm and cold waters are mixed together by the circular convergence of the subarctic gyre producing modified North Atlantic Waters (e.g., Reverdin et al., 1999). The different diatom assemblages (factors) serve as a proxy for these different water masses in the area (Fig. 3). The results of the quantitative SST reconstructions and the factor analysis are shown in Fig. 4a–d. Out of the eight different modern diatom assemblages that were established in Andersen et al. (2004), *factors 2, 3 and 7* are the primary contributors to the Holocene record of LO09-14 (Figs. 3, 4b–d). The North Atlantic Assemblage (*factor 2*), mirrors the influence of warm North Atlantic water masses over the site and fluctuations in this factor might indicate variability in the strength of the IC or the progressive cooling or warming

reconstruct details of the surface ocean variability. The downcore factors are then used in the temperature equations to estimate paleotemperatures. The estimations have a root mean square error of 1.25 °C, a coefficient of determination between observed and

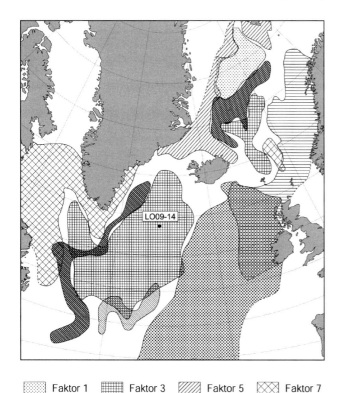

Fig. 3. The modern geographic distribution of the diatom assemblages and the study site LO09-14 (modified from Andersen et al., 2004). Factor 1 = Greenland Arctic Waters assemblage, Factor 2 = North Atlantic Current assemblage, Factor 3 = Subarctic Waters assemblage, Factor 4 = Norwegian-Atlantic Current assemblage, Factor 5 = Sea-ice assemblage, Factor 6 = Arctic Water assemblage, Factor 7 = East- and West-Greenland Current Assemblage, and Factor 8 = Transitional Waters assemblage.

of Atlantic water masses. *Factor 3* is the Sub-Arctic Assemblage and has its highest loadings in areas of mixture between Arctic and Atlantic waters, such as the present situation at the core site. High presence of *factor 3* would thereby indicate conditions similar to the modern. The last significant assemblage is the East- and West-Greenland Current Assemblage (*factor 7*), which reflect polar water originating from the EGC or the cold and warm water mixture of the subarctic waters from the area between the WGC and the LC.

3.2. Reconstruction of paleoceanographic conditions

The record of factor analysis and the reconstructed SSTs reveal highly variable surface water conditions through the last 11 kyr over the LO09-14 site (Fig. 4a–d). The general Holocene climate development of the site seem to be three-folded in trends and variability: During the early Holocene interval from 11 to 7.5 kyr, the surface conditions were highly variable with episodes of temperature fluctuations of 2–4 °C amplitudes and high amplitude fluctuations in the diatom assemblages.

This unstable period is followed by an interval of increasing SSTs to a Holocene optimum between 7.5 and 5 kyr. The final interval from 5 kyr to present reveals cooler and relatively more stable conditions compared to the period between 11–7.5 kyr.

3.2.1. Rhizosolenia borealis and Thalassiothrix longissima events

The time periods 11–9.5 kyr, 1.3–0.5 kyr and 2.3 kyr are characterized by several levels where the species *Rhizosolenia borealis* dominates the assemblages, reaching abundances between 50% and 90% of the total floral assemblage (hereafter referred to as *Rb*-events). Dense aggregations of the diatom genus *Rhizosolenia* sp. have been described along a strong convergence zone between warm and cold waters in the equatorial Pacific (Yoder et al., 1994). High concentrations of this species in the subpolar North Atlantic could thereby also imply responses to circulation conditions associated with an open-ocean front, i.e., the SAF. At 2.4 kyr there is also a level where the assemblage is dominated by the species *Thalassiothrix longissima* (Fig. 4a). Nearly monospecific-dominance of this species is not observed in modern analogues, and our interpretations are therefore again based on comparable counterparts from the Pacific Ocean, but also a thick diatom ooze reported from the subpolar North Atlantic during the last interglacial (Eem) (Bodén and Backman, 1996). The Eemian diatom ooze, consisting exclusively of *T. longissima*, is interpreted to be a result of a proximal convergence front (Bodén and Backman, 1996). Vast deposits of laminated diatom ooze in the equatorial Pacific cores formed during the Neogene by the rapid accumulation of the mat-forming diatom, *T. longissima*. These have also been associated with major frontal systems (Kemp and Baldauf, 1993). Based on the existing information on *Rhizosolenia* and *Thalassiothrix* mats we interpret the periods 11–9.5 kyr, 1.3–0.5 kyr and 2.3–2.4 kyr to be periods where there was exceptionally strong convergence in the surface waters of the northwest Atlantic Ocean.

3.2.2. A highly unstable Holocene climate

In the early Holocene (11–7.5 kyr) highly unstable conditions prevailed over the LO09-14 site on the Reykjanes ridge (Fig. 4a–d). From 11 kyr to 10 kyr surface conditions were relatively warm with SSTs between 11 and 13 °C. Several *Rb*-events occurred during this interval oscillating with warm water indicative diatoms (*factor 2*) and the EGC/WGC assemblage (*factor 7*) implying dynamic surface conditions with subpolar water advecting eastward and possibly the SAF migrating towards the site. A minor cold event around 10.4 precedes an abrupt and a drastic 4 °C SST drop at 9.8 kyr, which led to the coldest Holocene temperature of 8 °C in the record. This prominent

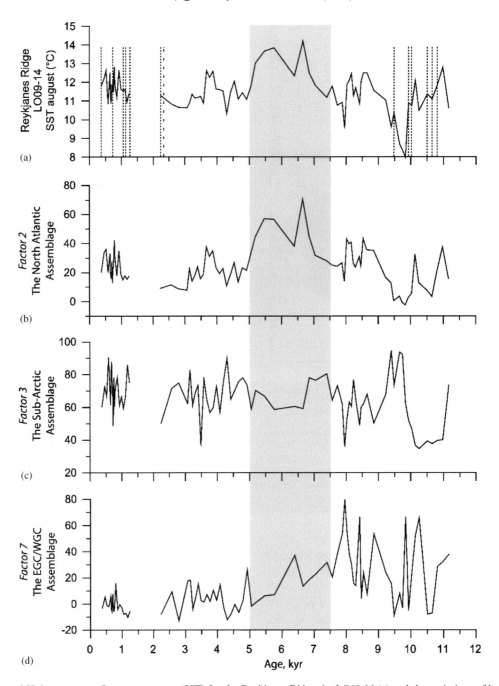

Fig. 4. Reconstructed Holocene sea surface temperatures (SST) for the Reykjanes Ridge site LO09-14 (a) and the variations of loadings of *factor 2* (b), *3* (c) and *7* (d). Dotted lines indicate *Rb*-events (*Rhizosolenia borealis* dominance) and the dashed line indicates the level of *Thalassiothrix longissima* dominance. The shaded area indicates the mid-Holocene optimum period.

cooling is associated with total absence of *factor 2* and a short-lasting peak of *factor 7* followed by the dominance of the Sub-Arctic assemblage (*factor 3*). *Rb*-events are bracketing the cooling with no prominent *Rhizosolenia* dominance occurring within the maximum cooling. The cooling is thereby not indicating an eastward migrating SAF, but rather a minimum in IC influence as evidenced by the total absence of *factor 2*. Following the cold event is an increase in SSTs towards temperatures around 12 °C between 9 and 8 kyr. The warming is mirrored by

the increasing contribution of *factor 2*, indicating a stronger IC influence or a progressive surface water warming. A 1.5 °C SST decrease centred at 8.3 kyr followed by another cooling at 7.9 kyr interrupts this general surface amelioration. Both these SST decreases are associated with high contribution of *factor 7* and minor decrease in *factor 2*. *Factor 7* displays a high amplitude variability, which has not been recorded since then, through the whole early Holocene. The record, thus, indicates surface conditions which were much

2161

more unstable and had a higher polar water imprint than at present. These pulses of *factor 7* might reflect a more zonal (eastward) circulation with polar and subpolar water expansion towards the east.

After 7.5 kyr a 2.5 °C rise over a 500-year period marks the start of a period of maximum Holocene temperatures (Fig. 4a). SSTs of 13–14 °C are reached between 7 and 5 kyr, which is 2–3 °C higher than the modern SSTs at this site. This warm interval displays a clear increase in the warm North Atlantic diatom assemblage (*factor 2*) and a progressive retreat of the East- and West-Greenland Current Assemblage (*factor 7*) towards modern values. This could imply a stronger IC flow over the site and/or a general warming of the surface water. The lack of *Rb*-events through this interval further suggests that the SAF had a more distal position to the study site or that the front was more hampered due to increased surface water mixing. A 1.5 °C decrease in SSTs centred at 6.4 kyr punctuate this ameliorated interval.

After 5 kyr, SSTs decreased and varied between 11 and 13 °C (Fig. 4a). Surface conditions became relatively stable compared to the early Holocene. Two cooling events around 4.7 and 4.3 kyr are followed by a warm peak centred around 3.7 kyr, which precedes a less constrained cooling about 2.8 kyr. These cooler surface conditions are reflected by the diminishing influence of warm Atlantic water (*factor 2*) and the re-establishment of *factor 3* as the dominant diatom assemblage over the site (Fig. 4b,c). In the interval 2.3–2.4 kyr there is one *Rb*-event, as well as a level of *T. longissima* ooze. Although much shorter in duration than the *T. longissima* ooze described from Eemian sediments (Bodén and Backman, 1996), the *T. longissima*-event seen at 2.4 kyr could be a miniature analogue with similar implications of a strong convergence front. From 2–1 kyr the record has a hiatus, due to intrusion of soupy sediments from the core liner during core sampling. The last 1 kyr of the LO09-14 record is characterized by SSTs varying between 11 and 12.5 °C. Prevailing *Rb*-events occur at this time interval, which may indicate the existence of a strongly developed convergence front in the area. To summarize the results; the subpolar North Atlantic seems to have experienced highly variable surface conditions throughout the entire Holocene, but especially during the early Holocene between 11 and 7.5 kyr. Repeated cooling events are centred at 10.4, 9.8, 8.3, 7.9, 6.4, 4.7, 4.3 and 2.8 kyr. A thermal optimum is recorded between 7.5 and 5 kyr.

4. Discussion and correlations

Holocene record of SSTs in the Nordic Seas display a development which is closely tied to the history of northern hemisphere high-latitude insolation, though with ca 2 kyr delay due most probably to the persistence of remnants of the Laurentide ice sheet into early Holocene and its effect on the oceanic and atmospheric circulation. In the Nordic Seas a Holocene climatic optimum is recorded between ca 9.5 and 6.5 kyr, though displaying a time transgressive behaviour in duration towards the west and north (Koç et al., 1993; Koç and Jansen, 1994; Eiriksson et al., 2000; Birks and Koç, 2002; Marchal et al., 2002). Our results show that the development of climate in the subpolar North Atlantic deviates from the development in the Nordic Seas considering both the timing of the Holocene climate optimum and the stability of the SSTs. Over the Reykjanes site the thermal optimum is not reached until 7.5–5 kyr, lagging the insolation maximum by 4000 years.

The differences observed between the early Holocene LO09-14 SST record and SST records further north indicate a decoupling in the surface regimes responsible for the observed variations (Fig. 5a–c). The sites on the North Iceland shelf and the East Greenland shelf reveal evidences for ameliorated surface conditions between 9.5 and 6.5 kyr to which a strong IC is invoked as a possible explanation (Eiriksson et al., 2000; Andersen et al., 2004). The presence of *factor 2* in LO09-14 during this time period shows that the Atlantic water inflow is also relatively high over the Reykjanes Ridge (Fig. 4b). However, the amount of cold water ejected into the convergence area south of Iceland was still high as indicated by the high presence of *factor 7* (Fig. 4d). This suggests that subpolar/subarctic water weakened the IC influence over the Reykjanes Ridge while the IC had a stronger impact further north. Since the two sites affected by the EGC or EGC-originating water show an early Holocene warming (Fig. 5b,c), the source of cold water over the Reykjanes Ridge probably comes from elsewhere. The answer could lie further south, namely in cold water outflow from the Davis Strait. At present most polar water outflow is through the Fram Strait and Denmark Strait (Aagaard and Carmack, 1989). However, during the early Holocene and until 7–6 kyr the outflow has been suggested to be larger from the Canadian Arctic due to wider channels as a result of isostatic depression (Williams et al., 1995). After the melting of the ice sheets the Canadian Arctic channels became shallower (Williams et al., 1995), establishing the present-day circulation. A sediment core from Orphan Knoll provides evidence for that the early Holocene was characterized by a relatively low salinity until ca 7.5 kyr in this area (Solignac et al., 2004). We, therefore, suggest that the cooler SSTs and the high variability observed in the LO09-14 early Holocene record might be due both to the flux of meltwater from remnants of the Laurentide ice sheet and to the higher polar water flux from the Canadian Arctic.

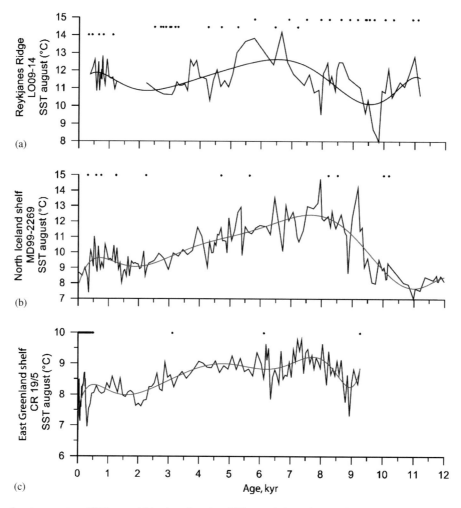

Fig. 5. LO09-14 sea surface temperature (SST) record (a), plotted against SST records from the North Iceland Shelf (b) and the East Greenland Shelf (c) (from Andersen et al., 2004). Dots mark AMS-dates and the black line depicts the Lead 210 measurements.

The different sources of cold water influence and varying warm water contribution over the Reykjanes Ridge site versus the North Iceland shelf site and the East Greenland shelf site is also visible in the temperature gradients (ΔT) between these sites (Fig. 6a,b). The modern SSTs reconstructed from the North Iceland shelf is 10 °C and 8 °C for the East Greenland shelf, such that the present ΔT to the Reykjanes Ridge is 2 °C and 4 °C, respectively. During the early Holocene when the two northerly sites experienced early Holocene climatic optimum reduced or negative temperature gradients are recorded between these sites and the Reykjanes Ridge site. As the thermal optimum was reached over the Reykjanes Ridge site, the transition towards colder surface conditions had already started further north over the EGC-influenced sites. Temperature differences thereby increased between the sites in the time interval 7–5 kyr. When the late Holocene cooling was initiated over the Reykjanes Ridge the SST decreased causing reduced ΔT. After 3 kyr the temperature gradients developed towards the present gradient.

A sequence of cold events is observed in the LO09-14 record throughout the Holocene which seem to be part of a larger, regional climate signal. These millennial scale coolings in core LO09-14 are centred at 10.4, 9.8, 8.3, 7.9, 6.4, 4.7, 4.3 and 2.8 kyr (Fig. 7a). These events correlate well with surface ocean perturbations noted as low concentrations of the coccolith *Emiliani huxleyi* from core MD95-2015 on the Gardar Drift (Giraudeau et al., 2000) (Figs. 1 and 7b). Both the LO09-14 and the MD95-2015 records show some similarities to the stacked record of hematite stained grains (HMG) from the North Atlantic showing that the sea ice events which transported the HMG to the North Atlantic was also accompanied by SST decrease (Fig. 7c) (Bond et al., 1997, 2001). Based on the LO09-14 SST record we can quantify some of these sea ice events to have caused 2–4 °C cooling of the surface waters. However, the correlation is rather poor in 8–5 kyr interval. This could be due either/both to the stacked nature of the HMG record and chronological uncertainties. The causes of Holocene millennial scale coolings are still under debate. But, both external forces like the sun's radiative output

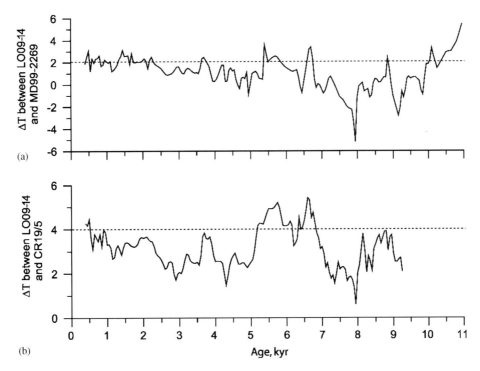

Fig. 6. ΔT between the interpolated SST curve of LO09-14 and interpolated SST curves from the North Iceland shelf (MD99-2269) (a) and the East Greenland shelf (CR19/5) (b). Dashed line depicts the modern ΔT between the sites.

(Bond et al., 2001), internal oscillations in the ocean circulation system (Campbell et al., 1998), and large-scale variations in atmospheric processes as the North Atlantic Oscillation have been suggested (Giraudeau et al., 2000).

Early Holocene coolings like the 8.2 kyr event, which might correspond to either the 8.3 or the 7.9 cooling in LO09-14, is often attributed to meltwater pulses caused by the decaying Laurentide and Scandinavian ice sheets (Bauch and Weinelt, 1997; Klitgaard-Kristensen et al., 1998; Barber et al., 1999). Frequent presence of similar freshwater episodes in the early Holocene could explain the larger temperature variability recorded in LO09-14 during this period compared to the late Holocene SST.

Deep-water production in the North Atlantic is important for the amount of heat transported north-ward and for the ocean ventilation. It is therefore considered as an amplifier or causality to millennial-scale climate variability during glacial and deglacial time (Broecker, 1990). Deep-water ventilation, however, also varied during the Holocene (Bianchi and McCave, 1999; Oppo et al., 2003). Marine archives record highly unstable deep-water conditions prior to the mid-Holocene optimum, which are attributed to remnants of ice sheets (Bianchi and McCave, 1999). As the decay of ice sheets stalled, millennial-scale variability still continued. The LO09-14 SST record, which shows evidence for a reduction in IC and an increased cold water influence prior to the mid-Holocene optimum in the subpolar North Atlantic, support changes in

circulation as a possible explanation to this variability. A record of carbon-isotope ($\delta^{13}C$) variations from ODP Site 980 in the subpolar northeastern Atlantic (Figs. 1 and 7d) shows reductions in the relative NADW contribution at 9.3, 8.0, 5.0 and 2.8 kyr (Oppo et al., 2003). The observed NADW reductions correlate well with the timing of the surface coolings recorded at the LO09-14 site. When NADW production is low the overturning circulation becomes weaker, hence heat flux transported to the north is decreased. As expected the decreasing NADW contribution from 6.5 to 5.0 kyr is reflected as a cooling from the mid-Holocene optimum to late-Holocene cooler SSTs at the LO09-14 site. Based on this close correlation, we argue that a surface-deepwater linkage also exists for the Holocene. The trigger for the NADW disturbances could, as discussed by Bond and co-authors (1997, 2001), be an increase in outflow of drift ice from the Fram Strait like the Great Salinity Anomaly in the 1960s and 1970s (Dickson et al., 1988). However, a study from the subarctic Nordic Seas does not reveal variability in the deep-water formation during the last 5 kyr (Bauch et al., 2001), adding to the complexity of the ventilation system. There is no simple one-to-one correlation between the amount of inflowing Atlantic water to the Nordic Seas and the amount of deep-water formation. Instead of participating in the deep-water formation most of the Atlantic water entering the Nordic Seas is modulated to sink to intermediate depths (Mauritzen, 1996).

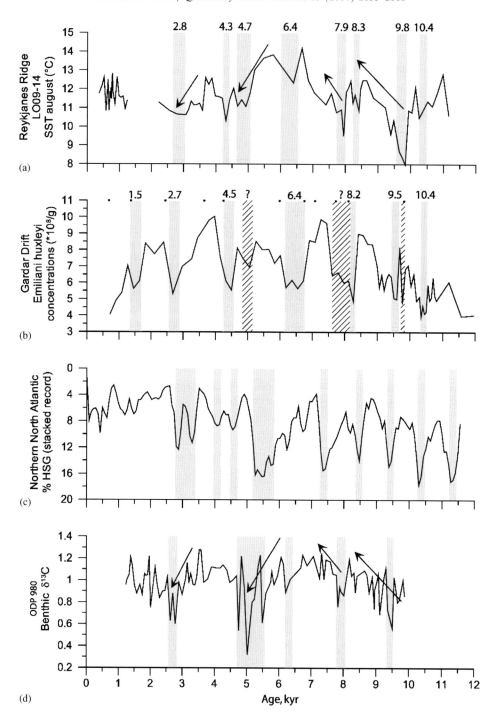

Fig. 7. Comparison of different Holocene climate records from the northern North Atlantic: LO09-14 sea surface temperature (SST) record from Reykjanes Ridge (a); *Emiliani huxleyi*-events in core MD95-2015 indicating surface perturbations on the Gardar Drift (Giraudeau et al., 2000) (b); Hematite-stained grains (HSG) indicating repeated IRD-events during the Holocene (Bond et al., 1997, 2001) (c); $\delta^{13}C$ from ODP Site 980 as a proxy for North Atlantic Deep Water production (Oppo et al., 2003) (d). The shaded areas show the Holocene cooling events and the hatched areas are possible cold events. The dots in (b) show the AMS-dates in the Gardar Drift record. All records are plotted against age models used by their authors.

Another aspect to the climate system in the subpolar North Atlantic is the recent postulations of deep-water formation in the Irminger Sea (Pickart et al., 2003a). The convection is driven by small-scale inter-annual atmospheric jets formed over the southern tip of Greenland. It is not yet captured in global weather models, but local effects are reported to be strong westerly winds, low air temperatures and heat loss along the 60 °N parallel (Pickart et al., 2003b). Similar storm events at a larger scale could possibly have defined a stronger configuration to the generally uncertain northernmost branch of the SAF, which could explain the recurring *Rb*-events between 1.3 and 0.5 kyr.

5. Conclusions

Our results from the Reykjanes Ridge site show that the Holocene climate developed differently in different regions of the North Atlantic due to different dominant forcing factors. It has been suggested that climate development in the Nordic Seas has been mainly driven by the insolation forcing (e.g., Koç et al., 1993; Koç and Jansen, 1994). Whereas, climate development in the subpolar North Atlantic in proximity of the SAF seems to have been mainly affected by (a) incursions of meltwater from the melting remnants of the Laurentide ice sheet (b) increased outflow of polar water from the Canadian Arctic and (c) variations in the production rate of NADW. Based on our results, we think the western branch of the NAD was reduced due to enhanced meltwater flux from Labrador Sea during the early Holocene and was instead compensated by a stronger eastern branch flowing into the Norwegian Sea. If this is true, then the climate optimum observed in the Nordic Seas during early Holocene is a combination of stronger NAD flux and higher insolation.

The warming in the subpolar North Atlantic, which had been suppressed until about 7 kyr due to the presence of cold and low salinity meltwater, occurs during the mid-Holocene when the northern hemisphere insolation was being strongly reduced. It also starts and ends rather abruptly indicating perhaps a change in atmospheric circulation rather than a gradual insolation forcing. An interesting question to investigate further is whether a long-term NAO/AO type of atmospheric circulation pattern as suggested by Rimbu et al. (2003) can explain the Holocene climate development in the high latitude North Atlantic.

Acknowledgements

This study was supported by VISTA (in collaboration with Statoil), the Norwegian Research Council through the NOClim and NORPAST projects, the European Commission and the Norwegian Polar Institute. We thank the "LO-14 group": J. McManus, A. Kuijpers, K. Emeis and I. Snowball for good collaboration, K.L. Knudsen and an anonymous reviewer for constructive comments. We also thank E. Jansen and P. de Menocal, coordinators of the IMAGES Holocene Working Group Workshop (Hafslo, August 2003) for a superbly organized meeting. [14]C datings were performed at Woods Hole Oceanographic Institute, the University of Aarhus, and at Leibnitz Labor für Altersbestimmung und Isotopenforschung, Christian-Albrechts-Universität, Kiel.

References

Aagaard, K., Carmack, E.C., 1989. The role of sea ice and other fresh water in the Arctic Circulation. Journal of Geophysical Research 94, 14485–14498.

Alley, R.B., Mayewski, P.A., Sowers, T., Stuiver, M., Taylor, K.C., Clark, P.U., 1997. Holocene climatic instability: a prominent, widespread event 8200 yr ago. Geology 25, 483–486.

Andersen, C., Koç, N., Jennings, A., Andrews, J.T., 2004. Nonuniform response of the major surface currents in the Nordic Seas to insolation forcing: implications for the Holocene climate variability. Paleoceanography 19, PA2003, doi: 10.1029/2002PA000873.

Barber, D.C., Dyke, A., Hillaire-Marcel, C., Jennings, A.E., Andrews, J.T., Kerwin, M.W., Bilodeau, G., McNeely, R., Southons, J., Morehead, M.D., Gagnon, J.-M., 1999. Forcing of the cold event of 8,200 years ago by catastrophic drainage of Laurentide lakes. Nature 400, 344–348.

Bauch, H.A., Weinelt, M.S., 1997. Surface water changes in the Norwegian Sea during last deglacial and Holocene times. Quaternary Science Reviews 16, 1115–1124.

Bauch, H.A., Erlenkeuser, H., Spielhagen, R.F., Struck, U., Matthiessen, J., Thiede, J., Heinemeier, J., 2001. A multiproxy reconstruction of the evolution of deep and surface waters in the subarctic Nordic seas over the last 30,000 yr. Quaternary Science Reviews 20, 659–678.

Bianchi, G.G., McCave, I.N., 1999. Holocene periodicity in North Atlantic climate and deep-ocean flow south of Iceland. Nature 397, 515–517.

Birks, C.J.A., Koç, N., 2002. A high-resolution diatom record of late-Quaternary sea-surface temperatures and oceanographic conditions from the eastern Norwegian Sea. Boreas 31, 323–344.

Boden, P., Backman, J., 1996. A laminated sediment sequence from the northern North Atlantic Ocean and its climatic record. Geology 24, 507–510.

Bond, G., Showers, W., Cheseby, M., Lotti, R., Almasi, P., deMenocal, P., Priore, P., Cullen, H., Hajdas, I., Bonani, G., 1997. A pervasive millennial-scale cycle in North Atlantic Holocene and glacial climates. Science 278, 1257–1266.

Bond, G., Kromer, B., Beer, J., Muscheler, R., Evans, M.N., Showers, W., Hoffmann, S., Lotti-Bond, R., Hajdas, I., Bonani, G., 2001. Persistent Solar Influence on North Atlantic Climate During the Holocene. Science 294, 2130–2136.

Broecker, W.S., 1990. Salinity history of the northern Atlantic during the last deglaciation. Paleoceanography 5, 459–467.

Bronk Ramsey, C., 1995. Radiocarbon Calibration and Analysis of Stratigraphy: the OxCal Program. Radiocarbon 37, 425–430.

Bronk Ramsey, C., 2001. Development of the Radiocarbon Program OxCal. Radiocarbon 43, 355–363.

Campbell, I.D., Campbell, C., Apps, M.J., Rutter, N.W., Bush, A.B.G., 1998. Late Holocene ~1500 yr climatic periodicities and their implications. Geology 26, 471–473.

Dansgaard, W., Johnsen, S.J., Clausen, H.B., Dahl-Jensen, D., Gundestrup, N.S., Hammer, C.U., Hvidberg, C.S., Steffensen, J.P., Sveinbjörnsdottir, A.E., Jouzer, J., Bond, G., 1993. Evidence for general instability of past climate from a 250-kyr ice-core record. Nature 364, 218–220.

Denton, G.H., Karlén, W., 1973. Holocene climatic variations—Their pattern and possible cause. Quaternary Research 3, 155–205.

Dickson, R.R., Meincke, J., Malmberg, S.A., Lee, A.J., 1988. The "Great Salinity Anomaly" in the Northern North Atlantic 1968–1982. Progress in Oceanography 20, 103–151.

Dietrich, G., 1969. Atlas of the hydrography of the Northern Atlantic Ocean based on the Polar Front Survey of the International Geophysical Year, winter and summer 1958. International Council for the Exploration of the Sea, Hydrographic Service, Copenhagen.

Eiriksson, J., Knudsen, K.L., Haflidason, H., Henriksen, P., 2000. Late-glacial and Holocene paleoceanography of the North Icelandic shelf. Journal of Quaternary Science 15, 23–42.

Fratantoni, D.M., 2001. North Atlantic surface circulation during the 1990's observed with satellite-tracked drifters. Journal of Geophysical Research 106, 22,067–22,093.

Giraudeau, J., Cremer, M., Manthé, S., Labeyrie, L., Bond, G., 2000. Coccolith evidence for instabilities in surface circulation south of Iceland during Holocene times. Earth and Planetary Science Letters 179, 257–268.

Hopkins, T.S., 1991. The GIN-Sea—A synthesis of its physical oceanography and literature review 1972–1985. Earth-Science Reviews 30, 175–318.

Hurdle, B.G., 1986. The Nordic Seas. Springer, New York, 777pp.

Jiang, H., Seidenkrantz, M.-S., Knudsen, K.L., Eiriksson, J., 2002. Late-Holocene summer sea-surface temperatures based on a diatom record from the north Icelandic shelf. The Holocene 12, 137–147.

Johannessen, O.M., 1986. Brief overview of the physical oceanography. In: Hurdle, B.G. (Ed.), The Nordic Seas, pp. 103–127.

Kemp, A., Baldauf, J., 1993. Vast Neogene laminated diatom mat deposits from the equatorial Pacific Ocean. Nature 362, 141–143.

Klitgaard-Kristensen, D., Sejrup, H.P., Haflidason, H., Johnsen, S., Spurk, M., 1998. A regional 8200 cal yr BP cooling event in northwest Europe, induced by final stages of the Laurentide ice-sheet deglaciation? Journal of Quaternary Science 13, 165–169.

Klitgaard-Kristensen, D., Sejrup, H.P., Haflidason, H., 2001. The last 18 kyr fluctuations in Norwegian Sea surface conditions and implications for the magnitude of climate change: evidence from the North Sea. Paleoceanography 16, 455–467.

Koç, N., Jansen, E., 1994. Response of the high-latitude Northern Hemisphere to orbital climate forcing: evidence from the Nordic Seas. Geology 22, 523–526.

Koç, N., Jansen, E., Haflidason, H., 1993. Paleoceanographic reconstructions of surface ocean conditions in the Greenland, Iceland and Norwegian Seas through the last 14 ka based on diatoms. Quaternary Science Reviews 12, 115–140.

Koç Karpuz, N., Jansen, E., 1992. A high-resolution diatom record of the last deglaciation from the SE Norwegian Sea: documentation of rapid climatic changes. Paleoceanography 7, 499–520.

Koç Karpuz, N., Schrader, H., 1990. Surface sediment diatom distribution and Holocene paleotemperature variations in the Greenland, Iceland and Norwegian Sea. Paleoceanography 5, 557–580.

Lehman, S.J., Keigwin, L.D., 1992. Sudden changes in North Atlantic circulation during the last deglaciation. Nature 356, 757–762.

Marchal, O., et al., 2002. Apparent long-term cooling of the sea surface in the northeast Atlantic and Mediterranean during the Holocene. Quaternary Science Reviews 21, 455–483.

Maurtizen, C., 1996. Production of dense overflow waters feeding the North Atlantic across the Greenland-Scotland Ridge. Part 2: an inverse model. Deep-Sea Research 43, 807–835.

Nesje, A., Dahl, S.O., Andersson, C., Matthews, J.A., 2000. The lacustrine sedimentary sequence in Sygneskardvatnet, western Norway: a continuous, high-resolution record of the Jostedalsbreen ice cap during the Holocene. Quaternary Science Reviews 19, 1047–1065.

O'Brien, S.R., Mayewski, P.A., Meeker, L.D., Meese, D.A., Twickler, M.S., Whitlow, S.I., 1995. Complexity of Holocene climate as reconstructed from a Greenland ice core. Science 270, 1962–1964.

Oppo, D.W., McManus, J.F., Cullen, J.L., 2003. Deepwater variability in the Holocene epoch. Nature 422, 277–278.

Pickart, R.S., Straneo, F., Moore, G.W.K., 2003a. Is labrador sea water formed in the irminger basin? Deep-Sea Research I 50, 23–52.

Pickart, R.S., Spall, M.A., Ribergaard, M.H., Moore, G.W.K., Milliff, R.F., 2003b. Deep convection in the Irminger Sea forced by the Greenland tip jet. Nature 424, 152–156.

Reverdin, G., Verbrugge, N., Valdimarsson, H., 1999. Upper ocean variability between Iceland and Newfoundland, 1993–1998. Journal of Geophysical Research 104, 29599–29611.

Rimbu, N., Lohmann, G., Kim, J.-H., Arz, W., Schneider, R., 2003. Arctic/North Atlantic Oscillation signature in Holocene sea surface temperature trends as obtained from alkenone data. Geophysical Research Letters 30 (6), 1280, doi:10.1029/2002GL016570.

Ruddiman, W.F., Glover, L.K., 1975. Subpolar North Atlantic circulation at 9300 yr BP: faunal evidence. Quaternary Research 5, 361–389.

Schrader, H.J., Gersonde, R., 1978. Diatoms and silicoflagellates in the eight metres section of the lower Pliocene of Capo Rossello. Utrecht Micropaleontological Bulletin 17, 129–176.

Solignac, S., de Vernal, A., Hillaire-Marcel, C., 2004. Holocene sea-surface conditions in the North Atlantic—contrasted trends and regimes in the western and eastern sectors (Labrador Sea vs. Iceland Basin). Quaternary Science Reviews 23, 319–334.

Stefánsson, U., 1962. North Icelandic waters. Journal Marine Research Institute 3, 1–269.

Stuiver, M., Reimer, P.J., Bard, E., Beck, J.W., Burr, G.S., Hughen, K.A., Kromer, B., McCormac, G., van der Plicht, J., Spurk, M., 1998. INTCAL98 Radiocarbon Age Calibration, 24000–0 cal BP. Radiocarbon 40, 1041–1083.

Williams, K.M., Andrews, J.T., Weiner, N.J., Mudie, P.J., 1995. Late Quaternary paleoceanography of the mid- to outer continental shelf, East Greenland. Arctic and Alpine Research 27, 352–363.

Yoder, J.A., Ackleson, S.G., Barber, R.T., Flament, P., Balch, W.M., 1994. A line in the sea. Nature 371, 689–692.

ELSEVIER

Quaternary Science Reviews 23 (2004) 2167–2181

QSR

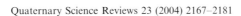

General circulation modelling of Holocene climate variability

Gavin A. Schmidt[a,b,*], Drew T. Shindell[a,b], Ron L. Miller[a,c],
Michael E. Mann[d], David Rind[a]

[a]*NASA Goddard Institute for Space Studies, New York, NY, USA*
[b]*Center for Climate Systems Research, Columbia University, New York, NY, USA*
[c]*Department of Applied Maths and Physics, Columbia University, New York, NY, USA*
[d]*Department of Environmental Sciences, University of Virginia, Charlottesville, VA, USA*

Abstract

Results from a series of Goddard Institute for Space Studies (GISS) General Circulation Models (GCMs) are used to assess climate variability in the pre-anthropogenic Holocene, the interval following the end of the last glacial beginning roughly 11.5 kyr BP. In particular, we focus on the forced aspects of this variability. The principle forcings are orbital, solar, volcanic and events (such as the 8.2 kyr BP event). Land use and greenhouse gases also play a small role. We discuss suitable comparisons to paleo-data and the appropriateness of model experimental design for single and multiple forcing runs using time-slices and transient simulations. As an example, we focus on the response to solar and volcanic forcing in the context of the Maunder Minimum, a cool period of the late 17th century and demonstrate that although (northern) hemispheric mean temperature changes can be reasonably simulated in most models, the details of regional patterns depend more heavily on the included physics. In particular, we highlight the role of the stratosphere as well as the importance of ocean and vegetation feedbacks.
© 2004 Elsevier Ltd. All rights reserved.

1. Introduction

There is an increasing community focus on Holocene climate variability that is driven partly by the increasing number of quality proxy paleo-records, but also by the obvious relevance for assessing future climate change (Renssen and Osborn, 2003). Climate models are becoming more sophisticated in how they can be forced and the quality of the physics contained therein. Better estimates of potential forcings are being produced (Bard et al., 2000; Crowley, 2000; Hansen et al., 2002) and better chronologies are allowing more complete and

integrated snapshots of past climate than were available previously (e.g. Harrison et al., 2003). It is therefore unsurprising that more climate model simulations are now being performed with relevance to Holocene climate variability (Gonzalez-Rouco et al., 2003; Shindell et al., 2003; Widmann and Tett, 2003; Goosse et al., 2004). However, due to inherent limitations in knowledge of past boundary conditions, the appropriate initial state of the ocean and the limited simulation lengths possible with state-of-the-art models, the experiments being performed are often abstractions, rather than a reconstruction of a particular time period. Analysis of such experiments can often be problematic for paleo-climatologists. We will therefore attempt to address the question: What can climate modelling tell us about Holocene variability?

Firstly, we discuss the principal forcings that might have played significant roles, the time history of these forcings that can be used by models, and what the modelled response to these forcings generally imply.

*Corresponding author. NASA Goddard Institute for Space Studies and Center for Climate Systems Research, Columbia University, 2880 Broadway, New York, NY 10025, USA. Tel.: +1-212-678-5627; fax: +1-212-678-5552.

E-mail addresses: gschmidt@giss.nasa.gov (G.A. Schmidt), dshindell@giss.nasa.gov (D.T. Shindell), rmiller@giss.nasa.gov (R.L. Miller), mann@virginia.edu (M.E. Mann), drind@giss.nasa.gov (D. Rind).

0277-3791/$ - see front matter © 2004 Elsevier Ltd. All rights reserved.
doi:10.1016/j.quascirev.2004.08.005

Modellers have an inbuilt bias toward forced climate change because the cause and effect are clear. Intrinsic variability in models or in the climate record is much more problematic due to the tight coupling of the physical (ocean, atmosphere, sea ice) and bio-geochemical cycles and the inevitable ambiguity regarding cause and effect that results. However, the characterisation of internal (or unforced) variability inevitably arises when trying to identify forced variability in the observed record. There are global mean, regional and local climate changes that will have occurred even in the absence of forcing, due for instance to variability in the ocean circulation on decadal or even centennial time-scales (e.g. Delworth et al., 1993). Distinguishing this kind of variability from the forced (and therefore more predictable) kind is a key research topic.

Secondly, we discuss time-slice experiments that have been, or can be, performed, and the kinds of paleo-data that are useful for validating model simulations. These kinds of experiments will focus on particular intervals where conditions are presumed to have been relatively stable (for timescales relevant for the particular model configuration) and attempt to model the impact of all particular forcings and relevant changes in boundary conditions. Due to the difficulty in producing time-slice paleo-reconstructions and the observed variability in the climate, only a few time-slices are likely to be examined in any detail; chiefly the mid-Holocene (8–6 kyr BP), the Maunder Minimum (late 17th century) and the pre-industrial (mid 19th century). Finally we discuss the simulations of transient climate change where coupled models of varying complexity are run with time-varying forcings to try and estimate the transient response of climate. These are principally being done for the last 500–1000 years.

2. Climate models

We will restrict the discussion to general circulation models (GCMs) of the climate. In particular, we will highlight model results using various flavours of the GISS GCM. These models, at minimum, have a reasonable simulation of the atmospheric circulation, radiative transfer and hydrologic cycle, but can differ markedly in the complexity of the ocean component and any associated bio-geochemical modules. Models of intermediate complexity have also been used to address these issues and are particularly useful for conceptual studies and for longer time-scales than can be accommodated in a GCM. For instance, these models have shown changes in the overturning circulation during the Holocene (Renssen et al., 2001; Goosse et al., 2004) and have been able to explore vegetation feedbacks more efficiently than GCMs (Claussen et al., 1999). However, many of the interesting issues for the Holocene relate to

spatial or seasonal patterns that cannot be resolved in such models, for instance, relating to the winter atmospheric circulation, El Niño/Southern Oscillation (ENSO) or monsoonal changes. GCMs too, have difficulty with some important aspects of climate variability (i.e. ENSO in coupled models, the stratospheric connection to the North Atlantic Oscillation (NAO)/Arctic Oscillation (AO) etc.), but many aspects of atmospheric variability are reasonably simulated.

Climate models can be run in two ways, in equilibrium experiments and in transient mode. For the first case, a change is generally made in the boundary conditions and the model is run long enough so that it reaches a statistical steady state. This takes between 5 and 10 years of simulation in an atmosphere-only model (AGCM), around 20–30 years in models with some kind of upper ocean component (such as an ocean mixed layer (AGCM-ML or Qflux model)) and around 500–1000 years for models with a deep ocean (a fully coupled ocean-atmosphere GCM) due to the enormous heat capacity and slow mixing of the oceans. Intermediate configurations sometimes include a mixed layer model with some diffusion into the deeper ocean. Depending on the question to be addressed and the time-scales involved, a suitable equilibrium experiment can usually be defined. Note however, that there is an implicit assumption that longer term processes that are not explicitly considered (ice sheets, carbon cycle, sea surface temperatures (for an AGCM), the ocean heat transports (for an AGCM-ML) etc.) are constant. This can be problematic, particularly for the ocean and land surface (including the vegetation), both of which influence, and are influenced by, the atmospheric state. However, the advantage of solid statistics (due to the averaging out of short-term 'weather' noise) is frequently paramount. These kinds of experiments are generally validated using time-slices and spatial patterns of change.

The second kind of experiment is a transient run, where starting from an initial condition, the forcings change in time and a climate history is simulated. This kind of experiment most resembles paleo-record time series but is plagued with difficulties for the modeller. Firstly, defining suitable initial conditions is extremely tricky and laborious. In particular, the oceans retain a 'memory' of the past few hundred years which can play a role in the subsequent climate evolution. This is likely to be most important in the early Holocene where the ocean is still reacting to deglaciation and also in subsequent periods when the ocean may not be in equilibrium and observations may place only a weak constraint on oceanic changes. Secondly, any one experiment will have a great deal of chaotic behaviour unique to that realization and so many simulations (i.e. a 5–10 member ensemble) need to be done to get good a signal-to-noise ratio, particularly for assessing regional changes. Finally, accurate estimates of changes in forcing through time are extremely sparse and come with high uncertainties. Thus, transient climate simulations are only now starting to appear for periods earlier than about 1850.

Some transient experiments can be performed using transient sea surface temperatures (SST) and sea ice as a 'forcing' to the atmospheric model. This is possible for periods where good estimates of global SST and sea ice are available (since about 1870 so far (i.e. Rayner et al., 2003)). Conceivably, the SST record could be extended back using calibrated networks of proxies (Rutherford et al., 2003), which could avoid the need for expensive ocean model calculations in these transient runs, but sea ice reconstructions will remain problematic prior to direct observations. However, the response of the atmosphere to varying SST does not necessarily reproduce the variations in the coupled system that gave rise to the SST changes in the first place (i.e. Kushnir et al., 2002).

The latest generation of GCMs come with optional interactive components for many bio-geochemical subsystems. Due to the natural delay in using state-of-the-art models for paleo-climate, very few of these have been applied to Holocene climate variability as yet. Preliminary results with dynamic vegetation (Braconnot et al., 1999) and interactive stratospheric ozone (Shindell et al., 2001a; Rind et al., 2004), are however available and are discussed briefly below. We discuss in the final section likely future experiments.

3. Forcings and response

For climate models, forcings can be simply defined as changes in the background conditions that are external to the model calculations. For instance, changes to insolation either due to solar variability or orbital changes are clearly an external forcing. Changes in atmospheric composition can be considered to be forcing, as long as that particular component is not a prognostic output from the model. For example, in a model that does not calculate atmospheric chemistry, methane concentrations can be considered a forcing, while in a model with a prognostic methane cycle, it would be a feedback to changes in emissions or temperature. We discuss all the principle forcings (including some more speculative ideas) and the possibility of having reasonably accurate time-histories of them. We consider particularly the changes in forcing prior to around 1850 (the pre-industrial) to avoid complications due to recent anthropogenic perturbations to the climate system.

Forcings can be usefully characterised by the instantaneous change in the radiation budget at the tropopause (the instantaneous forcing). The eventual equilibrium global temperature change is roughly proportional to the forcing, with the climate sensitivity as the constant of proportionality. Another quantity, the adjusted forcing where the stratosphere is allowed to adjust radiatively while the troposphere is fixed, is a slightly better predictor (Hansen et al., 1997a), but the typically small distinction will not be addressed here. The concept of instantaneous forcing though is not very useful either for forcings which are extremely regional in effects, or for predicting regional climate change. For instance, orbital radiative forcing is almost zero in the global mean, but regionally and seasonally is extremely large. For each forcing, we discuss some of the most useful paleo-records or analyses for validating climate models and some of the key results and continuing problems.

Climate sensitivity is calculated as a matter of course by the global mean response to a doubling of CO_2 using an AGCM with an ocean mixed layer. The instantaneous forcing from this is around $4\,W/m^2$ and the range of climate response to that is around $3 \pm 1\,°C$ (Houghton et al., 2001). In estimates that follow we will assume a canonical sensitivity of $0.75\,°C$ per W/m^2 unless otherwise stated. Some degree of polar amplification usually occurs (especially in winter), though the extent of this is dependent on the base state (e.g. ice extent) (Rind et al., 1995). Some forcings can have a greater impact on global mean temperature than an equivalent amount of CO_2, for instance, the black carbon (soot) impact on sea ice albedo (Hansen and Nazarenko, 2004), but the concept of equivalent radiative forcings is still a useful metric. Over the Holocene there is no strong evidence that climate sensitivity has changed substantially.

3.1. Well-mixed greenhouse gases

The most important greenhouse gas is water vapour. However, this has such a short lifetime in the troposphere (around 10 days) it can be considered a feedback to changes in other radiative forcings and is always modelled as such. The other principal natural greenhouse gases CO_2, CH_4 and N_2O have been measured in gas bubbles trapped in ice cores and have a past history that is rather well known. In fact the history of these gases over the pre-industrial Holocene is remarkably stable, with CO_2 variations of 20 ppmv around the pre-industrial value of 280 ppmv and N_2O values of around 260 ± 18 ppbv (Sowers et al., 2003). There is a small dip of around 4–10 ppmv in CO_2 from 1600–1800 CE in the Law Dome analysis (Etheridge et al., 1996) (although the timing of this dip is 300 years earlier in the Taylor Dome ice core (Indermuhle et al., 1999)). CH_4 is slightly more variable, with a peak value of 700 ppbv at 11 kyr BP, a minimum of around 575 ppbv at 5.5 kyr BP and up to 700 ppbv again by the pre-industrial period (Chappellaz et al., 1997; Houghton et al., 2001). There is a notable short term dip in CH_4 of at least 60 ppbv at around 8.2 kyr BP.

The radiative forcing from GHGs over the Holocene is therefore small, a maximum of $0.5 \, W/m^2$ from CO_2 and around $0.09 \, W/m^2$ from 5.5 kyr BP minimum to about 1850 for methane. The 17th to 18th century decrease in CO_2 is consistent with a $0.1–0.2 \, W/m^2$ negative forcing at that time. The 8.2 kyr dip in CH_4 gives a small $0.05 \, W/m^2$ and was probably not radiatively significant. The short lifetime of CH_4 in the atmosphere (8–10 years) and the relatively coarse sampling in the ice cores does not preclude shorter term excursions, but any such excursions are similarly unlikely to have had much climatic impact. There are additional indirect effects from CH_4 changes on stratospheric water vapour and on ozone, but these are smaller than the direct impact.

CO_2 and methane have effects on the global temperature proportional to their forcing and thus for the Holocene changes described above, the global annual mean changes will have been within the range of $\approx 0–0.5\,^{\circ}C$.

3.2. Orbital changes

The largest perturbation in radiative forcing since the disappearance of the Laurentian and Fenno-Scandinavian ice sheets is undoubtedly the change in insolation (incoming solar radiation) from the mid-Holocene to the present due principally to the change in precession (Fig. 1). At around 6 kyr BP Northern Hemisphere summers had a significantly larger insolation than at present ($>25 \, W/m^2$ over the season), while tropical areas received slightly less in the annual mean. The high accuracy to which this forcing is known (Berger, 1978) and the similarity of the ice sheet configuration to today, has made the 6 kyr BP period a favorite target for climate models, particularly through the Paleoclimate Modelling Intercomparison Project (PMIP) (Hewitt and Mitchell, 1996; Joussaume et al., 1999; Braconnot et al., 1999; Otto-Bleisner, 1999; Kitoh and Murakami, 2002; Liu et al., 2003a).

The most useful observational targets for this period are the records of Northern Hemisphere (NH) high latitude warmth and maps of hydrological changes across N. America, Africa and Asia (Cheddadi et al., 1997; Texier et al., 1997; Viau and Gajewski, 2001; Harrison et al., 2003) (Fig. 2). Some fossil coral analyses have indicated that there may have been reductions in the ENSO variability (Gagan et al., 1998; Liu et al., 2003b). No good estimates of NH mean or global temperature are available and so this period is principally a test of the regional climate change, particularly in the hydrologic cycle, produced by the models.

Fig. 2 illustrates the importance of allowing ocean feedbacks for forcings that have time-scales longer than the ocean response time (see also Braconnot et al., 1999; Otto-Bleisner, 1999). These results are from recent simulations with the GISS model where only orbital forcing is changed while using pre-industrial atmospheric composition. For the first experiment, the ocean temperatures are kept fixed at pre-industrial values (1876–1885 mean) (Rayner et al., 2003), while in the second experiment the Qflux ocean was used which allows for a thermodynamic response to the forcing (results shown are 5 year annual means after the models have equilibrated). An increase in Sahel rainfall is clear in both cases, consistent with the lake level records, but the effect is doubled in the Qflux experiment. In prescribed SST experiments the surface energy budget is out of balance and the evaporation and hydrologic cycle are not fully adjusted to the forcing (Miller and Tegen, 1998).

Other feedbacks (principally vegetation, see Section 3.5 also) are likely to have also been important for this period (Braconnot et al., 1999). Similarly, Eurasian warming is clearly enhanced when the ocean temperatures and sea ice are allowed to respond to the high latitude increase in insolation. Comparisons to the northern European reconstruction of coldest month temperature (Cheddadi et al., 1997) (not shown) indicate that the west-to-east transition from cooling to warming is not well captured by either experiment. Overall there is no significant global mean temperature change ($<-0.02\,^{\circ}C$). The reconstructed pattern of drying over North America is similar to that seen in during La Niña phases of ENSO (Liu et al., 2003b), and as in previous

Insolation difference 6000 BP - 2000CE (W/m^2)

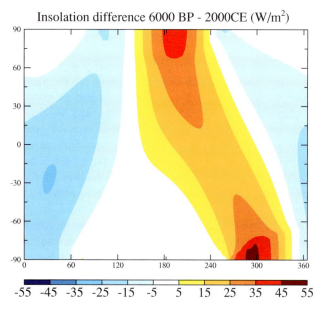

-55 -45 -35 -25 -15 -5   5   15 25 35 45 55

Fig. 1. The difference in orbital forcing at 6 kyr BP compared to 2000CE as a function of latitude and time of year (Julian days). The annual mean change in total global irradiance is extremely small, but in the high latitudes averages to about $5 \, W/m^2$ increase and about a $1 \, W/m^2$ decrease in the tropics (Berger, 1978).

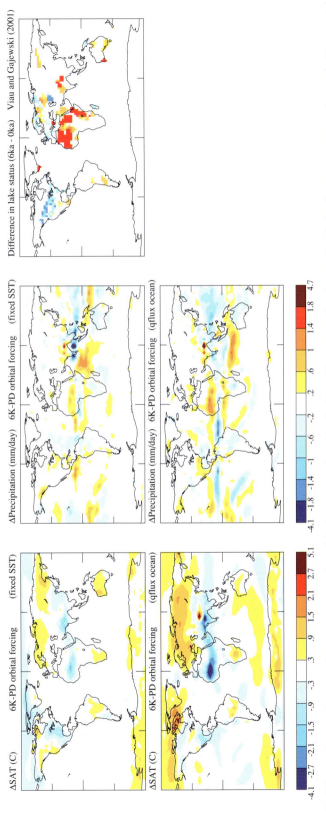

Fig. 2. The difference in annual mean surface temperature and precipitation patterns forced by the 6 kyr BP to present day change of insolation for two experiments (top row) with present day SST, (bottom row) with a Qflux ocean. The top right-hand figure shows reconstructed lake level differences that can be qualitatively compared to the changes in precipitation (Harrison et al., 2003; Viau and Gajewski, 2001) (red implies higher lake levels, blue lower).

studies, this pattern is not well simulated, conceivably because of the lack of change in ENSO with a purely thermodynamic ocean.

3.3. Volcanic aerosols

Explosive volcanic eruptions can deposit enormous amounts of sulphate aerosols into the stratosphere where they can linger long enough to cause significant radiative change. The aerosols are reflective and so increase the Earth's albedo (cooling the planet), while at the same time are locally absorbing (heating the lower stratosphere). Volcanic aerosols in the troposphere are swiftly rained out and are of less climatic importance due to the much shorter residence time (weeks rather than years). The global radiative forcing for an event such as the eruption of Mt. Pinatubo in 1991 reached a peak of -3 to $-4\,W/m^2$ (Hansen et al., 1996, 1997b). Note that the lower stratospheric warming impact of volcanic aerosols can only be simulated with a relatively sophisticated radiative transfer code and has been shown to be climatically important (see below) (Kirchner et al., 1999; Stenchikov et al., 2002; Shindell et al., 2001b, 2004).

Determining the time history of the volcanic forcing involves the timing of eruptions, the amount of sulphate released and the composition, particle size and the fraction of sulphate entering the stratosphere. Satellite and proxy measurements have shown that it is principally tropical volcanoes that have the most lasting effects (Pinatubo, Tambora, Toba, etc.) and that these lead to increased sulphate deposition in ice cores in both Antarctica and Greenland. These ice core records have therefore been used to estimate sulphate forcing back through the Holocene (Hammer et al., 1997; Udisti et al., 2000). The most reliable records are for the last 1000 years or so (Crowley, 2000; Robertson et al., 2001) where many of the peaks can be reliably calibrated to known eruptions (e.g. Tambora in 1815, Huaynaputina in 1600). Note that the radiative forcing from even well-observed eruptions can differ significantly depending on the data set used.

While the impact of a single eruption lasts at most a few years (the residence time for aerosols in the stratosphere), there are apparent changes in volcanic frequency on decadal and even centennial time scales in the Holocene. Thus, changes in volcanic forcing can impact longer term climate. Experiments with Energy Balance Models (EBMs) (i.e. Crowley, 2000) have shown that the hemispheric mean cooling in response to volcanic forcing (Mt. Pinatubo for instance) is well simulated by such models and is proportional to the climate sensitivity. This result is borne out in decadal-scale simulations with GCMs as well (Shindell et al., 2003). There is some evidence too for volcanic impacts

on ENSO frequency (Adams et al., 2003), though this has yet to be reproduced in models.

An interesting picture emerges when we consider the seasonal data. Based upon the instrumental period, Robock and Mao (1995) showed that there is a significant tendency toward winter warming over specific portions of the NH continents, particularly Eurasia, in the first and second winters following an eruption. Work with long historical archives (Fischer et al., 2004) and with seasonally resolved paleo-reconstructions (Shindell et al., 2004) has shown that this appears to be true over the last 500 years as well.

Volcanic aerosols from El Chichon (1982) and especially Pinatubo (1991) have been thoroughly observed and assuming that previous large eruptions had similar spatial and size distributions of aerosol and time decay, model simulations of the effects can easily be done. Fig. 3 shows a GCM study of the effects of the Krakatau. Santa Maria and Pinatubo eruptions from five ensemble members (so a total of 15 eruption cycles) (mean radiative forcing $-3.7\,W/m^2$ in the year following the eruption based on the Crowley (2000) dataset). Examining the mean temperature for the winter following the eruption demonstrates quite clearly the Eurasian warming/Mediterranean cooling pattern seen in the paleo-reconstructions and modern observations. This pattern of winter warmth is associated with an enhanced wintertime circulation, represented by a positive phase of the NAO/AO (Hurrell, 1995; Thompson and Wallace, 1998). Note that the wintertime NAO phase is highly variable and the changes seen in the figure in the mean are much smaller than the interannual variability and so are only statistically significant over many realizations. However, these regional patterns are almost completely absent in the decadal mean (Shindell et al., 2003). Over this longer term the patterns are of generalised cooling, with an enhancement toward the poles. This pattern of 'winter warming' is at least partially related to changes in temperature gradients and dynamics in the stratosphere, and models with only poor stratospheric resolution are less able to capture this phenomena (Robock et al., 1999).

3.4. Solar irradiance

Observations during the satellite era have shown definitively that solar irradiance can vary during the roughly 11-year solar cycle (e.g. Willson, 1997). During the solar maximum, the sun is particularly active and dark sunspots as well as bright faculae are more evident. Total solar irradiance (TSI) is about 0.1% stronger than during a solar minimum when there are very few sunspots. While the integrated solar irradiance varies little, the higher frequency bands (such as for UV radiation) can vary by 1–10% (Lean et al., 1995). The direct forcing from solar minimum to solar maximum is

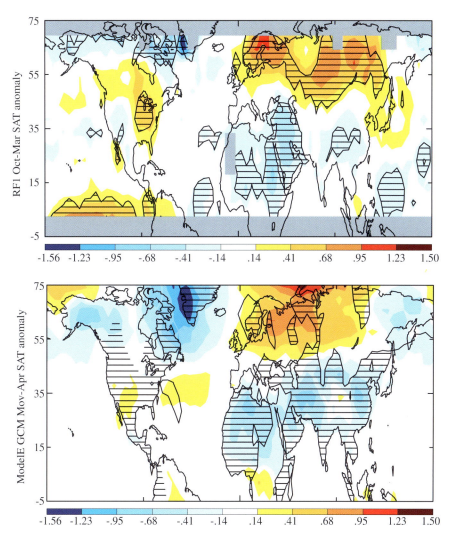

Fig. 3. (Top) The regional pattern of warming and cooling (°C) seen in seasonally resolved paleo-reconstructions in the winter following a major tropical eruption based on an average of 11 large and well identified historical eruptions (mean forcing $-3.8\,\text{W/m}^2$) and (bottom) the ensemble mean impact of increased volcanic aerosols seen in GCM simulations with a mean forcing of $-3.7\,\text{W/m}^2$ (Shindell et al., 2004). Significant (95% confidence) areas of change (with respect to winters with no volcanic forcing) are hatched.

about $0.24\,\text{W/m}^2$, after accounting for geometry and the Earth's albedo. It remains uncertain how longer-term solar variability relates to changes over the solar cycle.

Approximately in phase with the solar irradiance, the solar magnetic field also varies and this in turn modulates the amount of cosmogenic radiation hitting the terrestrial atmosphere. These high energy cosmic rays are the source for cosmogenic nuclides such as ^{14}C, ^{10}Be, ^{7}Be, ^{36}Cl and ^{3}H which thus have production rates that vary on solar cycle timescales (less production during a solar maximum) (Lal and Peters, 1967; Masarik and Beer, 1999). Paleo-records of these isotopes have therefore been used to estimate solar irradiance in the past (Stuiver and Quay, 1980; Bard et al., 2000). These records have generally been calibrated to the estimates of solar forcing at the Maunder Minimum, themselves highly uncertain (Bard et al., 1997; Lean et al., 2002) and so do not provide an

independent estimate of the forcing. Additionally, the amount of contamination in records of ^{10}Be due to climate impacts on its deposition have not yet been satisfactorily quantified (Salyk et al., 2003). However, millennial time series combining reconstructions based on the observed sunspot cycle with the cosmogenic isotopes do exist, with the caveat that significant uncertainties exist in the magnitude of long-term solar change (Bard et al., 1997).

Isolating the effects of solar forcing in the climate record is difficult, particularly in the instrumental record because of the number of other factors that change at the same time and the shortness of the record. Paleo-climate reconstructions (Mann et al., 1998) are, however, long enough to be used. Waple et al. (2002) showed that the presumed long-term solar variability appears to have a significant response in surface temperature patterns when filtered (>40 years) and lagged (by

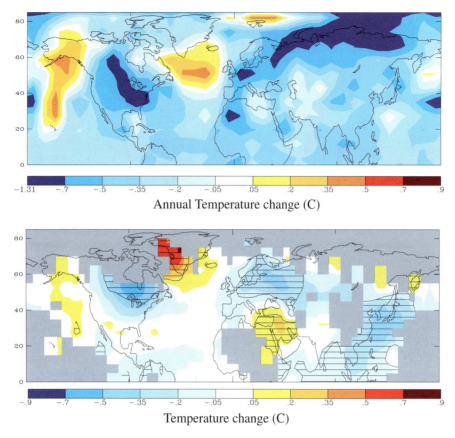

Fig. 4. (a) The temperature pattern seen in a model with interactive stratospheric ozone when forced by spectrally resolved solar forcing comparable to the scaling in lower panel (Shindell et al., 2001b). (b) The annual temperature pattern correlated to the solar irradiance reconstruction when smoothed to leave only variability > 40 years and lagged by 20 years (Waple et al., 2002).

10–30 years) (Fig. 4b). The lag is consistent with an ocean thermal response to the forcing. Note the approximate anti-correlation to the winter warming pattern for volcanic effects seen above, underscoring the spatial patterns of temperature change over oceanic and terrestrial regions that appears to be superposed on the global mean responses to certain climate forcings.

Models that are run with a TSI change generally produce minimal global mean temperature changes (around 0.2 °C at equilibrium for a 0.1% change), in line with their sensitivity to CO_2 increases (Hansen et al., 1997a). This is generally not sufficient to match observed variability over the solar cycle, particularly in the stratosphere. Stratospheric ozone, however, is particularly sensitive to changes in the UV portion of the spectrum and models that include a realistic strato-sphere and interactive ozone calculation tend to see much more realistic variability over the solar cycle (Haigh, 1996; Shindell et al., 1999). Over the solar cycle the ocean does not react significantly. However, for longer term variability, the ocean thermal response (at minimum) implied in the observations must be con-sidered.

The AGCM-ML results described in Shindell et al. (2001b) demonstrated a significant shift in regional

temperatures (Fig. 4a) which corresponded well with the paleo-data. These results are equilibrium results, i.e. the mixed layer ocean has had time to thermally adjust to the change in surface fluxes, while the ocean heat transport remained constant. The change of temperature seen here is mostly related to a shift in sea level pressure from mid- to high latitudes, consistent with a more negative phase of the NAO/AO. Interannual variability in these indices does not change appreciably. The comparison with the Waple et al. (2002) analysis is appropriate because both cases are looking only at the solar component of change (over a long time period) and the lag is appropriate for an ocean thermal response. Given the uncertainties in the magnitude of solar forcing, the climate sensitivity and the ability of the proxy data to capture long-term variability, the match is surprisingly good. As in the response to volcanic forcing, a large part ($\approx 50\%$) of this result was due to the stratospheric temperature gradient changes induced by the interactive ozone response.

The purported impacts of longer term (millennial) solar variability in the Holocene (Bond et al., 2001; Hu et al., 2003) will require both better forcing functions and fully dynamic ocean models to even start to address the issues adequately. Initial results suggest that ocean dynamics might well lead to hemispheric differences in response (Broecker et al., 1999; Goosse et al., 2004), as well as potentially important responses to any changes in the NAO/AO (Visbeck et al., 1998; Delworth and Dixon, 2000).

3.5. Land surface changes

Changes to the land surface impact the climate in a number of direct ways through changes in the albedo, evapo-transpiration and runoff, but also indirectly through changes to biogenic emissions of volatile organic compounds (which impact atmospheric oxidation and hence methane and tropospheric ozone), organic aerosol formation (which interact directly with the radiation and indirectly with cloud formation) and the potential for dust emission (which decreases as vegetation expands). Other impacts are possible through changes in biomass burning (a source of CO, black carbon etc.) linked to aridity and changing vegetation types, and the effects of wetland expansion or contraction on methane emissions. Vegetation change estimates exist for the last 3 centuries (i.e. Ramankutty and Foley, 1999) and through projects such as BIOME6000 for the mid-Holocene (Prentice and Webb III, 1998).

At present, no climate models have included the full range of effects. However, preliminary results with changes to vegetation distributions (Braconnot et al., 1999) have shown that this is probably a key factor in modelling the mid-Holocene and that feedbacks between climate and vegetation are likely to be important in more recent periods (Bauer et al., 2003).

3.6. Other forcings

There are a large number of additional radiative forcings that may have been of some importance during the Holocene. These include dust, tropospheric and stratospheric ozone, natural sulphates, organic aerosols, stratospheric water vapour and minor greenhouse gases (such as SO_2). Unfortunately, these fields are not well-mixed (i.e. there is significant spatial heterogeneity) and there are very few direct measurements.

The dust record is somewhat constrained due to a few ocean sediment records downwind of the major source regions. During the African Humid Period (8 kyr BP to 5.5 kyr BP), Atlantic dust deposition was reduced by about one-third, compared to the present day (deMenocal et al., 2000). Currently, African and Arabian sources contribute $-0.12 \, W/m^2$ to the global forcing at the top of the atmosphere (Miller et al., 2004), corresponding to a forcing of $0.04 \, W/m^2$ during the African Humid Period. However, this forcing could be up to four times higher since the present-day dust load calculated by Miller et al. (2004) is at the low end of estimates and the assumed particle absorption is

possibly too high. On the other hand, Prospero and Lamb (2003) suggest that transport of Sahelian dust across the Atlantic may have been unusually large during the satellite era used to constrain dust models, compared to the pre-industrial value reflected in the sediment cores. The potential for African dust to contribute significantly to the climate forcing at 6 kyr BP is thus still uncertain.

Another example is the potential for greenhouse gas forcing by volcanic emissions of SO_2 that has been hypothesised to have been a factor in the response to non-explosive eruptions such as Laki in 1783 (Highwood and Stevenson, 2003).

In order to assess the global importance of these ancillary forcings, models must be run to produce fields that are consistent with the appropriate boundary conditions and available emission histories. This is an active research area and given the uncertainties involved and the sparsity of good validating data, results from such simulations are likely to remain speculative for some time to come.

4. Time-slices

Time-slice experiments can be very useful in matching model results to multiple sources of paleo-data with reasonable time control (particularly ice cores, tree rings, etc.). However, a number of points need to be borne in mind. Firstly, the period of time for which the forcings are assumed fixed determines which model configurations are most appropriate. For example, the response of a short-lived forcing (up to a couple of decades) may be calculated with a mixed layer ocean whereas the ocean dynamic response would be crucial to longer forcings. Secondly, all known relevant forcings must be included and thirdly, the data to which the models are compared must be tightly controlled chronologically and with as widespread coverage as possible. The response will depend on the duration of the forcing as in the example above distinguishing the ocean response to a single volcano versus several decades of with unusually high volcanic activity. Similarly, the dynamic ocean response can depend on the time-scale of the forcing (Visbeck et al., 1998). One must therefore be extremely vary of extrapolating results to longer periods.

Below, we highlight the Maunder Minimum ($\approx$ 1650–1710) period. We refer the readers interested in the 6 kyr BP time slice to the more comprehensive discussion of the PMIP experiments in Joussaume et al. (1999) and references cited therein. The PMIP2 project is slated to coordinate simulations for the early Holocene (10 kyr BP), but we are not aware of any further organised efforts to look at other Holocene time-slices.

4.1. Maunder minimum

The Maunder Minimum (1650–1710) has become the standard period for assessing variability in pre-industrial climate. It was a period during which very few sunspots were observed and has been recognised as one of the coolest intervals of the so-called Little Ice Age (LIA) (Eddy, 1976). Unlike the LIA, the Maunder Minimum is a fairly well defined chronological period and since it is well within the historical and instrumental period (for a few long records), there is a relative wealth of data available for comparison. The time period involved is certainly long enough for the ocean thermal response to be important and possibly for ocean dynamics, particularly in the North Atlantic, to start to respond. Therefore the experiments described below using a mixed layer ocean are at the margin of appropriateness for this period.

The data comparison that we use is the difference in the 1660–1680 mean (which allows time for the ocean thermal response to be felt) compared to a century later (1770–1790). There is a significant difference in both the estimated solar forcing and volcanic forcing for these two periods and yet sizable anthropogenic modification of atmospheric composition has not yet taken place. Land use (probably) and greenhouse gas changes are minimal between these two periods. Fig. 5 shows the comparison in the multi-decadal mean temperatures. The spatial pattern is consistent with the results from the solar-only cases (Fig. 4), although the scale of change is larger in the model (hemispheric mean change of 0.5 °C, compared to 0.2–0.4 °C in the observations) (Shindell et al., 2003). This could be because the sensitivity of this particular model (1 °C per W/m^2) is a little high (compared to the canonical 0.75 °C per W/m^2), or because the forcing is too large (i.e. the solar change is over-estimated in the reconstruction of Lean et al. (2002)), or due to uncertainties in the amplitude of low-frequency variability in paleo-climate reconstructions (Esper et al., 2002; Mann and Hughes, 2002). Given these uncertainties the comparison is reasonable, but this does highlight some of the problems in performing time-slice experiments as compared to single factor experiments described above.

5. Transient simulations

For a successful transient simulation, good records for all the relevant forcings, a reasonable initial condition and computer resources to run multiple long runs must be available. These restrictions have meant that simulations have generally been done only for the last 500–1000 years with coupled models of any

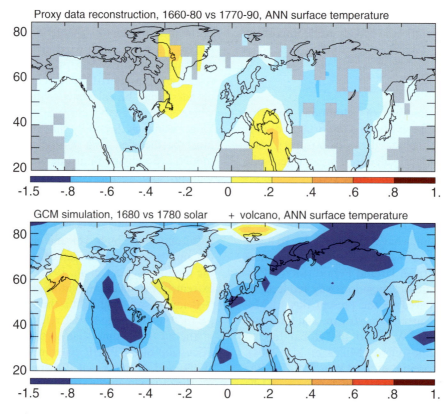

Fig. 5. Difference in annual mean temperature (°C) between 1660 and 1680 and a century later (a) from Mann et al. (1998) spatial reconstructions and (b) from a combination of the solar and volcanic forcing results from a GCM (Shindell et al., 2003).

complexity (Gonzalez-Rouco et al., 2003; Widmann and Tett, 2003; Goosse et al., 2004; Rind et al., 2004). The length of the simulations implies that the runs are almost always done with a relatively old version of a GCM, and so some components of the forcing or the physics are somewhat less than state-of-the-art. The target data set for comparison are predominantly the various NH temperature reconstructions (Briffa et al., 1998; Mann et al., 1999; Esper et al., 2002; Luterbacher et al., 2004, etc.).

Other Holocene transient climate events that have been modelled thus far are the 8.2 kyr BP event (Renssen et al., 2001) and the termination of the African Humid Period (Claussen et al., 1999; Renssen et al., 2003) albeit with simplified models. The 8.2 kyr event is likely to be a good target for future modelling attempts because the signal is quite strong and there is a relatively well constrained forcing function (from the final draining of Lake Agassiz or possibly the final collapse of the Hudson Bay Dome, around 5–15 Sv yrs of freshwater) (von Grafenstein et al., 1998; Barber et al., 1999).

The principle modelling issue is finding a suitable (ocean) initial condition to perturb. However, we will not focus on these events here.

5.1. The last 500 years

Given the emphasis in this review on climate changes during the past few centuries, we focus on transient simulations of a period (1550–1800) that encompasses the Maunder Minimum but that does not include either the spin-up phase of the simulation or the very large radiative perturbations that start in the 19th century. We calculate the anomalies in the NH mean surface temperature so that the mean over this period is zero for both the target paleo-reconstruction and the model results. This gives comparable results for the natural, unforced variability for this period and minimises distortions introduced from varying climate sensitivities to increasing GHG toward the end of the simulation.

Fig. 6 shows the NH mean surface temperature anomaly for a meta-ensemble of 6 runs (Waple et al., 2002) reconstruction. We use the term meta-ensemble to distinguish from a standard ensemble where only the initial conditions are different. In these cases, the depth of the deep ocean layer, the ocean initial conditions (cold or warm starts) and magnitude of the volcanic forcing vary. These runs are relatively coarse resolution (8° × 10°), 9 layers in the vertical and have an ocean mixed layer with diffusion into a deep ocean with solar, volcanic and GHG forcing (Robertson et al., 2001). A number of points can be made: (i) The short term impact of volcanic eruptions is clear in the model ensemble mean and in the observations (despite some uncertainty over timing and importance). The magnitude of the model response is dependent on the radiative forcing function and the model sensitivity, thus a match between the reconstruction and the models could be

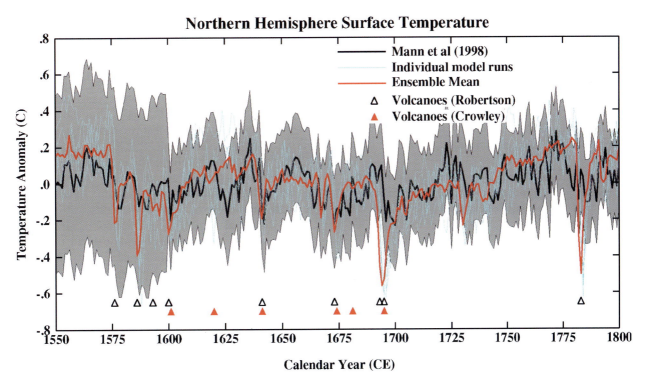

Fig. 6. 1550–1800 comparison between a proxy reconstruction (with error bars) and a "meta-ensemble" of six model runs. The years with significant volcanic forcing in the simulation are highlighted, together with significant volcanoes from the latest Crowley (2000) and Robertson et al. (2001) compilations.

fortuitous. It seems clear however, that the response to Laki (1783–1784) is over-estimated, as is the response to Tambora in 1816 (not shown), (ii) the background variability in NH mean SAT is similar in the proxy record ($\sigma = 0.1\,°C$) and individual model results ($\sigma = 0.14$–$0.2\,°C$), the higher values are associated with two runs with relatively shallow oceans (1300 m) compared to 5000 m in the other runs and (iii) the long-term (decadal and longer) variability is reasonably captured in the models. However, while the NH mean temperature series are well correlated to the proxy reconstruction ($r^2 = 0.1$–0.26 (individual runs), 0.23 (for the ensemble mean) ± 0.12 for 95% significance levels), the spatial patterns of variability (not shown) are not. This is a clear demonstration that while it appears to be relatively easy to get a reasonable estimate of the forced changes in hemispheric mean temperature (as with energy balance or intermediate models (Crowley, 2000; Bauer et al., 2003; Goosse et al., 2004)), the regional patterns of change are much more sensitive to the quality of the simulation and the physics contained therein (i.e. compare Shindell et al., 2001a; Waple et al., 2002). In particular, the improvements to the horizontal resolution, the inclusion of a reasonably resolved stratosphere, ocean dynamics, and interactive ozone are all likely to impact the regional changes.

6. Future directions

Over the next few years, computing power will increase, the physics included in models will become more sophisticated and longer simulations will be performed. At the same time, estimates of potential climate forcings will likely improve and more complete records of paleo-climate will become available. Forward modelling of proxy data (such as ice core isotopes or biomes) has the potential to further improve the match to the data (and/or highlight continuing problems in the models).

It can be seen from the results discussed in this paper that databases of paleo-climate change that can be used to isolate forcing mechanisms (such as for long term solar variability, or volcanic forcing), or provide time-slice reconstructions are extremely useful for modelling purposes. Since most paleo-records are single location time-series, this implies that more work will be needed to integrate different records into spatial networks such as the Mann et al. (1998) and Luterbacher et al. (2004) reconstructions, or the hydrological databases of Harrison et al. (2003). This task is quite difficult, due in part to problems of chronology, differences in time resolution, confounding non-climatic influences, biases in spatial sampling etc. However, this will be increasingly necessary in order to provide a solid framework for validating climate models and testing hypotheses. Some

individual records (for instance, from ice cores) are also very useful targets, but matches to these are not sufficient to validate global climate changes without support from the spatial networks discussed above.

It is clear that the successes in modelling Holocene variability have relied on extending the range of feedbacks that are permitted in the models. For instance, ozone for the solar case, SST for almost all cases, vegetation dynamics for the mid-Holocene, and well-resolved stratospheric dynamics for capturing some of the atmospheric circulation changes. Further feedbacks will continue to be added into the models, such as interactive dust and other aerosols, carbon cycles and tropospheric chemistry. The addition of these bio-geochemical modules in fully coupled models should allow for a fuller assessment, for instance, of whether GHG changes from around 8 kyr BP can conceivably be anthropogenic (as recently speculated (Ruddiman, 2003)), or whether they were likely just responding to changes in climate (Gerber et al., 2003).

For quasi-steady states longer than a few decades, but shorter than a few millennia, ocean dynamics cannot be expected to be in equilibrium with the forcing. Thus for multi-decadal to centennial changes, coupled transient simulations are required with all the extra work that this requires (Widmann and Tett, 2003; Gonzalez-Rouco et al., 2003). This is the timescale most frequently seen in Holocene paleo-data, and there is still some way to go in being able to validate models at this frequency band.

One thing that it is extremely unlikely is that models will suddenly start to provide exact matches to the time-series of any particular climate record. There is such a degree of weather 'noise' or intrinsic variability that no particular climate simulation, however sophisticated, will follow the same semi-random path that the actual climate took. Ensembles over many such simulations may average over much of this 'noise' but these averages by necessity smooth over many short term or spatial variations, which are none the less recorded in any particular proxy. Therefore while we are likely to have reasonable estimates of the changes in global or hemispheric mean temperature, accurate simulations of regional climate are likely to remain elusive. Simulations of hydrological variables (precipitation, drought) or extreme events are more problematic given their much greater variability in time and space. Some information can be gained from down-scaling large scale circulation behaviour (such as the NAO/AO, or ENSO variability) to the regional or even local scale (i.e. Thompson and Wallace, 2001), but that is likely to be the exception. Regional modelling, or zoomed models (with locally higher resolution) may play some role as well.

So to answer the question posed at the beginning of this paper; Climate modelling has hinted that, for the Holocene (i) although hemispheric mean temperature changes appear to be robust responses to forcings, the spatial patterns of response can be quite complicated and dependent on the physics contained within the models, (ii) modes of variability such as the NAO/AO (and potentially ENSO) can be affected by forcings, and (iii) the kinds of changes that have occurred are still posing challenges to our ability to model them. By further focusing on the variability of this period, we may therefore be able to increase our understanding of Holocene climate and improve our confidence that models are capable of modelling more recent and future climate changes.

Acknowledgements

Model development is funded by NASA Climate Modeling Development grants to GISS. Thanks to Rick Healy for providing some of the model output, and especially to Eystein Jansen for organising the 2003 Holocene workshop in Bergen. Comments from Hans Renssen and an anonymous reviewer greatly helped improve the clarity of the manuscript.

References

Adams, J.B., Mann, M.E., Ammann, C.M., 2003. Proxy evidence for an El Niño-like response to volcanic forcing. Nature 426, 274–278.

Barber, D.C., Dyke, A., Hillaire-Marcel, C., Jennings, A.E., Andrews, J.T., Kerwin, M.W., Bilodeau, G., McNeely, R., Southon, J., Morehead, M.D., Gagnon, J.-M., 1999. Forcing of the cold event of 8,200 years ago by catastrophic drainage of Laurentide lakes. Nature 400, 344–348.

Bard, E., Rostek, F., Sonzogni, C., 1997. Interhemispheric synchrony of the last deglaciation inferred from alkenone paleothermometry. Nature 385, 707–710.

Bard, E., Raisbeck, G.M., Yiou, F., Jouzel, J., 2000. Solar irradiance during the last 1200 years based on cosmogenic nuclides. Tellus 52B, 985–992.

Bauer, E., Claussen, M., Brovkin, V., 2003. Assessing climate forcings of the earth system for the past millennium. Geophysical Research Letters 30.

Berger, A., 1978. Long-term variations of daily insolation and Quaternary climatic changes. Journal of Atmospheric Sciences 35, 2362–2367.

Bond, G.C., Kromer, B., Beer, J., Muscheler, R., Evans, M.N., Showers, W., Hoffmann, S., Lotti-Bond, R., Hajdas, I., Bonani, G., 2001. Persistent solar influence on the North Atlantic climate during the Holocene. Science 294, 2130–2136.

Braconnot, P., Joussaume, S., Marti, O., de Noblet, N., 1999. Synergistic feedbacks from ocean and vegetation on the African monsoon response to mid-Holocene insolation. Geophysical Research Letters 26, 2481–2484.

Briffa, K.R., Jones, P.D., Schweingruber, F.H., Osborn, T.J., 1998. Influence of volcanic eruptions on Northern Hemisphere summer temperature over the past 600 years. Nature 393, 350–354.

Broecker, W., Sutherland, S., Peng, T.H., 1999. A possible 20th-century slow-down of Southern Ocean deep water formation. Science 286, 1132–1135.

Chappellaz, J., Blunier, T., Kints, S., Dällenbach, A., Barnola, J.-M., Schwander, J., Raynaud, D., Stauffer, B., 1997. Changes in the atmospheric CH_4 gradient between Greenland and Antarctica

during the Holocene. Journal of Geophysical Research 102, 15987–15999.

Cheddadi, R., Yu, G., Guiot, J., Harrison, S.P., Prentice, I.C., 1997. The climate of Europe 6000 years ago. Climate Dynamics 13, 1–9.

Claussen, M., Kubatzki, C., Brovkin, V., Ganopolski, A., 1999. Simulation of an abrupt change in Saharan vegetation in the mid-Holocene. Geophysical Research Letters 26, 2037–2040.

Crowley, T.J., 2000. Causes of climate change over the past 1000 years. Science 289, 270–277.

Delworth, T.L., Dixon, K.W., 2000. Implications of the recent trend in the Arctic/North Atlantic Oscillation for the North Atlantic thermohaline circulation. Journal of Climate 13, 3721–3727.

Delworth, T.L., Manabe, S., Stouffer, R.J., 1993. Interdecadal variations of the thermohaline circulation in a coupled ocean-atmosphere model. Journal of Climate 6, 1993–2011.

deMenocal, P., Ortiz, J., Guilderson, T., Sarnthein, M., 2000. Coherent high and low-latitude climate variability during the Holocene warm period. Science 288, 2198–2202.

Eddy, J.A., 1976. The Maunder Minimum. Science 192, 1189–1202.

Esper, J., Cook, E.R., Schweingruber, F.H., 2002. Low-frequency signals in long tree-ring chronologies for reconstructing past temperature variability. Science 295, 2250–2253.

Etheridge, D.M., Steele, L.P., Langenfelds, R.L., Francey, R.J., Barnola, J.M., Morgan, V.I., 1996. Natural and anthropogenic changes in atmospheric CO_2 over the last 1000 years from air in Antarctic ice and firn. Journal of Geophysical Research 101, 4115–4128.

Fischer, E., Luterbacher, J., Wanner, H., 2004. Atmospheric circulation changes and European precipitation anomalies following major tropical volcanic eruptions over the last 500 years. Geophysical Research Abstract 6, 04218.

Gagan, M.K., Ayliffe, L.K., Hopley, D., Cali, J.A., Mortimer, G.E., Chappell, J., McCulloch, M.T., Head, M.J., 1998. Temperature and surface-ocean water balance of the mid-Holocene Tropical West Pacific. Science 279, 1014–1018.

Gerber, S., Joos, F., Bruegger, P.P., Stocker, T.F., Mann, M.E., Sitch, S., 2003. Constraining temperature variations over the last millennium by comparing simulated and observed atmospheric CO_2. Climate Dynamics 20, 281–299.

Gonzalez-Rouco, F., von Storch, H., Zorita, E., 2003. Deep soil temperature as proxy for surface air-temperature in a coupled model simulation of the last thousand years. Geophysical Research Letters 30, 2116.

Goosse, H., Masson-Delmotte, V., Renssen, H., Delmotte, M., Fichefet, T., Morgan, V., van Ommen, T., Khim, B.K., Stenni, B., 2004. A late medieval warm period in the Southern Ocean as a delayed response to external forcing? Geophysical Research Letters 31, L06203.

Haigh, J.D., 1996. The impact of solar variability on climate. Science 272, 981–984.

Hammer, C.U., Clausen Jr., B., Langway Jr., C.C., 1997. 50,000 years of recorded global volcanism. Climate Change 35, 1–15.

Hansen, J., Nazarenko, L., 2004. Soot climate forcing via snow and ice albedos. Proceedings of the National Academy of Science 101, 423–428.

Hansen, J., Ruedy, R., Sato, M., Reynolds, R., 1996. Global surface air temperature in 1995: Return to pre-Pinatubo level. Geophysical Research Letters 23, 1665–1668.

Hansen, J., Sato, M., Ruedy, R., 1997a. Radiative forcing and climate response. Journal of Geophysical Research 102, 6831–6864.

Hansen, J.E., Sato, M., Ruedy, R., Lacis, A., Asamoah, K., Beckford, K., Borenstein, S., Brown, E., Cairns, B., Carlson, B., Curran, B., de Castro, S., Druyan, L., Etwarrow, P., Ferede, T., Fox, M., Gaffen, D., Glascoe, J., Gordon, H., Hollandsworth, S., Jiang, X., Johnson, C., Lawrence, N., Lean, J., Lerner, J., Lo, K., Logan, J., Luckett, A., McCormick, M., McPeters, R., Miller, R., Minnis, P.,

Ramberran, I., Russell, G., Russell, P., Stone, P., Tegen, I., Thomas, S., Thomason, L., Thompson, A., Wilder, J., Willson, R., Zawodny, J., 1997b. Forcings and chaos in interannual to decadal climate change. Journal of Geophysical Research 102, 25679–25720.

Hansen, J.E., Sato, M., Nazarenko, L., Ruedy, R., Lacis, A., Koch, D., Tegen, I., Hall, T., Shindell, D., Santer, B., Stone, P., Novakov, T., Thomason, L., Wang, R., Wang, Y., Jacob, D.J., Hollandsworth, S., Bishop, L., Logan, J., Thompson, A., Stolarski, R., Lean, J., Willson, R., Levitus, S., Antonov, J., Rayner, N., Parker, D., Christy, J., 2002. Climate forcings in Goddard Institute for Space Studies SI2000 simulations. Journal of Geophysical Research 107, 4347.

Harrison, S.P., Kutzbach, J.E., Liu, Z., Bartlein, P.J., Otto-Bliesner, B., Muhs, D., Prentice, I.C., Thompson, R.S., 2003. Mid-Holocene climates of the Americas: a dynamical response to changed seasonality. Climate Dynamics 20, 663–688.

Hewitt, C.D., Mitchell, J.F.B., 1996. GCM simulations of the climate of 6 kyr BP: mean changes and changes and interdecadal variability. Journal of Climate 9, 3505–3529.

Highwood, E.J., Stevenson, D.S., 2003. Atmospheric impact of the 1783–1784 Laki eruption: Part II Climatic effect of sulphate aerosol. Atmospheric Chemistry and Physics 3, 1177–1189.

Houghton, J.T., Ding, Y., Griggs, D.J., Nouger, M., van der Linden, P.J., Dai, X., Maskell, K., Johnson, C.A., 2001. Climate Change 2001: The Scientific Basis. Cambridge University Press, New York.

Hu, F.S., Kaufman, D., Yoneji, S., Nelson, D., Shemesh, A., Huang, Y., Tian, J., Bond, G., Clegg, B., Brown, T., 2003. Cyclic variation and solar forcing of Holocene climate in the Alaskan subarctic. Science 301, 1890–1893.

Hurrell, J.W., 1995. Decadal trends in the North Atlantic oscillation: regional temperatures and precipitation. Science 269, 676–679.

Indermuhle, A., Stocker, T., Joos, F., Fischer, H., Smith, H.J., Wahlen, M., Deck, B., Mastroianni, D., Tschumi, J., Blunier, T., Meyer, R., Stauffer, B., 1999. Holocene carbon-cycle dynamics based on CO_2 trapped in ice at Taylor Dome, Antarctica. Nature 398, 121–126.

Joussaume, S., Taylor, K.E., Braconnot, P., Mitchell, J.F.B., Kutzbach, J.E., Harrison, S.P., Prentice, I.C., Broccoli, A.J., Abe-Ouchi, A., Bartlein, P.J., Bonfils, C., Dong, B., Guiot, J., Herterich, K., Hewitt, C.D., Jolly, D., Kim, J.W., Kislov, A., Kitoh, A., Loutre, M., Masson, V., McAvaney, B., McFarlane, N., de Noblet, N., Peltier, W.R., Peterschmitt, J.Y., Pollard, D., Rind, D., Royer, J.F., Schlesinger, M.E., Syktus, J., Thompson, S., Valdes, P., Vettoretti, G., Webb, R.S., Wyputta, U., 1999. Monsoon changes for 6000 years ago: results of 18 simulations from the Paleoclimate Modeling Intercomparison Project (PMIP). Geophysical Research Letters 26, 859–862.

Kirchner, I., Stechnikov, G.L., Graf, H.-F., Robock, A., Antuna, J.C., 1999. Climate model simulation of winter warming and summer cooling following the 1991 Mount Pinatubo volcanic eruption. Journal of Geophysical Research 104, 19039–19055.

Kitoh, A., Murakami, S., 2002. Tropical Pacific climate at the mid-Holocene and the last glacial maximum simulated by a coupled ocean-atmosphere GCM. Paleoceanography 17, 1047.

Kushnir, Y., Robinson, W.A., Blade, I., Hall, N.M.J., Peng, S., Sutton, R., 2002. Atmospheric GCM response to extratropical SST anomalies: synthesis and evaluation. Journal of Climate 15, 2233–2256.

Lal, D., Peters, B., 1967. Cosmic ray produced radioactivity on the Earth. Handbook of Physics 46, 551–612.

Lean, J., Beer, J., Bradley, R., 1995. Reconstruction of solar irradiance since 1610: implications for climate change. Geophysical Research Letters 22, 3195–3198.

Lean, J., Wang, Y.-M., Sheeley Jr., N.R., 2002. The effect of increasing solar activity on the Sun's total and open magnetic flux during multiple cycles: implications for solar forcing of climate. Geophysical Research Letters 29, 224.

Liu, Z., Brady, E., Lynch-Stieglitz, J., 2003a. Global ocean response to orbital forcing in the Holocene. Paleoceanography 18, 1041.

Liu, Z., Brady, E., Lynch-Stieglitz, J., 2003b. Modeling climate shift of El Niño variability in the Holocene. Geophysical Research Letters 27, 2265–2268.

Luterbacher, J., Dietrich, D., Xoplaki, E., Grosjean, M., Wanner, H., 2004. European seasonal and annual temperature variability, trends, and extremes since 1500. Science 303, 1499–1503.

Mann, M.E., Hughes, M.K., 2002. Tree-ring chronologies and climate variability. Science 296, 848.

Mann, M.E., Bradley, R.S., Hughes, M.K., 1998. Global-scale temperature patterns and climate forcing over the past six centuries. Nature 392, 779–787.

Mann, M.E., Bradley, R.S., Hughes, M.K., 1999. Northern hemisphere temperatures during past millenium: inferences, uncertainties and limitations. Geophysical Research Letters 26, 759–762.

Masarik, J., Beer, J., 1999. Simulation of particle fluxes and cosmogenic nuclide production in the Earth's atmosphere. Journal of Geophysical Research 102, 12099–12111.

Miller, R.L., Tegen, I., 1998. Climate response to soil dust aerosols. Journal of Climate 11, 3247–3267.

Miller, R.L., Tegen, I., Perlwitz, J., 2004. Surface radiative forcing by soil dust aerosols and the hydrologic cycle. Journal of Geophysical Research 109, D04203.

Otto-Bliesner, B.L., 1999. El Niño/La Niña and Sahel precipitation during the middle Holocene. Geophysical Research Letters 26, 87–90.

Prentice, I.C., Webb III, T., 1998. BIOME 6000: reconstructing global mid-Holocene vegetation patterns from palaeoecological records. Journal of Biogeography 25, 997–1005.

Prospero, J.M., Lamb, P.J., 2003. African droughts and dust transport to the Caribbean: climate change implications. Science 302, 1024–1027.

Ramankutty, N., Foley, J.A., 1999. Estimating historical changes in global land cover: croplands from 1700 to 1992. Glob. Biogeochem. Cycles 1, 997–1027.

Rayner, N.A., Parker, D.E., Horton, E.B., Folland, C.K., Alexander, L.V., Rowell, D.P., Kent, E.C., Kaplan, A., 2003. Global analyses of SST, sea ice and night marine air temperature since the late nineteenth century. Journal of Geophysical Research 108.

Renssen, H., Osborn, T., 2003. Holocene climate variability investigated using data-model comparisons. CLIVAR Exchanges 26, 1–3.

Renssen, H., Goosse, H., Fichefet, T., Campin, J.-M., 2001. The 8.2 kyr BP event simulated by a global atmosphere-sea-ice-ocean model. Geophysical Research Letters 28, 1567–1570.

Renssen, H., Brovkin, V., Fichefet, T., Goosse, H., 2003. Holocene climate instability during the termination of the African Humid Period. Geophysical Research Letters 30, 1184.

Rind, D., Healy, R., Parkinson, C., Martinson, D., 1995. The role of sea ice in $2 \times CO_2$ climate model sensitivity. Part I: the total influence of sea ice thickness and extent. Journal of Climate 8, 449–463.

Rind, D., Shindell, D., Perlwitz, J., Lerner, J., Lonergan, P., Lean, J., McLinden, C., 2004. The relative importance of solar and anthropogenic forcing of climate change between the Maunder Minimum and the present. Journal of Climate 17, 906–929.

Robertson, A., Overpeck, J., Rind, D., Mosley-Thompson, E., Zielinski, G., Lean, J., Koch, D., Penner, J., Tegen, I., Healy, R., 2001. Hypothesized climate forcing time series for the last 500 years. Journal of Geophysical Research 106, 14783–14804.

Robock, A., Mao, J., 1995. The volcanic signal in surface temperature observations. Journal of Climate 8, 1086–1103.

Robock, A., Stechnikov, G.L., Ramachandran, S., Ramaswamy, V., 1999. Winter warming following volcanic eruptions: observations and climate model simulations of forced Arctic Oscillation patterns. EOS 80, 232.

Ruddiman, W.F., 2003. The anthropogenic greenhouse era began thousands of years ago. Climate Change 61, 261–293.

Rutherford, S., Mann, M.E., Delworth, T.L., Stouffer, R., 2003. Climate field reconstruction under stationary and nonstationary forcing. Journal of Climate 16, 462–479.

Salyk, C., Schmidt, G.A., Koch, D., Veeder, C., 2003. Climate and solar effects on ^{10}Be deposition in ice cores. EOS Trans. AGU 84 (46), Abstract PP52A-0949.

Shindell, D.T., Rind, D., Balachandran, N., Lean, J., Lonergan, P., 1999. Solar cycle variability, ozone, and climate. Science 284, 305–308.

Shindell, D.T., Schmidt, G.A., Mann, M.E., Rind, D., Waple, A., 2001a. Solar forcing of regional climate change during the Maunder Minimum. Science 294, 2149–2152.

Shindell, D.T., Schmidt, G.A., Miller, R.L., Rind, D., 2001b. Northern hemisphere winter climate response to greenhouse gas, ozone, solar and volcanic forcing. Journal of Geophysical Research 106, 7193–7210.

Shindell, D.T., Schmidt, G.A., Miller, R.L., Mann, M.E., 2003. Volcanic and solar forcing of climate changes during the preindustrial era. Journal of Climate 16, 4094–4107.

Shindell, D.T., Schmidt, G.A., Mann, M.E., Faluvegi, G., 2004. Dynamic winter climate response to large tropical volcanic eruptions since 1600. Journal of Geophysical Research 109, D05104.

Sowers, T., Alley, R.B., Jubenville, J., 2003. Ice core records of atmospheric N_2O covering the last 106,000 years. Science 301, 945–948.

Stenchikov, G., Robock, A., Ramaswamy, V., Schwarzkopf, M.D., Hamilton, K., Ramachandran, S., 2002. Arctic Oscillation response to the 1991 Mount Pinatubo eruption: effects of volcanic aerosols and ozone depletion. Journal of Geophysical Research 107, 4803.

Stuiver, M., Quay, P.D., 1980. Changes in atmospheric C-14 attributed to a variable sun. Science 207, 11–19.

Texier, D., de Noblet, N., Harrison, S.P., Haxeltine, A., Jolly, D., Joussaume, S., Laarif, F., Prentice, I.C., Tarasov, P., 1997. Quantifying the role of biosphere-atmosphere feedbacks in climate change: coupled model simulation for 6000 years BP and comparison with palaeodata for northern Eurasia and northern Africa. Climate Dynamics 13, 865–882.

Thompson, D.W.J., Wallace, J.M., 1998. The Arctic Oscillation signature in the wintertime geopotential height and temperature fields. Geophysical Research Letters 25, 1297–1300.

Thompson, D.W.J., Wallace, J.M., 2001. Regional climate impacts of the Northern Hemisphere annular mode. Science 293, 85–89.

Udisti, R., Becagli, S., Castellano, E., Mulvaney, R., Schwander, J., Torcini, S., Wolff, E., 2000. Holocene electrical and chemical measurements from the EPICA-Dome C ice core. Ann. Glaciol. 30, 20–26.

Viau, A.E., Gajewski, K., 2001. Holocene variations in the global hydrological cycle quantified by objective gridding of lake level databases. Journal of Geophysical Research 106, 31703–31716.

Visbeck, M., Hurrell, M., Polvani, L., Cullen, H., 1998. An ocean model's response to North Atlantic Oscillation-like wind forcing. Geophysical Research Letters 25, 4521–4524.

von Grafenstein, U., Erlenkeuser, H., Müller, J., Jouzel, J., Johnsen, S., 1998. The cold event 8200 years ago documented in oxygen isotope records of precipitation in Europe and Greenland. Climate Dynamics 14, 73–81.

Waple, A.M., Mann, M.E., Bradley, R.S., 2002. Long-term patterns of solar irradiance forcing in model experiments and proxy-based surface temperature reconstructions. Climate Dynamics 18, 563–578.

Widmann, M., Tett, S.F.B., 2003. Simulating the climate of the Last Millennium. Pages Newsletter 11, 21–23.

Willson, R.C., 1997. Total solar irradiance trend during solar cycles 21 and 22. Science 277, 1963–1965.

ELSEVIER

Quaternary Science Reviews 23 (2004) 2183–2205

QSR

Holocene millennial-scale summer temperature variability inferred from sediment parameters in a non-glacial mountain lake: Danntjørn, Jotunheimen, central southern Norway

Atle Nesje[a,b,*], Svein Olaf Dahl[b,c], Øyvind Lie[c]

[a]*Department of Earth Science, University of Bergen, Allégt. 41, N-5007 Bergen, Norway*
[b]*Department of Geography, University of Bergen, Breiviksveien 40, N-5045 Bergen, Norway*
[c]*Bjerknes Centre for Climate Research, Allégt. 55, N-5007 Bergen, Norway*

Abstract

Holocene changes in summer temperature in the Jotunheimen area have been inferred from variations in several sediment parameters from a composite (combined from two overlapping sediment cores) lake-core record (mean time resolution of ~15 years) from Lake Danntjørn, a small mountain lake located close to the modern pine-tree limit in Jotunheimen, central southern Norway. Loss-on-ignition analysis at contiguous 0.5-cm intervals throughout the cores revealed that the variations in residue after ignition of the sediments at 550 °C in the upper 260 cm (~8500 calendar years) were not associated with variations in magnetic susceptibility. In the basal, deglacial and early Holocene sequence of the lake sediments, however, the two sediment parameters fluctuated mainly in phase. In addition, the biogenic silica (BSi) concentration is relatively high. Thus, in the early part of the Holocene, the variations in the residue were probably primarily driven by input of minerogenic, detrital material (inwash and/or windblown) and biological productivity (including BSi from diatoms) in the lake itself and in the lake surroundings. From approximately 8500 calendar years BP, variations in the residue probably primarily reflect variations in low-minerogenic (low magnetic susceptibility with little variations) and diatom (BSi) productivity, as demonstrated with analysis of BSi concentration between 260 and 233 cm (~8500–7760 calendar years BP). The highest residue values (probably mainly reflecting high diatom productivity) occurred at 7800–7200, 7000, 6700, 6200, 5400–5050, 4300, 2500, 1800, 1550, 900, 700, and <100 calendar years BP. The most significant <8000 calendar years BP residue minima (mainly low diatom productivity) were recorded at 7100, 6800, 6600, 6000–5700, 5000–4400, 4200, 3800, 3600, 3200, 3000, 2750, 2300, 1900, 1700, 1070, 740, 600, 380, and 80 calendar years BP (BP = AD 2000). The lake record from Jotunheimen is discussed in the context of Holocene terrestrial and marine records from the North Atlantic region and possible natural climate forcing factors.

© 2004 Elsevier Ltd. All rights reserved.

1. Introduction

Continental climate records from Scandinavia capture Holocene insolation changes as well as changes in North Atlantic Ocean thermohaline and atmospheric circulation. In order to make reliable and likely predictions of future climate, and to separate natural from human-induced climate variability, it is important to know the rate and magnitude of past climate changes (e.g. Alverson et al., 2003). Because instrumental meteorological records commonly are too short to cover the entire climate variability, the climate history for earlier periods has to be reconstructed from indirect (proxy) indicators. Previously, climate reconstructions from biological sedimentary remains were based on single indicator species or assemblages of taxa and the results were usually interpreted in a qualitative and descriptive manner. Numerical techniques and approaches are now available that allow quantitative reconstructions

*Corresponding author. Department of Earth Science, University of Bergen, Allégt 41, N-5007 Bergen, Norway. Tel.: +47-555-83502; fax: +47-555-84330.

E-mail address: atle.nesje@geo.uib.no (A. Nesje).

0277-3791/$ - see front matter © 2004 Elsevier Ltd. All rights reserved.
doi:10.1016/j.quascirev.2004.08.015

from floral and faunal assemblages (ter Braak and Juggins, 1993; Birks, 1995, 1998, 2003). Holocene climate variations in Europe and the north Atlantic region have been reconstructed from a number of proxies and archives (Table 1).

Mountain lakes respond quickly to the environmental impact of climate changes. Lakes close to the modern (or past) tree lines or at ecotone boundaries are commonly more sensitive to such changes than lowland lakes (e.g. Körner, 1998). Such lakes are therefore used as sensors of modern and past environmental changes and climate variations. Climate variability observed from instrumental records (e.g. Schindler et al., 1996) over the last few decades has been recorded as changes in lake ecosystems inferred from sediment records. Climate variations influence lakes in different ways and the direct and indirect linkages between climate and lake sediments need to be understood in order to realise the potential of using lake sediments to obtain records of past climate variability. The key processes are those affecting radiation balance and water balance. Radiation determines light and temperature regimes that are modulated by winter lake ice and snow cover and by wind. These factors then influence depth, duration and intensity of water stratification affecting chemical and biological processes (primary production, nutrient cycling, oxygen consumption and pH).

Lake sediments commonly accumulate quite rapidly and sediment cores can be sub-sampled at mm-interval to provide data with decadal to sub-decadal time resolution. Climate reconstructions from lake sediments rely on relationships between the sediment record and the climate. The use of biological proxies may involve the use of transfer functions (e.g. Birks, 2003). Large diameter (~110 mm) cores together with non-destructive techniques, or techniques that require only small samples, make it possible to carry out multi-proxy studies. Climate reconstructions using a multi-proxy approach are becoming increasingly common (e.g. Ammann et al., 2000; Birks et al., 2000; Lotter and Birks, 2003). Reliable inter-core correlations may be carried out using simple physical sediment parameters (e.g. dry weight, water content, loss-on-ignition, bulk density, magnetic susceptibility).

Recent research on lake and marine cores has demonstrated that the climate of the Holocene has been more variable than previously recognised. Organisms in ecotonal environments may be responsive to climatic variations because they are close to their physiological limits of their distribution (Lotter and Birks, 2003). Most commonly, summer temperature, the length of the growing season, the depth/duration of the snow cover, and wind strength/exposure are the most controlling physiographic factors. It is only in the past few decades that lacustrine records from near ecotones have been the focus of detailed palaeoclimatological studies (e.g. Battarbee et al., 2002). Mountain lakes are particularly sensitive recorders of past and present climate change as the occurrence and composition of aquatic and

Table 1
Examples of Holocene climate reconstructions in Europe and the North Atlantic region from different proxies and archives

Ice cores: Dansgaard et al. (1993), Grootes et al. (1993), O'Brian et al. (1995), Stuiver et al. (1995), Alley et al. (1997), Dahl-Jensen et al. (1998), Johnsen et al. (2001), Fisher and Koerner, (2003).

Alpine tree-limit fluctuations: Kullman (1981, 1995), Eronen et al. (1993), Berglund et al. (1996), Karlén and Kuylenstierna, (1996).

Glacier variations: Karlén (1976, 1988), Nesje et al. (1991, 1994, 1995, 2000, 2001), Dahl and Nesje (1992, 1994, 1996), Karlén and Matthews (1992), Matthews and Karlén, 1992, Leeman and Niessen, (1994), Karlén et al. (1995), Snowball and Sandgren (1996), Svendsen and Mangerud (1997), Matthews et al. (2000), Snyder et al. (2000), Barnekow and Sandgren (2001), Dahl et al. (2002), Seierstad et al. (2002), Winkler et al. (2003).

Pollen, chronomids and diatoms: Barber et al. (1999a b), Barnekow (2000), Rosén et al. (2001), Seppä and Birks, (2001,2002), Bigler et al. (2002), Korhola et al. (2002), Seppä et al. (2002), Bigler and Hall (2003), Davis et al. (2003), Hannon et al. (2003), Heiri et al. (2003), Heikkilä and Seppä, (2003).

Plant macrofossils: Birks (1991), Birks and Ammann (2000), Birks et al. (2000), Hannon et al. (2003).

Tree rings: Briffa et al. (1992), Schweingruber and Briffa (1996), Briffa et al. (1998, 2002), Kalela-Brundin (1999), Kirchhefer (2001), Grudd et al. (2002); Helama et al. (2002), Thun (2002).

Stable-isotope records from tree rings: See review by McCarroll and Loader (2004).

Stable-isotope records from lake sediments: Grafenstein et al. (1998), Shemesh et al. (2001), Hammarlund et al. (2002, 2003), Leng and Marshall (2004), Rosqvist et al. (2004).

Physical sediment parameters (LOI and mineral magnetic properties) in lake sediments: Snowball et al. (1999), Willemse and Törnquist (1999), Nesje and Dahl (2001), Battarbee et al. (2001, 2002), Kaplan et al. (2002), Lotter and Birks (2003), Rubensdotter and Rosqvist (2003).

Peatlands: See review by Chambers and Charman (2004).

Speleothems: Lauritzen (1996), Lauritzen and Lundberg (1999).

Marine proxies (e.g. stable isotopes, faunal variations, physical sediment parameters): Lehman et al. (1991), Koc et al. (1993), Haflidason et al. (1995), Bond et al. (1997, 2001), Fronval and Jansen (1997), Klitgaard-Kristensen et al. (1998, 2001), Barber et al. (1999a b), Bianchi and McCave (1999); Grøsfjeld et al. (1999), Hald et al. (2001), Mikalsen et al. (2001), Birks and Koc (2002), Husum and Hald (2002), Andersson, et al (2003), Oppo et al. (2003), Risebrobakken et al. (2003), Sarnthein et al. (2003), Andrews and Giraudeau (2003), Solignac et al. (2004).

Marine molluscs: Salvigsen et al. (1992), Hjort et al. (1995), Salvigsen (2002).

Historical instrumental evidence: Pfister et al. (1999), O'Sullivan et al. (2002), Nordli et al. (2003).

terrestrial biota are directly or indirectly related to climate (e.g. Lotter et al., 1997; Battarbee, 2000). The length of lake ice-free period and the summer temperature are commonly the main controlling factors of the primary production in the water column in addition to the mixing regime and the amount of oxygen depletion (Livingstone, 1997). Catchment soils and vegetation have also a strong influence on water chemistry (e.g. Birks et al., 2000).

The main aim of this study was: (a) to reconstruct the Holocene environmental variability from several sediment parameters in a non-glacial mountain lake close to the modern pine-tree limit in Jotunheimen, central southern Norway, and (b) to discuss the inferred Holocene temperature variability record from Jotunheimen in the context of other terrestrial and marine records in the North Atlantic realm and with the respect to possible natural climate forcing factors.

2. Lakes as climate archive

2.1. Introduction

Commonly, lake sediments consist of three components (in varying proportion): *organic matter, nonecarbonate clastic material*, and one or more *carbonate minerals*. In lake sediments, the organic content reflect autochthonous production from plants and input of eroded organic and minerogenic material from the catchment. The organic material in lake sediments may be produced in the lake itself, washed in from rivers in the catchment, and blown in from the lake surroundings. The minerogenic material in lakes may originate from lake-shore erosion and wave action, lake-ice erosion along the shallow shores, transport by glacier meltwater streams, fluvial erosion of deposits along the rivers in the catchment, redeposition of material by slumping and turbidity currents, and windblown material (local and regional sources). Slope processes, such as gully inwash, debris flows, and 'wet' snow avalanches may also contribute to minerogenic and/or organic input to lakes due to erosion of soils and clastic material in the catchment. In proglacial lakes, high variability of minerogenic input to the lacustrine system is generally higher than the internal variability/production of organic material. In such lakes, changes of the organic/minerogenic content mainly reflect allochthonous, glacially produced minerogenic influx. If the dominant process(es) of the minerogenic material input to a lake is (are) identified, specific lake catchment processes may be reconstructed in great detail from lacustrine sediments. Mapping and monitoring of processes in the catchment are therefore recommended in order to better understand variations in the sediment parameters.

2.2. Organic matter

The organic matter content of lake sediments provides a variety of indicators that can be used to reconstruct palaeoenvironmental variations of lakes and their catchments. Organic matter originates from organic matter components (lipids, carbohydrates, proteins, etc.) produced by organisms that have lived in and around the lake (e.g. Meyers and Teranes, 2001). Organic matter derived from the residues of plants is gradually altered into humus through physical fragmentation, faunal and microfaunal interactions, mineralisation and other processes of humus formation. Measurement of the *organic content* of soils and sediments is either by loss-on-ignition (LOI) or determination of *organic carbon* (OC) content by the wet oxidation technique. The *inorganic carbon* fraction may consist of coal, charcoal, and carbonates. *Total organic carbon* (TOC) is commonly measured in the analysis of lake and marine sediment cores by means of an LECO carbon-carbonate determination.

LOI is the most commonly applied method for estimating the organic content in lake sediments (e.g. Dean, 1974; Håkanson and Jansson, 1983; Heiri et al., 2001 and references therein). LOI has been shown to be a remarkable 'composite' proxy of environmental change in high latitude lakes (Willemse and Törnquist, 1999; Battarbee et al., 2001, 2002; Nesje and Dahl, 2001; Kaplan et al., 2002). The interpretation of the LOI signal may, however, be complex as the residue after ignition at 550 °C may consist of varying inputs of inorganic mineral matter, carbonate and biogenic silica (diatoms) (e.g., Battarbee et al., 2002). The LOI values are also influenced by varying input and preservation of organic matter. The LOI signal may also be controlled by sedimentation rate changes of mineral matter (most relevant in proglacial lakes). Both inorganic and organic matter can be produced within the lake as well as in the catchment. Small mountain lakes located on crystalline bedrock, with sparse vegetation and thin soils in the catchment, few or small inlets, and restricted aquatic macrophyte flora, are commonly well suited because the factors influencing on the LOI signal are fewer.

Differential thermal analysis (DTA) thermograms show that when a dried, powdered sample containing organic material and calcium carbonate is heated in a muffle furnace, the organic material begins to ignite at about 200 °C and is completely ignited by the time the furnace temperature has reached approximately 550 °C. The high correlation ($r = > 0.95$) between ignition loss organic matter and percent organic carbon determined chromatographically shows that the LOI method is a measure of the amount of organic matter in a sample [the weight LOI is 2.13 ± 0.4 times the organic carbon content (Dean, 1974)]. Snowball and Sandgren (1996) showed that LOI was 2–3% higher than TOC, probably due to crystalline water in clayey material.

According to Dean (1974) evolution of CO_2 from the calcium carbonate will begin at about 800 °C and proceed rapidly so that most of the CO_2 has been evolved by the time the furnace has reached 850 °C. If any dolomite is present in the sample, it will evolve CO_2 at a lower temperature than calcite (at approximately 700–750 °C). Sutherland (1998) suggested that structural water may be lost by metal oxides at temperatures as low as 280–400 °C. Most clays contain up to 5% lattice OH water which is not removed until heated to 550–1000 °C. Therefore, the 550–950 °C ignition loss would contain a significant amount of lattice water in samples that are low in carbonate and high in clay. For example, lake sediments containing no carbonate but with high clay content usually show a 2–4% LOI between 550 and 950 °C. Sediments containing 100% clay would presumably yield up to 5% weight loss between 550 and 950 °C.

Inorganic carbon may be lost at temperatures between 425 and 520 °C in minerals such as siderite, magnesite and rhodochrosite (Weliky et al., 1983; Sutherland, 1998). There is also a possible loss of volatile salts at 550 °C (Bengtsson and Enell, 1986). The LOI method does not, however, indicate which carbonate minerals that are present. Different carbonate minerals will evolve CO_2 at slightly different temperatures, but they cannot be separated by this method, thus giving total carbonate.

Diatoms are photosynthetic algae that secrete biogenic silica (SiO_2 nH_2O) as an internal shell. Biogenic silica (BSi) reflects the sedimentary abundance of diatoms, which are single-celled algae that commonly dominate lake primary productivity (Wetzel, 2001).

The isotopic composition of autigenic and biogenic carbonates and diatom silica are used as palaeoclimatic proxies from lake sediments (e.g. Leng and Marshall, 2004). The oxygen isotopic composition of BSi depends on the ambient water temperature and the isotopic composition of the lake water when the shell is secreted (e.g. Shemesh et al., 1992). In lacustrine environments stratigraphic changes in $\partial^{18}O$ values are commonly attributed to changes in temperature or precipitation/ evaporation ratio, whereas carbon and nitrogen isotopes are used to infer changes in carbon, nutrient cycling and productivity within lakes and their catchments.

2.3. Mineral matter

Magnetic susceptibility of lacustrine sediments is a useful indicator of erosion and transport of clastic sediments in lake catchments (Snowball and Thompson, 1990; Snowball et al., 1999). Commonly magnetic susceptibility reflects the concentration of magnetic minerals (e.g. Thompson and Oldfield, 1986). Cold climates without a stabilising vegetation cover and/or glaciers in the catchment cause high susceptibility due to increased erosion and deposition of minerogenic sediments (e.g. Stockhausen and Zolitschka, 1999; Nesje et al., 2000, 2001). Mineral magnetic measurements have been used, in combination with other physical sediment parameters, as an indicator of glacier activity (Sandgren and Risberg, 1990; Nesje et al., 1991, 1994, 2000, 2001; Karlén and Matthews, 1992; Matthews and Karlén, 1992; Snowball, 1993; Dahl and Nesje, 1994, 1996; Sohlenius, 1996; Matthews et al., 2000). Increased glacier activity in the lake catchment associated with increased erosion and input of clastic sediments increases the minerogenic content, also reflected in the LOI content. Periods of insignificant and reduced glacier activity in the lake catchment are characterised by low production of glacially derived clastic material. Increased magnetic susceptibility has therefore been related to the amount of allochthonous clastic material transport into lakes (e.g. Thompson et al., 1975). High magnetic susceptibility values may, however, also reflect increased input of minerogenic material into lakes associated with surface runoff during rainstorms, floods and mass movement events (snow avalanches and debris flows) from adjacent valley sides (Karlén and Matthews, 1992; Nesje et al., 1991, 1995, 2000, 2001; Dahl et al., 2003; Sletten et al., 2003).

3. Study area and previous research

Lake Danntjørn (8°99′E, 61°35′N), at an altitude of ~950 m, located in Sjodalen, eastern Jotunheimen, southern Norway, is a small mountain lake (maximum water depth of 4.1 m) located on top of a low-relief, NW/SE trending ridge just below the modern pine tree limit (~1000 m a.s.l.) in Sjodalen (Figs. 1 and 2A, B). The deepest part of the lake (the coring site) is located in the SE part of the lake. The lake has no major inlet streams and the outlet is in the SE end. Commonly the lake is ice free from May/June to September/October. Data on local hydrology, pH, summer lake water temperature and nutrient status are not available.

Mountain regions, such as Jotunheimen in central southern Norway, show large elevational differences over short horizontal distances. Jotunheimen is the highest mountain range in northern Europe and contains approximately 300 glaciers (Østrem et al., 1998). Most of Jotunheimen lies in the alpine zone, above the birch (*Betula pubescens*) altitudinal tree line at approximately 1000 m. Several mountains rise to more than 2000 m. Jotunheimen is located about 150 km from the western coast of southern Norway and the central part of Jotunheimen marks the main water divide in southern Norway. The climate in Jotunheimen is transitional between maritime western Norway and the

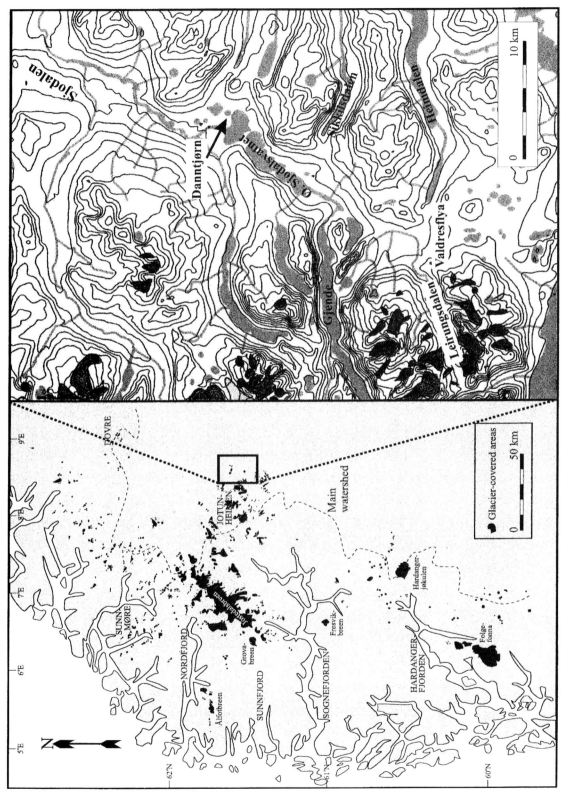

Fig. 1. Location map of southern Norway (left) and the Jotunheimen area (right). The contour interval is 100 m. The location of Danntjørn is indicated. Glaciers and lakes on the right-hand map are indicated by dark and light shading, respectively.

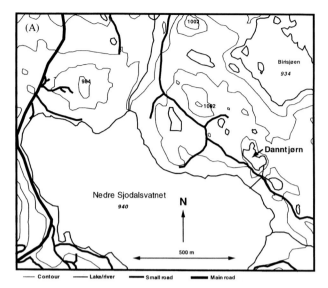

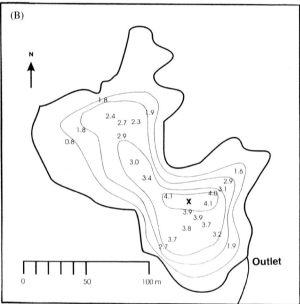

Fig. 2. (A) Map of the Danntjørn area. (B) Bathymetric map (contour interval 1 m) of Lake Danntjørn with the individual measurements (in metres) and the coring site indicated (x).

more continental eastern part of southern Norway. The meteorological station Sognefjell (1413 m a.s.l.) in western Jotunheimen had a AD1978–89 mean annual temperature of −3.1 °C, a July mean of 5.7 °C, a January mean of −10.7 °C, a mean summer (1 May–30 September) temperature of 3.3 °C and a mean winter (1 October–30 April) temperature of −7.7 °C (Aune, 1993). The mean annual, winter, and summer precipitation were 860, 488 and 372 mm, respectively (Førland, 1993). The lowland region just east of Jotunheimen (Skjåk) is among the driest region in Norway with a mean annual precipitation of approximately 300 mm. The precipitation, however, increases by 8–10% per 100 m altitudinal rise (Haakensen, 1989; Dahl and Nesje, 1992; Laumann and Reeh, 1993), attain-

ing > 1000 mm over much of the high mountains. The lower altitudinal permafrost boundary is approximately 1460 m in eastern Jotunheimen and rises westward (Isaksen et al., 2002). The bedrock in Jotunheimen is predominately metamorphosed rocks of Precambrian age in the Caledonian nappe (charnockitic to anorthositic rocks). Along the valley bottom of Sjodalen, in which Lake Danntjørn is located, there is a belt of phyllite and sandstone of late Precambrian age (Sigmond et al., 1984).

Geographical and glaciological research in Jotunheimen has a history going back to the late 19th/early 20th centuries (e.g. Øyen, 1893, 1908). In western Jotunheimen, pioneering glacier mass-balance studies were carried out by Ahlmann (1922, 1940). In eastern Jotunheimen, glacial geomorphological processes were studied by Lewis (1960). Ice-cored moraines were studied by Østrem, 1964,1965). Measurements and mapping of glacier-front variations were initiated by Hoel, Werenskiold and Liestøl (Hoel and Werenskiold, 1962) and annual glacier-front variations and glacier mass-balance data are published annually by Norges vassdrags- og Energidirektorat (NVE). The glacier mass balance record for Storbreen in central Jotunheimen, started in 1947, is the second longest and continuous record (after Storglaciären in Sweden) in the world (Andreassen and Østrem, 1999). In addition, Matthews (1974) used the Storbreen glacier foreland to contribute to the development of the statistical basis for lichenometric dating of 'Little Ice Age' moraines. Recently, Matthews et al. (1997) reconstructed the Holocene colluvial history in Leirdalen, Matthews et al. (2000) used lake sediments from Bøvertunsvatnet and Dalsvatnet to reconstruct upstream glacier variations in western Jotunheimen, Barnett et al. (2001) reconstructed Holocene climatic change and tree-line response in Leirdalen, central Jotunheimen, and Sandvold et al. (2001) reconstructed the Holocene glacial and colluvial activity in Leirungsdalen.

4. Methodology

In summer 2000, two long, overlapping cores were retrieved from Lake Danntjørn from the deepest (~4.1 m depth) and flat-bottomed part of the lake (Fig. 2B). The long, 110-mm diameter cores were retrieved from a raft using a piston corer with a 110 mm diameter core tube constructed to obtain up to 6 m of sediments (Nesje, 1992). The upper core (core I) was 234 cm long, whereas the lower core (core II, top at 208 cm below lake bottom and basal part at 360 cm below lake bottom) was 152 cm long. The cores were brought unopened back to the laboratory where they were stored in a cold room until opening. After opening, the sediment layer closest to the tube wall was removed and the sediment surface was cleaned carefully. Lithofacies and sedimentological structures and textures were described before the cores were subject to magnetic susceptibility measurements and sampled contiguously for LOI at 0.5 cm intervals. In the summer of 2003, a 36-cm-long surface core was retrieved from a Zodiac rubber boat by a modified Kajak-gravity corer (Renberg, 1991; manufactured by HTH Teknik, Luleå, Sweden) from the same part of the lake as cores I and II. The Renberg gravity corer is designed to retrieve undisturbed cores from the water-sediment interface and downward in the upper part of the sediments.

The samples for LOI were dried overnight at 105 °C in ceramic crucibles before the dry weight was measured (normally 1–3 g). The water content was calculated in percent of the dry weight. In the furnace, the samples were subjected to gradually rising temperatures for half an hour and ignited at 550 °C for one hour. The crucibles were then put into a desiccator to cool for approximately half an hour and weighed at room temperature. The weight LOI was calculated in per cent of dry weight.

Magnetic susceptibility was carried out at 0.5 mm intervals on the cleaned core face of the two long sediment cores. The magnetic susceptibility measurements (10^{-6} SI) of the two long Danntjørn cores were carried out by a Bartington MS2B sensor.

Core II was sampled contiguously (1-cm slices) between 311 and 355 cm ($n = 45$; corresponding to

232.5–277.5 cm in the composite record) for biogenic silica (BSi) content. BSi was extracted with 10% Na_2CO_3 and determined with spectrophotometer following Mortlock and Froelich (1989).

For radiocarbon dating, birch (*Betula* sp.) bark fragments were wet sieved (1-mm mesh) from the lake sediments. Accelerator mass spectrometry (AMS) radiocarbon dating (all 11 samples were bark fragments of birch) was carried out by the Poznan Radiocarbon Laboratory in Poland, following standard procedures for AMS radiocarbon dating. Calibration of the radiocarbon ages to calendar years BP (BP = AD 1950) was done using the calibration program CALIB 4.1.2 for atmospheric samples (Stuiver et al., 1998). When more than one possible intercept age, the median value was used. The age/depth model for Danntjørn lake record was based on linear interpolation between the intercept calibrated ages obtained from the different levels.

5. Results

5.1. Age/depth model

The depths in Table 2 refer to the depth below the sediment surface in the composite (core I and II combined, Fig. 5A–C) Lake Danntjørn sediment record, but the depths given in brackets refer to the original depths in core Danntjørn II (Fig. 5B). The date obtained

Table 2
Radiocarbon AMS dates from Lake Danntjørn

Depth (cm)	Lab. ref.	^{14}C age	Calendar year BP	Calendar year age range	Sedimentation rate (mm/year)	Time resolution (year/0.5 cm)
57	Poz-932	2790 ± 30	2915	2945–2850		
					0.44	12
91	Poz-930	3450 ± 45	3695	3825–3640		
					0.39	13
119	Poz-931	3920 ± 35	4410	4420–4295		
152	Poz-936	2760 ± 35*	2850	2920–2785	0.31	16
157.5 (235)	Poz-939	4980 ± 40	5670	5745–5655		
					0.27	18
180	Poz-933	5720 ± 40	6495	6550–6415		
					0.58	8
198.5 (276)	Poz-935	6060 ± 40	6815	6970–6805		
					0.35	15
207	Poz-942	6200 ± 40	7060	7210–7010		
					0.23	21
210.5 (288)	Poz-941	6260 ± 40	7210	7250–7095		
					0.41	12
258.5 (336)	Poz-937	7555 ± 45	8375	8390–8345		
					0.13	38
273.5 (351)	Poz-938	8550 ± 50	9535	9545–9500		

The depths refer to the composite (cores I and II combined) LOI record. Original depths in core II in brackets. For original depths in cores I and II, see also Fig. 5A, B. Dating marked* is not used in the age/depth model in Fig. 3. The ^{14}C and calibrated age ranges are given with ±1 sigma. All radiocarbon dates were obtained on birch (*Betula*) bark fragments. Calibration of the radiocarbon ages to calendar years BP (BP = AD1950) was done using the calibration program CALIB 4.1.2 for atmospheric samples (Stuiver et al., 1998). When more than one possible intercept age, the median value was used. Sedimentation rates and time resolution between the radiocarbon-dated levels are indicated.

from 152 cm was apparently too young (the dated material may have been dragged downward during coring) and was therefore not used in the age/depth model based on linear interpolation (as recommended by R. Telford, pers. comm.) between the intercept calendar ages obtained from the dated levels (Fig. 3). Apparently from the age/depth curve, the topmost lake sediments were not sampled by the Nesje corer (core I), because the top sediments were extremely fluid and therefore difficult to recover. The age/depth curve was extrapolated from the three uppermost radiocarbon ages further upward in the core, indicating that approximately 60 cm (~1500 years) were lacking at the top. A surface core was retrieved in order to sample the time span lacking in Lake Danntjørn core I. Unfortunately, the entire time span could not be sampled (see below). The sedimentation rate (and time resolution) in the Lake Danntjørn sediment record was rather uniform (mean 0.37 mm/yr), corresponding to a mean time resolution of ~13 yr per 0.5 cm sample thickness in the upper 258 cm of the core (Table 2, Fig. 4). Below this level, however, the sedimentation rate was apparently significantly lower (0.13 mm/yr) corresponding to a time resolution in the order of 38 yr per 0.5 cm sample thickness.

5.2. Physical sediment parameters

The loss-on-ignition (LOI) records (contiguous samples at 0.5-cm intervals) from Lake Danntjørn cores I and II are shown in Fig. 5A, B. The LOI records from

the Danntjørn cores I and II show the same main features (a–k in Fig. 5A, B) over the overlapping 77 cm sequence. Cores I and II from Lake Danntjørn were therefore combined into a composite LOI record (Fig. 5C). The tie points (based on LOI, magnetic susceptibility and age/depth models based on radiocarbon dates from the two cores) between the two cores were at 137 cm in core I and at 213.5 cm in core II (the vertical lines in Fig. 5A, B). The 0.5-cm interval magnetic susceptibility record (10^{-6} SI) from cores I and II were spliced at the same levels as the LOI records, and plotted together with the record of weight percent residue after ignition at 550 °C on a depth scale (Fig. 6). Except for the basal part, from approximately 250 cm and downward, where the two curves are mainly in phase, there is no apparent relationship between the residue curve and the magnetic susceptibility curve. The dry weight, LOI and residue after ignition at 550 °C in the 36-cm long surface Lake Danntjørn core are shown in Fig. 7.

5.3. Biogenic silica (BSi) concentration

The BSi concentration was measured contiguously every cm between 311 and 355 cm (45 samples) in the lower part of core Danntjørn II, corresponding to 233–278 cm in the composite Lake Danntjørn record (Fig. 8A). The BSi content shows a marked transition at 260 cm (Fig. 8, upper panel). Below this level, the BSi values are apparently relatively high (~120 mg/g), whereas above this level, BSi concentration apparently is considerably lower (~50 mg/g). Independent LOI analysis was carried on the same samples as subject to BSi analysis (1-cm intervals) and they are highly correlated ($r = 0.93$) with the primary LOI analysis over the same depth interval. Below 260 cm the BSi concentration is apparently higher than above. When flux-corrected, however, the BSi concentration is somewhat lower below 260 cm (before ~8500 calendar years BP) than above (Fig. 8, lower panel). The BSi data show a strong negative correlation ($r = -0.81$) between LOI and BSi above 260 cm (<8500 calendar years BP) in the composite record, indicating that the sediments are diluted by BSi (diatoms giving low magnetic susceptibility and high residue/low LOI) resulting in a negative correlation between LOI (organic matter) and BSi (Fig. 9A, B).

6. Discussion

Standardised (individual values subtracted from the mean value for entire record and divided by the standard deviation) residue values after ignition at 550 °C in the surface core show similar features as a tree-ring width series (standardised according to the

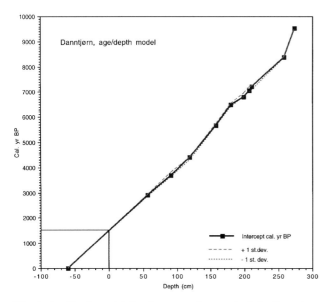

Fig. 3. Age/depth model for the composite core in Lake Danntjørn obtained by linear interpolation between the dated levels (Table 2). Lines indicating ±1 standard deviation age ranges of the radiocarbon dates are shown. By extrapolating the age/depth curve from the lowest three radiocarbon dates, apparently 60 cm of the topmost sediments in the lake, representing ~1500 calendar years, are missing (frame in lower left-hand corner).

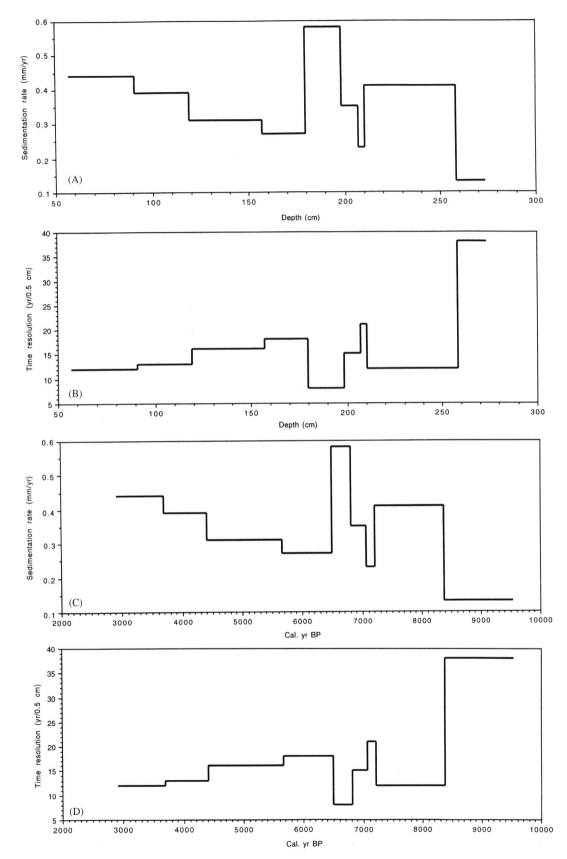

Fig. 4. Sedimentation rate (mm/year) and time resolution (year/0.5 cm; sampling interval is 0.5 cm) between the dated levels in the Danntjørn composite record (Table 2) on depth (A, B) and calendar years BP (C, D) scales.

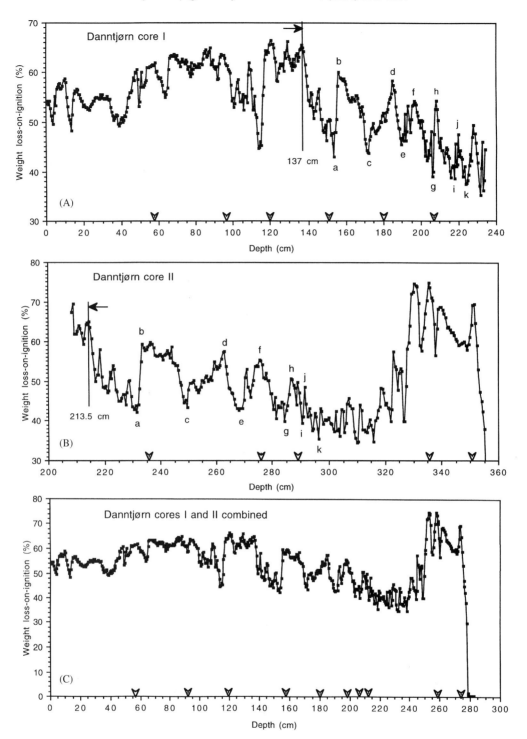

Fig. 5. (A) and (B)-Weight LOI in cores I and II, respectively, retrieved from Danntjørn. Common features recognised in both cores in the overlapping part (77 cm) are labelled a–k. (C)-Composite LOI record from Danntjørn when combining the upper 137 cm in core A and from 213.5 cm and downward in core II. The arrows indicate the location of the AMS radiocarbon dates in cores I and II and when combined in the composite record.

same procedure) from the Trøndelag area (Thun, 2002; Nesje et al., in preperation) about 200 km north of Jotunheimen (Fig. 10). This indicates that variations in the residue after ignition (probably mainly BSi from diatoms, as demonstrated in the depth interval 233–260 cm in the Lake Danntjørn composite sediment

record) at 550 °C (not the LOI) are mainly controlled by summer temperature variations.

Fig. 11 shows the residue of the composite Lake Danntjørn sediment record and the surface core together with the composite magnetic susceptibility record according to the age/depth model. Apparently,

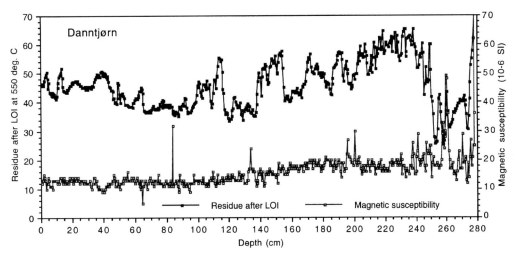

Fig. 6. Residue after LOI at 550 °C and the magnetic susceptibility (10^{-6} SI) in the composite Danntjørn core plotted on a depth scale.

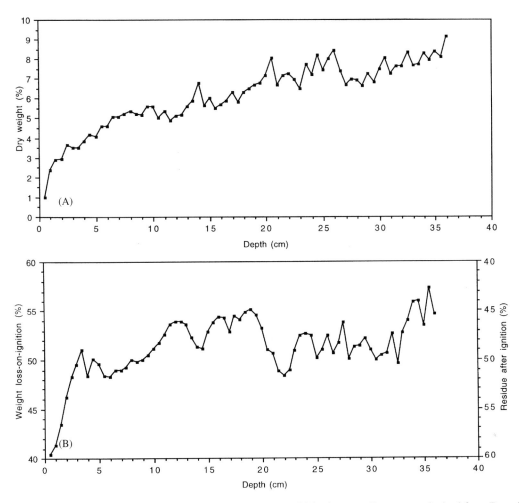

Fig. 7. (A)-Dry weight, (B)-Weight LOI and residue after LOI at 550 °C in the top sediment core obtained from Danntjørn.

the first organic sedimentation in the lake occurred ~9900 calendar years BP. In the basal part of the core, up to about 8150 calendar years BP, the residue and

magnetic susceptibility curves fluctuate mainly in phase. In this part of the core, the material also consists of significant (also when flux-corrected) amounts of BSi

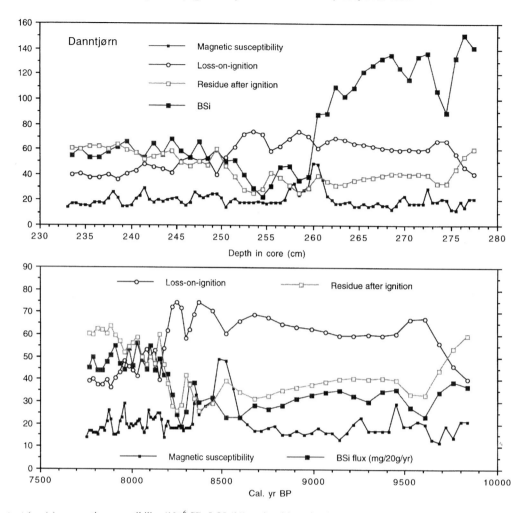

Fig. 8. BSi content (mg/g), magnetic susceptibility (10^{-6} SI), LOI (%) and residue after ignition at 550 °C (%) between 233 and 278 cm in the Lake Danntjørn composite record plotted on a depth scale (upper panel) and on a calendar years BP time scale (lower panel). Note that the BSi content has been flux-corrected in the lower panel.

(Fig. 8), indicating that the residue mainly consists of detrital, minerogenic material and diatoms. Two significant maxima/minima in the residue/LOI curves at 260 and 255 cm, the first of which also represented by a significant magnetic susceptibility peak, date to 8500 and 8300 calendar years BP, respectively. Subsequent to approximately 8150 calendar years BP, the residue and magnetic susceptibility curves are decoupled because the residue after ignition most probably mainly consists of BSi, as demonstrated up to 233 cm (~7760 calendar years BP). The highest residue values occurred at 7800–7200, 7000, 6700, 6200, 5400–5050, 4300, 2500, 1800, 1550, 900, 700, and <100 calendar years BP. The most significant <8000 calendar years BP residue minima are recorded at 7100, 6800, 6600, 6000–5700, 5000–4400, 4200, 3800, 3600, 3200, 3000, 2750, 2300, 1900, 1700, 1070, 740, 600, 380, and 80 calendar years BP.

Several climatic implications follow from the temporal pattern of LOI variations shown in Fig. 12A. The Danntjørn lake record indicates major millennial-scale

and minor centennial-scale events. These may relate <1400–1600-year periodicities/quasi-periodicities identified in other climate archives in the North Atlantic region and other regions (Bond et al., 1997, 2001; Stuiver et al., 1997; Campbell et al., 1998; Bianchi and McCave, 1999; Chapman and Shackleton, 2000; O'Sullivan et al., 2002; Gupta et al., 2003; Risebrobakken et al., 2003; Sarnthein et al., 2003).

6.1. 10,000–8500 calendar years BP

Between 10,000 and ~8500 calendar years BP, when the LOI in Danntjørn mainly reflects lake productivity (residue mainly in phase with magnetic susceptibility), organic content was high, indicating high summer temperatures (Fig. 12A). This is also seen in reconstructed summer temperatures from altitudinal variations in the pine-tree limit in the Scandes Mountains (Dahl and Nesje, 1996 and references therein), chironomid-inferred mean July temperature reconstructed from the Holebudalen site, northern Setesdalen, southern

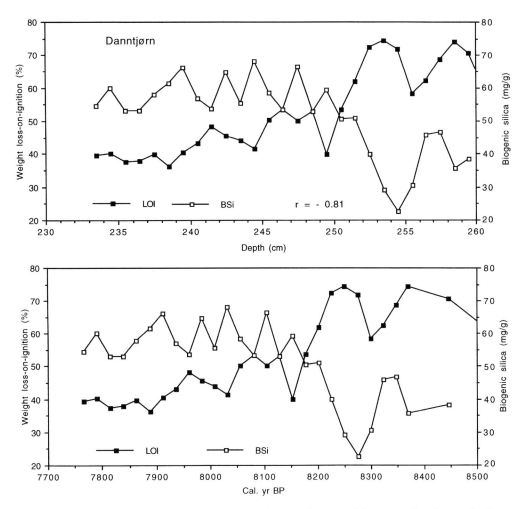

Fig. 9. LOI and BSi content between 233 and 260 cm in the composite Lake Danntjørn record (upper panel) and on a calendar years BP time scale (lower panel).

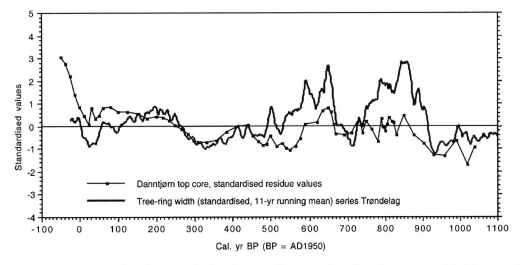

Fig. 10. The residue after LOI at 550 °C (standardised values) in the top sediment core from Danntjørn wiggle-matched to a tree-ring width series (standardised values) from Trøndelag (Thun, 2002; Nesje et al., in preparation) on a calendar years BP time scale.

Norway (Brooks, 2003), SST (August) (diatom-inferred) in the eastern Norwegian Sea (Vøring Plateau; Jansen and Koç, 2000; Birks and Koc, 2002; Koç and Jansen, 2002; Andersen, 2003; Andersen et al., 2004), and generally little (except for a peak around 9500 calendar years BP) drift ice in the North Atlantic (Bond et al.,

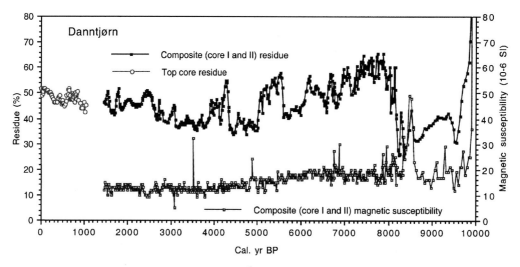

Fig. 11. Residue after LOI at 550 °C, the magnetic susceptibility (10⁻⁶ SI) in the composite Lake Danntjørn core and residue values in the top sediment core plotted on a calendar years BP time scale.

2001) (Fig. 12). Oxygen isotope records from planktonic foraminifera in the eastern Norwegian Sea (Vøring Plateau; Risebrobakken et al., 2003) do, however, not show such an early Holocene thermal maximum (Fig. 12).

6.2. The 8200 calendar years BP event

The two significant LOI minima/residue and magnetic susceptibility maxima present at 260 and 255 cm in the composite stratigraphy (Figs. 5 and 6), corresponding to ages of ~8400 and 8300 calendar years BP, are probably the same widespread Northern Hemisphere event as recorded in the GRIP (Dansgaard et al., 1993) and GISP2 (Grootes et al., 1993) Greenland ice cores, in lacustrine and proglacial lake records (Karlén, 1976, 1988; Karlén et al., 1995; von Grafenstein et al., 1998; Nesje et al., 2000, 2001; Nesje and Dahl, 2001), marine records (Bond et al., 1997; Klitgaard-Kristensen et al., 1998) and in speleothemes (Baldini et al., 2002. This climate oscillation was in south Norway termed the 'Finse Event' after the type-site at Finse north of the Hardangerjøkulen ice cap (Dahl and Nesje, 1994, 1996). Barber et al. (1999a,b) and Clarke et al. (2003, 2004) suggested that this '8200 event' was triggered by a catastrophic drainage episode from glacial lakes south of the Laurentide ice sheet. The slight offset in the timing of the '8.2 ka'/'Finse event' in Lake Danntjørn may indicate that the age model at the base of the core is slightly skewed (oldest ¹⁴C dates slightly too old).

The period subsequent to the Finse Event/'8.2 ka event' until about 7200 calendar years BP, exhibits the highest residue (high BSi concentration) values during the entire Holocene in Danntjørn, indicating high summer temperatures during that time interval. This is also observed in reconstructed summer tempera-

tures from altitudinal variations in the pine-tree limit in the Scandes Mountains (Dahl and Nesje, 1996 and references therein), chironomid-inferred mean July temperature reconstructed from the Holebudalen site, northern Setesdalen (Brooks, 2003), southern Norway, August sea-surface temperatures (SST) (diatom-inferred) in the eastern Norwegian Sea (Vøring Plateau; Jansen and Koç, 2000; Birks and Koc, 2002; Koç and Jansen, 2002; Andersen, 2003; Andersen et al., 2004) (Fig. 12B–D). Oxygen isotope records from planktonic foraminifera in the eastern Norwegian Sea (Vøring Plateau; Risebrobakken et al., 2003) did not indicate high temperatures during this time span (Fig. 12E) and this period was characterised by significant drift ice in the North Atlantic (Bond et al., 2001; Fig. 12F).

6.3. 7200–5600 calendar years BP

The Danntjørn record indicates variable but generally falling temperatures between 7200 and 6000 calendar years BP. Summer temperatures reconstructed from altitudinal variations in the pine-tree limit in the Scandes Mountains, chironomid-inferred mean July temperature reconstructed from the Holebudalen site, northern Setesdalen, August SST (diatom-inferred) in the eastern Norwegian Sea (Vøring Plateau) all show an almost similar development (Fig. 12B–D). Oxygen isotope records from planktonic foraminifera in the eastern Norwegian Sea (Vøring Plateau) indicate no temperature trend during this time span (Fig. 12E). This period was characterised by variable, but generally increasing, drift ice in the North Atlantic (Fig. 12F).

An inferred significant summer temperature minimum around 6000 calendar years BP is also seen (with small time offsets) in a summer temperature record reconstructed from altitudinal variations in the pine-tree limit in the Scandes Mountains, chironomid-inferred mean July temperature reconstructed from the Hole-budalen site, northern Setesdalen, southern Norway, SSTs inferred from diatoms in the eastern Norwegian Sea and a significant drift ice peak in the North Atlantic (Fig. 12).

6.4. 5600–5000 calendar years BP

In Danntjørn, high residue values, indicating high summer temperatures, were recorded between 5600 and 5000 calendar years BP. This interval was also characterised by high summer temperatures reconstructed from altitudinal variations in the pine-tree limit in the Scandes Mountains, in chironomid-inferred mean July temperature (slightly delayed probably due to dating inaccuracy) reconstructed from the Holebudalen site, northern Setesdalen, southern Norway, August SSTs in the eastern Norwegian Sea (Vøring Plateau), and a significant drift ice minimum in the North Atlantic. The oxygen isotope records from planktonic foraminifera in the eastern Norwegian Sea, however, do not indicate high temperatures during this time span.

6.5. 5000–4400 calendar years BP

Low residue values are recorded in Danntjørn between 5000 and 4400 calendar years BP, indicating low summer temperatures. This interval was also characterised by a period of low summer temperatures reconstructed from altitudinal variations in the pine-tree limit in the Scandes Mountains. This was also recorded in chironomid-inferred mean July temperature (slightly delayed probably due to dating inaccuracy) reconstructed from the Holebudalen site, northern Setesdalen, southern Norway, SSTs (diatom-inferred) in the eastern Norwegian Sea and presence of drift ice in the North Atlantic. The oxygen isotope records from planktonic foraminifera in the eastern Norwegian Sea, however, do not indicate low temperatures during this time span, despite some oxygen isotope minima in a couple of levels.

6.6. 4400–3900 calendar years BP

In Danntjørn, high residue values were recorded between 4400 and 3900 calendar years BP (peak value at

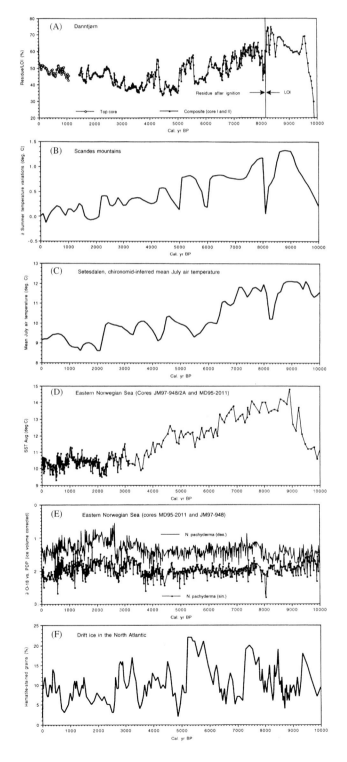

Fig. 12. (A) Holocene variations in the residue (after ~8150 calendar years BP) and LOI (before ~8150 calendar years BP) in the composite Lake Danntjørn core and in the top sediment core obtained from Danntjørn. (B) Summer temperature variations reconstructed from variations in the pine-tree limit in the Scandes Mountains, Sweden (Dahl and Nesje, 1996 and references therein). (C) Chironomid-inferred mean July air temperature variations from the Holebudalen in Setesdalen, southern Norway. The error lines of ±1.1 °C with respect to the central line given in the original paper are not indicated (adapted from Brooks, 2003). (D) Diatom-inferred SST (August) variations in the eastern Norwegian Sea (Vøring Plateau) (Jansen and Koç, 2000; Birks and Koc, 2002; Koç and Jansen, 2002; Andersen, 2003; Andersen et al., 2004). (E) Stabile oxygen isotopes (ice-volume corrected) in N. pachyderma (dex.) and N. pachyderma (sin.) from cores MD95-2011 and JM97-948 obtained from the eastern Norwegian Sea (Vøring Plateau) (Risebrobakken et al., 2003). (F) Variations in the occurrence of hematite-stained grains as an indicator of drift sea ice in the northern Atlantic (Bond et al., 2001).

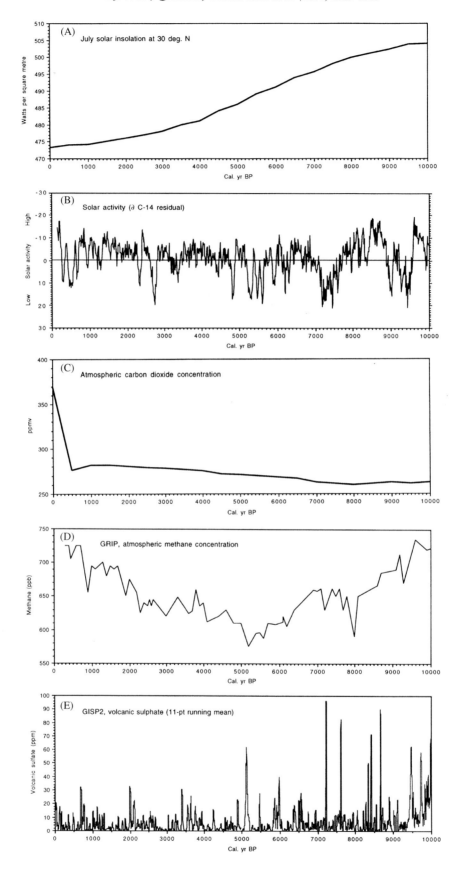

4300 calendar years BP). This interval was also characterised by a period of high summer temperatures reconstructed from altitudinal variations in the pine-tree limit in the Scandes Mountains, high mean July temperature (slightly delayed probably due to dating inaccuracy) reconstructed from chironomids at the Holebudalen site, northern Setesdalen, August SSTs in the eastern Norwegian Sea. The drift ice record from the North Atlantic shows, however, some drift ice, though decreasing, during this time interval. Again, the oxygen isotope records from planktonic foraminifera in the eastern Norwegian Sea do not indicate high temperatures during this time span.

6.7. 3900 calendar years BP to the present

After the inferred temperature drop at 3900 calendar years BP in the Danntjørn record, low residue values prevailed until ca 2700 calendar years BP, followed by high values ~2400 calendar years BP. A similar development was observed in the record of summer temperatures reconstructed from altitudinal variations in the pine-tree limit in the Scandes Mountains, chironomid-inferred mean July temperatures at the Holebudalen site, northern Setesdalen, southern Norway, August SSTs in the eastern Norwegian Sea. The drift ice record from the North Atlantic also indicates significant drift ice during this time interval. The oxygen isotope records from planktonic foraminifera in the eastern Norwegian Sea show a more similar development than in the lower part of the core.

A residue minimum in the Danntjørn record ~1700–1600 calendar years BP is also recorded in the Scandes temperature record, chironomid-inferred mean July temperatures (though slightly delayed) at the Holebudalen site, northern Setesdalen, in the oxygen isotope records from planktonic foraminifera in the eastern Norwegian Sea, and as a period of significant drift ice in the North Atlantic.

For the remaining part of the Holocene, the records compared in Fig. 12, show mainly a similar temperature development. The summer temperature records discussed above, except the oxygen isotope records from planktonic foraminifera from the Vøring Plateau. The reason for this discrepancy may be that the planktonic foraminifera mainly reflect the isotopic composition of

the water masses below the thermocline, thus mainly reflecting a mean annual temperature signal than 'pure' summer temperature.

7. Possible climate forcing factors

There is growing evidence of millennial-scale variability of Holocene climate, at periodicities of ~2500 and 950 years possibly caused by changes in solar flux (e.g. O'Brian et al., 1995) and ~1500 years possibly related to an internal oscillation of the climate system (e.g. Bond et al., 1997).

A comparison between possible climate forcing factors (Fig. 13) and the Holocene records in Fig. 12, suggests that the early Holocene thermal maximum observed in most records was caused by increased summer solar insolation to the northern hemisphere. The sub-orbital millennial to decennial climate variability observed in the Holocene climate reconstructions was most probably a combined effect of solar activity changes, periods of increased volcanic aerosols, and internal feedback mechanisms in the earth's climate system. Internal modes of the climate system variability (e.g. the North Atlantic Oscillation; Hurrell et al., 2003) may also have been responsible for some of the observed climate variability. In addition to external climate forcing factors, abrupt, catastrophic lake drainage from ice-marginal lakes during the termination of the last ice age had significant, regional climate impact (Barber et al., 1999a b). The best example of this is perhaps the ~'8.2 ka event' detected in many palaeoclimatic archives that resulted from sudden glacial lake drainage at the margin of the Laurentide ice sheet (e.g. Clarke et al., 2003, 2004).

Most palaeoclimatic studies that discuss potential climate forcing factors have simply made correlations by curve matching (in the time domain) or by finding spectral peaks that correspond to similar frequencies (in the frequency domain) of one or more potential forcing factors (e.g. Bond et al., 2001). By the use of realistic climate models (coupled ocean-atmosphere models, preferentially also incorporating hydrological and vegetation feedbacks), it should be possible to combine high-resolution palaeoclimatic reconstruction, like those shown in Fig. 8, with climate models in order to test

---

Fig. 13. Possible climate forcing mechanisms for Holocene climate variability: (A) Holocene variations in July solar insolation at 30°N. Adapted from Berger and Loutre (1996). (B) Holocene variations in solar activity as inferred from residual 14C variations. Adapted from Stuiver et al. (1997). (C) Holocene variations in atmospheric carbon dioxide concentration in ice cores from Byrd Station, Taylor Dome and Siple Station (adapted from Neftel et al., 1988; Staffelbach et al., 1991; Marchal et al., 1999; Indermühle et al., 1999) and from the Mauna Loa Observatory, Hawaii (Keeling and Whorf, continuously updated; http://cdiac.esd.ornl.gov/ftp/maunaloa-co2/maunaloa.co2). (D) Holocene variations in atmospheric methane (CH_4) concentration from the Greenland GRIP ice core. Adapted from Blunier et al. (1995). (E) The volcanic sulphate record (11-pt. running mean) from the Greenland GISP2 ice core. Adapted from Zielinski et al. (1994) and Zielinski and Mershon (1997).

the complex interactions between the different forcing factors to help explaining the primary forcing factors behind the observed Holocene climate development.

8. Conclusions

(1) Holocene summer temperature variations have been inferred from several sediment parameters in a lacustrine sediment record obtained from Lake Danntjørn, a small mountain lake located close to the modern pine-tree limit in Jotunheimen, central southern Norway.

(2) In the lake sediments from the deglaciation and early Holocene (below ca 260 cm; ~8500 calendar years BP), variations in residue after ignition at 550 °C and magnetic susceptibility fluctuated mainly in phase (positively correlated). The BSi was also relatively high in this part of the core. Thus, in the early part of the Holocene, the variations in the residue were primarily driven by input of detrital, minerogenic sediments, BSi (diatom) and terrestrial/lacustrine biological productivity in the lake itself and in the lake surroundings.

(3) LOI analysis throughout Lake Danntjørn cores I and II revealed that the variations in residue after ignition of the sediments at 550 °C in the upper 260 cm (<8500 calendar years) were not associated with variations in magnetic susceptibility, but most probably diatom productivity (BSi), as demonstrated between 8500 and 7760 calendar years BP.

(4) From approximately 8500 calendar years BP up to the present, variations in the residue most likely primarily reflected variations in low-minerogenic (low magnetic susceptibility with little variations), diatom (BSi) productivity. The summer temperature variability has been inferred from variations in residue after ignition at 550 °C in the composite Lake Danntjørn lake sediment record, which has a mean time resolution of ~15 years per 0.5 cm sample thickness.

(5) The highest residue values (probably mainly reflect high diatom productivity due to high summer temperatures) occurred at 7800–7200, 7000, 6700, 6200, 5400–5050, 4300, 2500, 1800, 1550, 900, 700, and <100 calendar years BP. The most significant <8000 calendar years BP residue minima (probably mainly reflect low diatom productivity due to low summer temperatures) were recorded at 7100, 6800, 6600, 6000–5700, 5000–4400, 4200, 3800, 3600, 3200, 3000, 2750, 2300, 1900, 1700, 1070, 740, 600, 380, and 80 calendar years BP.

(6) The inferred summer temperature record from Jotunheimen exhibits many similarities with other terrestrial Holocene summer temperature reconstructions and temperature reconstructions based on surface ocean proxies, whereas oxygen-isotope records from planktonic foraminifera, which mainly live below the thermocline at a few hundred metres water depth, from the eastern Norwegian Sea (Risebrobakken et al., 2003) deviate significantly from the other proxies; especially, during the first-half of the Holocene.

(7) A comparison between the most likely climate forcing factors and the Holocene records presented in this paper, may suggest that the early Holocene thermal maximum observed in most records may have been caused by increased summer solar insolation to the Northern Hemisphere. However, abrupt and high-amplitude climate oscillations prior to ~8000 calendar years BP (including the 8.2 ka event) may have been triggered by abrupt, catastrophic lake drainage from ice-marginal lakes during the termination of the last ice age (Barber et al., 1999a b; Clarke et al., 2003, 2004; Nesje et al., 2004). Sub-orbital, millennial to decennial climate variability observed in the Holocene climate reconstructions, on the other hand, may have been a combined effect of changes in solar activity, periods of increased volcanic aerosols, and internal feedback mechanisms in the Earth's climate system.

Acknowledgements

Anne-Grete Bøe Pytte assisted with the summer coring in 2000. This project was part of the NOR-PAST-1 project funded by the Norwegian Research Council. The magnetic susceptibility measurements were carried out by Øyvind Paasche in Reidar Løvlie's laboratory at the Department of Earth Science, University of Bergen. The radiocarbon dates were provided by Tomasz Goslar at the Poznan Dating Laboratory in Poland through an agreement with the NORPEC Strategic University Programme led by John Birks, Department of Biology, University of Bergen. Richard Telford produced the age/depth model. To these persons and institutions we offer our sincere thanks. We express our gratitude to the two referees G. Rosqvist and D. Kaufmann, whose comments and suggestions greatly improved the clarity of the manuscript, the latter also providing the biogenic silica data from the Danntjørn lake sediments. This is publication Nr. A53 from the Bjerknes Centre for Climate Research.

References

Ahlmann, H.W., 1922. Glaciers in Jotunheimen and their physiography. Geografiska Annaler 4, 1–57.

Ahlmann, H.W., 1940. The Styggedal glacier in Jotunheimen, Norway. Geografiska Annaler 22, 95–130.

Alley, R.B., Mayewski, P.A., Sowers, T., Stuiver, M., Taylor, K.C., Clark, P.U., 1997. Holocene climate instability: a prominent, widespread event 8200 yr ago. Geology 25, 483–486.

Alverson, K.D., Bradley, R.S., Pedersen, T.F., 2003. Paleoclimate, Global Change and the Future, Global Change—The IGBP Series, 221pp.

Ammann, B., Birks, H.J.B., Brooks, S.J., et al., 2000. Quantification of biotic responses to rapid climatic changes around the Younger Dryas: a synthesis. Palaeogeography, Palaeoclimatology, Palaeoecology 159, 313–347.

Andersen, C., 2003. Surface ocean climate development and heat flux variability in the Nordic Seas and the subpolar North Atlantic during the Holocene. Dr. Scientiarum Thesis, Department of Earth Science, University of Bergen, Norway and Norwegian Polar Institute, Tromsø, Norway.

Andersen, C., Koç, N., Jennings, A., Andrews, T., 2004. Nonuniform response of the major surface currents of the Nordic Seas to insolation forcing: implications for the Holocene climate variability. Paleoceanography 19, PA2003.

Andersson, C., Risebrobakken, B., Jansen, E., Dahl, S.O., 2003. Late Holocene surface ocean conditions of the Norwegian Sea (Vøring Plateau). Paleoceanography 18, 1044.

Andreassen, L.M., Østrem, G., 1999. Storbresymposiet: 50 år med massebalansemålinger. Oslo: Norges vassdrags- og energidirektorat [Document 5, 1999].

Andrews, J.T., Giraudeau, J., 2003. Multiproxy records showing significant Holocene environmental variability: the inner N. Iceland shelf (Húnaflói). Quaternary Science Reviews 22, 175–193.

Aune, B., 1993. Temperaturnormaler. Det norske meteorologiske institutt. Rapport nr. 02/93, 63pp.

Baldini, J.U.L., McDermott, F., Fairchild, I.J., 2002. Structure of the 8200-year cold event revealed by a speleothem trace element record. Science 296, 2203–2206.

Barber, D.C., Dyke, A., Hillaire-Marcel, C., Jennings, A.E., Andrews, J.T., Kerwin, M.W., Bilodeau, G., McNeely, R., Southon, J., Morehead, M.D., Gagnon, J.-M., 1999a. Forcing of the cold event of 8200 years ago by catastrophic drainage of Laurentide lakes. Nature 400, 344–348.

Barber, K.E., Battarbee, R.W., Brooks, S., Eglinton, G., Haworth, E.Y., Oldfield, F., Stevenson, A.C., Thompson, R., Appleby, P.G., Austin, W.N., Cameron, N.G., Ficken, K.J., Golding, P., Harkness, D.D., Holmes, J., Hutchinson, R., Lishman, J.P., Maddy., D., Pinder, L.C.V., Rose, N., Stoneman, R., 1999b. Proxy records of climate change in the UK over the last two millennia: documented change and sedimentary records from lakes and bogs. Journal of the Geological Society of London 156, 369–380.

Barnekow, L., 2000. Holocene regional and local vegetation history and lake-level changes in the Torneträsk area, northern Sweden. Journal of Paleolimnology 23, 399–420.

Barnekow, L., Sandgren, P., 2001. Palaeoclimate and tree-line changes during the Holocene based on pollen and plant macrofossil records from six lakes at different altitudes in northern Sweden. Review of Palaeobotany and Palynology 117, 109–118.

Barnett, C.T., Dumayne-Peaty, L., Matthews, J.A., 2001. Holocene climatic change and tree-line response in Leirdalen, central Jotunheimen, south central Norway. Review of Palaeobotany and Palynology 117, 119–137.

Battarbee, R.W., 2000. Palaeolimnological approaches to climate change, with special regard to the biological record. Quaternary Science Reviews 19, 107–124.

Battarbee, R.W., Cameron, N.G., Golding, P., Brooks, S.J., Switsur, R., Harkness, D., Appleby, P., Oldfield, F., Thompson, R., Monteith, D.T., McGovern, A., 2001. Evidence for Holocene climate variability from the sediments of a Scottish remote mountain lake. Journal of Quaternary Science 16, 339–346.

Battarbee, R.W., Grytnes, J.-A., Thompson, R., Appleby, P.G., Catalan, J., Korhola, A., Birks, H.J.B., Heegaard, E., Lami, A., 2002. Comparing palaeolimnological and instrumental evidence of climate change for remote mountain lakes over the last 200 years. Journal of Paleolimnology 28, 161–179.

Bengtsson, L., Enell, M., 1986. Chemical analysis. In: Berglund, B.E. (Ed.), Handbook of Holocene Palaeoecology and Palaeohydrology. Wiley, Chisester, pp. 423–451.

Berger, A., Loutre, M.-F., 1996. Modeling the climate response to astronomical and CO_2 forcings. Geophysical External Climate and Environment, Comptes Rendus Academy of Science Paris 323 (IIa), 1.

Berglund, B.E., Barnekow, L., Hammarlund, D., Sandgren, P., Snowball, I., 1996. Holocene forest dynamics and climate changes in the Abisko area, northern Sweden—the Sonesson model of vegetation history reconsidered and confirmed. Ecological Bulletins 45, 15–30.

Bianchi, G.G., McCave, I.N., 1999. Holocene periodicity in North Atlantic climate and deep ocean flow south of Iceland. Nature 397, 515–517.

Bigler, C., Hall, R.I., 2003. Diatoms as quantitative indicators of July temperature: a validation attempt at century-scale with meteorological data from northern Sweden. Palaeography, Palaeoclimatology, Palaeoecology 189, 147–160.

Bigler, C., Larocque, I., Peglar, S.M., Birks, H.J.B., Hall, R.I., 2002. Quantitative multiproxy assessment of long-term patterns of Holocene environmental change from a small lake near Abisko, northern Sweden. The Holocene 12, 481–496.

Birks, C.J.A., Koc, N., 2002. A high-resolution diatom record of late Quaternary sea-surface temperatures and oceanographic conditions from the eastern Norwegian Sea. Boreas 31, 323–344.

Birks, H.H., 1991. Holocene vegetational history and climatic change in west Spitsbergen—plant macrofossils from Skardtjørna, an Arctic lake. The Holocene 1, 209–218.

Birks, H.H., Ammann, B., 2000. Two terrestrial records of rapid climatic change during the glacial-Holocene transition (14,000–9000 calendar years B.P.) from Europe. Proceedings of the National Academy of Science 97, 1390–1394.

Birks, H.H., Battarbee, R.W., Birks, H.J.B., 2000. The development of the aquatic ecosystem at Kråkenes Lake, western Norway, during the late-glacial and early Holocene—a synthesis. Journal of Paleolimnology 23, 91–114.

Birks, H.J.B., 1995. Quantitative palaeoenvironmental reconstructions. In: Maddy, D., Brew, J.S. (Eds.), Statistical Modelling of Quaternary Science Data. Quaternary Research Association, Cambridge, pp. 161–254.

Birks, H.J.B., 1998. Numerical tools in quantitative palaeolimnology—progress, potentialities, and problems. Journal of Paleolimnology 20, 301–332.

Birks, H.J.B., 2003. Quantitative palaeoenvironmental reconstructions from Holocene biological data. In: Mackay, A., Battarbee, R.W., Birks, H.J.B., Oldfield, F. (Eds.), Global Change in the Holocene. Arnold, London, pp. 107–122.

Blunier, T., Chappellaz, J., Schwander, J., Stauffer, J., Raynaud, D., 1995. Variations in atmospheric methane concentrations during the Holocene Epoch. Nature 374, 46.

Bond, G., Kromer, B., Beer, J., Muscheler, R., Evans, M.N., Showers, W., Hoffman, S., Lotti-Bond, R., Hajdas, I., Bonani, G., 2001. Persistent solar influence on North Atlantic climate during the Holocene. Science 294, 2130–2136.

Bond, G., Showers, W., Chesby, M., Lotti, R., Almasi, P., deMenocal, P., Priore, P., Cullen, H., Hajadas, I., Bonani, G., 1997. A pervasive millennial-scale cycle in North Atlantic Holocene and glacial climates. Science 278, 1257–1266.

Briffa, K.R., Jones, P.D., Bartholin, T.S., Eckstein, D., Schweingruber, F.H., Karlèn, W., Zetterberg, P., Eronen, M., 1992.

Fennoscandian summers from AD 500: temperature changes on short and long timescales. Climate Dynamics 7, 111–119.

Briffa, K.R., Jones, P.D., Schweingruber, F.H., Osborn, T.J., 1998. Influence of volcanic eruptions on Northern Hemisphere summer temperature over the past 600 years. Nature 393, 450–455.

Briffa, K.R., Osborn, T.J., Schweingruber, F.H., Jones, P.D., Shiyatov, S.G., Vaganov, E.A., 2002. Tree-ring width and density data around the Northern Hemisphere: Part 1. Local and regional climate signals. The Holocene 12, 737–757.

Brooks, S.J., 2003. Chironomid analysis to interpret and quantify Holocene climate change. In: Mackay, A., Battarbee, R.W., Birks, H.J.B., Oldfield, F. (Eds.), Global Change in the Holocene. Arnold, London, pp. 328–341.

Campbell, I.D., Campbell, C., Apps, M.J., Rutter, N.W., Bush, A.B.G., 1998. Late Holocene ~1500 yr climatic periodicities and their implications. Geology 26, 471–473.

Chambers, F.M., Charman, D.J., 2004. Holocene environmental change: contributions from the peatland archive. The Holocene 14, 1–6.

Chapman, M.R., Shackleton, N.J., 2000. Evidence of 550 year and 1000 year cyclicities in North Atlantic circulation patterns during the Holocene. The Holocene 10, 287–291.

Clarke, G.K.C., Leverington, D., Teller, J., Dyke, A., 2003. Super-lakes, megafloods, and abrupt climate change. Science 301, 922–923.

Clarke, G.K.C., Leverington, D.W., Teller, J., Dyke, A.S., 2004. Paleohydraulics of the last outburst flood from glacial Lake Agassiz and the 8200 BP cold event. Quaternary Science Reviews 23, 389–407.

Dahl, S.O., Bakke, J., Lie, Ø., Nesje, A., 2003. Reconstruction of former glacier equilibrium-line altitudes based on proglacial sites: an evaluation of approaches and selection of sites. Quaternary Science Reviews 22, 275–287.

Dahl, S.O., Nesje, A., 1992. Paleoclimatic implications based on equilibrium-line altitude depressions of reconstructed Younger Dryas and Holocene cirque glaciers in inner Nordfjord, western Norway. Palaeogeography, Palaeoclimatology, Palaeoecology 94, 87–97.

Dahl, S.O., Nesje, A., 1994. Holocene glacier fluctuations at Hardangerjøkulen, central-southern Norway: a high resolution composite chronology from lacustrine and terrestrial deposits. The Holocene 4, 269–277.

Dahl, S.O., Nesje, A., 1996. A new approach to calculating Holocene winter precipitation by combining glacier equilibrium-line altitudes and pine-tree limits: a case study from Hardangerjøkulen, central southern Norway. The Holocene 6, 381–398.

Dahl, S.O., Nesje, A., Lie, Ø., Fjordheim, K., Matthews, J.A., 2002. Timing, equilibrium-line altitudes and climatic implications of two early Holocene glacier readvances during the Erdalen Event at Jostedalsbreen, western Norway. The Holocene 12, 17–25.

Dahl-Jensen, D., Mosegaard, K., Gundestrup, N., et al., 1998. Past temperatures directly from the Greenland ice sheet. Science 282, 268–271.

Dansgaard, W., Johnsen, S.J., Clausen, H.B., Dahl-Jensen, D., Gundestrup, N.S., Hammer, C.U., Hvidberg, C.S., Steffensen, J.P., Sveinbjörnsdottir, A.E., Jouzel, J., Bond, G., 1993. Evidence for general instability of past climate from a 250-kyr ice-core record. Nature 364, 218–220.

Davis, B.A.S., Brewer, S., Stevenson, A.C., Guiot, J., and Data contributors, 2003. The temperature of Europe during the Holocene reconstructed from pollen data. Quaternary Science Reviews 22, 1701–1716.

Dean, W.E., 1974. Determination of carbonate and organic matter in calcareus sediments and sedimentary rocks by loss on ignition: comparison with other methods. Journal of Sedimentary Petrology 44, 242–248.

Eronen, M., Frenzel, B., Vorren, K.-D., (Eds.) 1993. Oscillations of the Alpine and Polar Tree Limits in the Holocene. Paleoclimate Research 9, 1–234.

Fisher, D.A., Koerner, R.M., 2003. Holocene ice-core climate history: a multi-variable approach. In: Alverson, K.D., Bradley, R.S., Pedersen, T.F. (Eds.), Paleoclimate, Global Change and the Future, Global Change—The IGBP Series, pp. 281–293.

Fronval, T., Jansen, E., 1997. Eemian and early Weichselian (140–60 ka) paleoceanography and paleoclimate in the Nordic seas with comparisons to Holocene conditions. Paleoceanography 12, 443–462.

Førland, E.J., 1993. Nedbørsnormaler. Det norske meteorologiske institutt, Rapport nr. 39/93 Klima, 63pp.

von rafenstein, U., von Erlenkeuser, H., Müller, J., Jouzel, J., Johnsen, S., 1998. The cold event 8200 years ago documented in oxygen isotope records of precipitation in Europe and Greenland. Climate Dynamics 14, 73–81.

Grootes, P.M., Stuiver, M., White, J.W.C., Johnsen, S., Jouzel, J., 1993. Comparison of oxygen isotope records from the GISP2 and GRIP Greenland ice cores. Nature 366, 552–554.

Grudd, H., Briffa, K.R., Karlén, W., Bartholin, T.S., Jones, P.D., Kromer, B., 2002. A 7400-year tree-ring chronology in northern Swedish Lapland: natural climatic variability expressed on annual to millennial timescales. The Holocene 12, 657–665.

Grøsfjeld, K., Larsen, E., Sejrup, H.P., de Vernal, A., Flatebø, T., Vestbø, M., Haflidason, H., Aarseth, I., 1999. Dinoflagellate cysts reflecting surface-water conditions in Voldafjorden, western Norway durin ghe last 11 300 years. Boreas 28, 403–415.

Gupta, A.K., Anderson, D.M., Overpeck, J.T., 2003. Abrupt changes in the Asian southwest monsoon during the Holocene and their links to the North Atlantic Ocean. Nature 421, 354–357.

Haakensen, N., 1989. Akkumulasjon på breene i Sør-Norge vinteren 1988–89. Været 13, 91–94.

Haflidason, H., Sejrup, H.P., Klitgaard-Kristensen, D., Johnsen, S., 1995. Coupled response of late glacial climatic shifts of northwest Europe reflected in Greenland ice cores: evidence from the northern North Sea. Geology 23, 1059–1062.

Hald, M., Dahlgren, T., Olsen, T.-E., Lebesbye, E., 2001. Late Holocene palaeooceanography in Van Mijenfjorden, Svalbard. Polar Research 20, 23–35.

Hammarlund, D., Barnekow, L., Birks, H.J.B., Buchardt, B., Edwards, T.W.D., 2002. Holocene changes in the oxygen-isotope stratigraphy of lacustrine carbonates from northern Sweden. The Holocene 12, 339–351.

Hammarlund, D., Björck, S., Buchardt, B., Israelson, C., Thomsen, C.T., 2003. Rapid hydrological changes during the Holocene revealed by stable isotope records from lacustrine carbonates from Lake Igelsjön, southern Sweden. Quaternary Science Reviews 22, 353–370.

Hannon, G.E., Bradshaw, R.H.W., Wastegård, S., 2003. Rapid vegetation change during the early Holocene in the Faroe Islands detected in terrestrial and aquatic ecosystems. Journal of Quaternary Science 18, 615–619.

Heikkilä, M., Seppä, H., 2003. A 11,000 yr palaeotemperature reconstruction from the southern boreal zone in Finland. Quaternary Science Reviews 22, 541–554.

Heiri, O., Lotter, A.F., Lemcke, G., 2001. Loss on ignition as a method for estimating organic and cabonate content in sediments: reprodicibility and comparability of results. Journal of Paleolimnolgy 25, 101–110.

Heiri, O., Lotter, A.F., Hausmann, S., Kienast, F., 2003. A chironomid-based Holocene summer air temperature reconstruction from the Swiss Alps. The Holocene 13, 477–484.

Helama, S., Lindholm, M., Timonen, M., Merilänen, J., Eronen, M., 2002. The supra-long Scots pine tree-ring record for Finnish Lapland: Part 2. Interannual to centennial variability in summer temperatures for 7500 years. The Holocene 12, 681–687.

Hjort, C., Mangerud, J., Adrielsson, L., Bondevik, S., Landvik, J.Y., Salvigsen, O., 1995. Radiocarbon dated common mussels *Mytilus edulis* from eastern Svalbard and the Holocene marine climatic optimum. Polar Research 14, 239–243.

Hoel, A., Werenskiold, W., 1962. Glaciers and Snowfields in Norway. Norsk Polarinstitutts Skrifter 126, 1–242.

Hurrell, J.W., Kushnir, Y., Ottesen, G., Visbeck, M., 2003. The North Atlantic oscillation—climatic significance and environmental impact. Geophysical Monograph 134, 279pp.

Husum, K., Hald, M., 2002. Early Holocene cooling events in Malangenfjord and the adjoining shelf, North-East Norwegian Sea. Polar Research 21, 267–274.

Håkanson, L., Jansson, M., 1983. Principles of Lake Sedimentology. Springer, Berlin.

Isaksen, K., Hauck, C., Gudevang, E., Ødegård, R.S., Sollid, J.L., 2002. Mountain permafrost distribution in Dovrefjell and Jotunheimen, southern Norway, based on BTS and DC resistivity tomography data. Norsk Geografisk Tidsskrift 56, 122–136.

Indermühle, A., Stocker, T.F., Joos, F., Fischer, H., Smith, H.J., Wahlen, M., Deck, B., Mastroianni, D., Tschumi, J., Blunier, T., Meyer, R., Stauffer, B., 1999. Holocene carbon-cycle dynamics based on CO_2 trapped in ice at Taylor Dome, Antarctica. Nature 398, 121–126.

Jansen, E., Koç, N., 2000. Century to decadal scale records of Norwegian Sea surface temperature variations of the past 2 millennia. PAGES/CLIVAR Newsletter 8 (1), 13–14.

Johnsen, S.J., Dahl-Jensen, D., Gundestrup, N., Steffensen, J.P., lausen, H.B., Miller, H., Masson-Delmotte, V., Sveinbjørnsdottir, A.E., 2001. Oxygen isotope and paleotemperature records from six Greenland ice-core stations: Camp Century, Dye-3, GRIP, GISP2, Renland and NorthGRIP. Journal of Quaternary Science 16, 299–307.

Kalela-Brundin, M., 1999. Climatic information from tree-rings of *Pinus sylvestris* L. and reconstruction of summer temperatures back to AD 1500 in Femundsmarks, eastern Norway, using partial least squares regression (PLS) analysis. The Holocene 9, 59–77.

Kaplan, M.R., Wolfe, A.P., Miller, G.H., 2002. Holocene environmental variability in southern Greenland inferred from lake sediments. Quaternary Research 58, 149–159.

Karlén, W., 1976. Lacustrine sediments and tree-limit variations as indicators of Holocene climatic fluctuations in Lapland: Northern Sweden. Geografiska Annaler 58A, 1–34.

Karlén, W., 1988. Scandinavian glacier and climatic fluctuations during the Holocene. Quaternary Science Reviews 7, 199–209.

Karlén, W., Bodin, A., Kuylenstierna, J., Näslund, J.-O., 1995. Climate of northern Sweden during the Holocene. Journal of Coastal Research Special Issue 17: Holocene Cycles: Climate, Sea Levels, and Sedimentation, 49–54.

Karlén, W., Kuylenstierna, J., 1996. On solar forcing of Holocene climate: evidence from Scandinavia. The Holocene 6, 359–365.

Karlén, W., Matthews, J.A., 1992. Reconstructing Holocene glacier variations from glacial lake sediments: studies from Nordvestlandet and Jostedalsbreen-Jotunheimen, southern Norway. Geografiska Annaler 74, 327–348.

Kirchhefer, A.J., 2001. Reconstruction of summer temperatures from tree-rings of Scots pine (*Pinus sylvestris* L.) in coastal northern Norway. The Holocene 11, 41–52.

Klitgaard-Kristensen, D., Sejrup, H.-P., Haflidason, H., 2001. The last 18 kyr fluctuations in Norwegian Sea surface conditions and implications for the magnitude of climatic change: evidence from the North Sea. Paleoceanography 16, 455–467.

Klitgaard-Kristensen, D., Sejrup, H.-P., Haflidason, H., Johnsen, S., Spurk, M., 1998. A regional 8200 calendar years BP cooling event in northwest Europe, induced by final stages of the Laurentide ice-sheet deglaciation? Journal of Quaternary Science 13, 165–169.

Koç, N., Jansen, E., 2002. Holocene climate evolution of the North Atlantic Ocean and the Nordic Seas—a synthesis of new results. In: Wefer, G., Berger, W.H., Behre, K.E., Jansen, E. (Eds.), Climate and History in the North Atlantic Realm. Springer, Berlin, Heidelberg, pp. 163–165.

Koc, N., Jansen, E., Haflidason, H., 1993. Paleoceanographic reconstruction of surface ocean conditions in the Greenland, Iceland and Norwegian Seas through the last 14 ka based on diatoms. Quaternary Science Reviews 12, 115–140.

Korhola, A., Vasko, K., Toivonen, H.T.T., Olander, H., 2002. Holocene temperature changes in northern Fennoscandia reconstructed from chironomids using Bayesian modelling. Quaternary Science Reviews 21, 1841–1860.

Kullman, L., 1981. Recent tree-limit dynamics of Scots pine (*Pinus sylvestris* L.) in the southern Swedish Scandes. Wahlenbergia 8, 1–67.

Kullman, L., 1995. Holocene tree-limit and climate history from the Scandes Mountains, Sweden. Ecology 768, 2490–2502.

Körner, C., 1998. A re-assessment of high elevation treeline positions and their explanations. Oecologia 115, 445–459.

Laumann, T., Reeh, N., 1993. Sensitivity to climate change of the mass balance of glaciers in southern Norway. Journal of Glaciology 39, 656–665.

Lauritzen, S.-E., 1996. Calibration of speleothem stable isotopes against historical records: a Holocene temperature curve for north Norway. In: Climate Change: the Karst Record. Karst Waters Institute Special Publication 2, Charles Town, West Virginia, pp. 78–80.

Lauritzen, S.-E., Lundberg, J., 1999. Calibration of the speleothem delta function: an absolute temperature record for the Holocene in northern Norway. The Holocene 9, 659–669.

Leeman, A., Niessen, F., 1994. Holocene glacial activity and climatic variations in the Swiss Alps: reconstructing a continuous record from proglacial lake sediments. The Holocene 4, 259–268.

Lehman, S.J., Jones, G.A., Keigwin, L.D., Andersen, E.S., Butenko, G., Østmo, S.-R., 1991. Initiation of Fennoscandian ice-sheet retreat during the last deglaciation. Nature 349, 513–516.

Leng, M.L., Marshall, J.D., 2004. Palaeoclimatic interpretation of stable isotope data from lake sediment archives. Quaternary Science Reviews 23, 811–831.

Lewis, W.V., (Ed.), 1960. Norwegian Cirque Glaciers, RGS Research Series No. 4, Royal Geographical Society, London.

Livingstone, D.M., 1997. Break-up dates of Alpine lakes as proxy data for local and regional mean surface air temperatures. Climatic Change 37, 407–439.

Lotter, A.F., Birks, H.J.B., 2003. Holocene sediments of Sägistalsee, a small lake at the present-day tree-line in the Swiss Alps. Journal of Paleolimnology 30, 253–260.

Lotter, A.F., Birks, H.J.B., Hofmann, W., Marchetto, A., 1997. Modern diatom, cladocera, chronomid, and chrysophyte cyst assemblages as quantitative indicators for the reconstruction of past environmental conditions in the Alps. I. Climate. Journal of Paleolimnology 18, 395–420.

Marchal, O., Stocker, T.F., Joos, F., Indermühle, A., Blunier, T., Tschumi, J., 1999. Modelling the concentration of atmospheric CO_2 during the Younger Dryas climate event. Climate Dynamics 15, 341–354.

Matthews, J.A., 1974. Families of the lichenometric dating curves from the Storbreen gletschervorveld, Jotunheimen, Norway. Norsk Geografisk Tidsskrift 28, 215–235.

Matthews, J.A., Dahl, S.O., Berrisford, M.S., Nesje, A., Dresser, P.Q., Dumayne-Peaty, L., 1997. A preliminary history of Holocene colluvial (debris-flow) activity, Leirdalen, Jotunheimen. Journal of Quaternary Science 12, 117–129.

Matthews, J.A., Karlén, W., 1992. Asynchronous Neoglaciation and Holocene climatic change reconstructed from Norwegian glacio-lacustrine sedimentary sequences. Geology 20, 991–994.

Matthews, J.A., Dahl, S.O., Nesje, A., Berrisford, M.S., Andersson, C., 2000. Holocene glacier variations in central Jotunheimen, southern Norway based on distal glaciolacustrine sediment cores. Quaternary Science Reviews 19, 1625–1647.

McCarroll, D., Loader, N.J., 2004. Stable isotopes in tree rings. Quaternary Science Reviews 23, 771–801.

Meyers, P.A., Teranes, J.L., 2001. Sediment organic matter. In: Last, W.M., Smol, J.P. (Eds.), Tracking Environmental Change Using Lake Sediments. Physical and Geochemical Methods, vol. 2. Kluwer Academic Publishers, Dordrecht, The Netherlands, pp. 239–269.

Mikalsen, G., Serjup, H.P., Aarseth, I., 2001. Late-Holocene changes in ocean circulation and climate: foraminiferal and isotopic evidence from Sulafjord, western Norway. The Holocene 11, 437–466.

Mortlock, R.A., Froelich, P.N., 1989. A simple method for the rapid determination of biogenic opal in pelagic marine sediments. Deep-Sea Research 36, 1415–1426.

Neftel, A., Oeschger, H., Staffelbach, T., Stauffer, B., 1988. CO_2 records in the Byrd ice core 50,000–5000 years BP. Nature 331, 609–611.

Nesje, A., 1992. A piston corer for lacustrine and marine sediments. Arctic and Alpine Research 24, 257–259.

Nesje, A., Dahl, S.O., 2001. The Greenland 8200 calendar years BP event detected in loss-on-ignition profiles in Norwegian lacustrine sediment sequences. Journal of Quaternary Science 16, 155–166.

Nesje, A., Dahl, S.O., Andersson, C., Matthews, J.A., 2000. The lacustrine sedimentary sequence in Sygneskardvatnet, western Norway: a continuous, high-resolution record of the Jostedalsbreen ice cap during the Holocene. Quaternary Science Rewiews 19, 1047–1065.

Nesje, A., Dahl, S.O., Bakke, J., 2004. Were abrupt Lateglacial and early Holocene climatic changes in northwest Europe linked to freshwater outbursts to the North Atlantic and Arctic Oceans? The Holocene 14, 299–310.

Nesje, A., Dahl, S.O., Løvlie, R., 1995. Late Holocene glacier and avalanche activity in the Ålfoten area, western Norway: evidence from a lacustrine sedimentary record. Norsk Geologisk Tidsskrift 75, 120–126.

Nesje, A., Dahl, S.O., Løvlie, R., Sulebak, J., 1994. Holocene glacier activity at the southern side of Hardangerjøkulen, central southern Norway; evidence from lacustrine sediments. The Holocene 4, 377–382.

Nesje, A., Kvamme, M., Rye, N., Løvlie, R., 1991. Holocene glacial and climate history of the Jostedalsbreen region, western Norway; evidence from lake sediments and terrestrial deposits. Quaternary Science Reviews 10, 87–114.

Nesje, A., Matthews, J.A., Dahl, S.O., Berrisford, M.S., Andersson, C., 2001. Holocene glacier fluctuations of Flatebreen and winter precipitation changes in the Jostedalsbreen region, western Norway, based on glaciolacustrine records. The Holocene 11, 267–280.

Nesje, A., Thun, T., Bjune, A., Larsen, J., in preparation. Organic content in lake sediments as a proxy of summer temperature: a comparison between loss-on-ignition in sediments from a small mountain lake in western Norway and a tree-ring width series from Trøndelag, mid Norway.

Nordli, P.Ø., Lie, Ø., Nesje, A., Dahl, S.O., 2003. Spring–summer temperature reconstruction in western Norway 1734–2003: a data-synthesis approach. International Journal of Climatology 23, 1821–1841.

O'Brian, S.R., Mayewski, P.A., Meeker, L.D., Meese, D.A., Twickler, M.S., Whitlow, S.I., 1995. Complexity of Holocene climate as reconstructed from a Greenland ice core. Science 270, 1962–1964.

Oppo, D., McManus, J.F., Cullen, J.L., 2003. Deepwater variability in the Holocene epoch. Nature 422, 277–278.

O'Sullivan, P.E., Moyeed, R., Cooper, M.C., Nicholson, M.J., 2002. Comparison between instrumental, observational and high resolution proxy sedimentary records of Late Holocene climatic change—a discussion of possibilities. Quaternary International 88, 27–44.

Østrem, G., 1964. Ice-cored moraines in Scandinavia. Geografiska Annaler 46 (A), 282–337.

Østrem, G., 1965. Problems of dating ice-cored moraines. Geografiska Annaler 47 (A), 1–38.

Østrem, G., Dale Selvig, K., Tandberg, K., 1998. Atlas over breer i Sør-Norge. Norges vassdrags-og energiverk. Vassdragsdirektoratet. Meddelselse nr. 61 fra Hydrologisk avdeling.

Øyen, P.A., 1893. Isbræstudier i Jotunheimen. Nyt Magazin for Naturvidenskaberne 34, 26–27.

Øyen, P.A., 1908. Bidrag til vore bræegnes glacialgeologi. Nyt Magazin for Naturvidenskaberne 46, 301–379.

Pfister, C., Brázdil, R., Glaser, R. (Eds.), 1999. Climatic Variability in Sixteenth Century Europe and its Social Dimension, Kluwer, Dordrect.

Renberg, I., 1991. The HON-Kajak sediment corer. Journal of Paleolimnology 6, 167–170.

Risebrobakken, B., Jansen, E., Andersson, C., Mjelde, E., Hevrøy, K., 2003. A high-resolution study of Holocene paleoclimatic and paleocenographic changes in the Nordic Seas. Paleoceanography 18, 1017.

Rosén, P., Segerström, U., Eriksson, L., Renberg, I., Birks, H.J.B., 2001. Holocene climatic change reconstructed from diatoms, chironomids, pollen and near-infrared spectroscopy at an alpine lake (Sjuodjijaure) in northern Sweden. The Holocene 11, 551–562.

Rosqvist, G., Jonsson, C., Yam, R., Karlén, W., Shemesh, A., 2004. Diatom oxygen isotopes in pro-glacial lake sediments from northern Sweden: a 5000 year record of atmospheric circulation. Quaternary Science Reviews 23, 851–859.

Rubensdotter, L., Rosqvist, G., 2003. The effect of geomorphological setting on Holocene lake sediment variability, northern Swedish Lapland. Journal of Quaternary Science 18, 757–767.

Salvigsen, O., 2002. Radiocarbon-dated *Mytilus edulis* and *Modiolus modiolus* from northern Svalbard: climatic implications. Norsk Geografisk Tidsskrift 56, 56–61.

Salvigsen, O., Forman, S.L., Miller, G.H., 1992. Thermophilous molluscs on Svalbard during the Holocen and their paleoclimatic implications. Polar Research 11, 1–10.

Sandgren, P., Risberg, J., 1990. Magnetic mineralogy of the lakes in lake Ådran, eastern Sweden, and an interpretation of early Holocene water level changes. Boreas 19, 57–68.

Sandvold, S., Lie, Ø., Nesje, A., Dahl, S.O., 2001. Holocene glacial and colluvial activity in Leirungsdalen, eastern Jotunheimen, south central Norway. Norsk Geologisk Tidsskrift 81, 25–40.

Sarnthein, M., van Kreveld, S., Erlenkeuser, H., Grootes, P.M., Kucera, M., Pflaumann, U., Schulz, M., 2003. Centennial-to-millennial-scale periodicities of Holocene climate and sediment injections off the western Barents shelf, 75oN. Boreas 32, 447–461.

Schindler, D.W., Bayley, S.E., Parker, B.R., Beaty, K.G., Cruikshak, D.R., Fee, E.J., Schindler, E.U., Stainton, M.P., 1996. The effects of climate warming on the properties of boreal lakes and streams at the Experimental Lakes Area, northwestern Ontario. Limnology and Oceanography 41, 1004–1017.

Schweingruber, F.H., Briffa, K.R., 1996. Tree-ring density networks for climate reconstruction. In: Jones, P.D., Bradley, R.S., Jouzel, J. (Eds.), Climatic Variations and Forcing Mechanisms of the Last 2000 years, NATO ASI Series, vol. 141. Springer, Berlin, pp. 43–66.

Seierstad, J., Nesje, A., Dahl, S.O., Riis Simonsen, J., 2002. Holocene glacier fluctuations of Grovabreen and Holocene snow-avalanche activity reconstructed from lake sediments in Grøningstølsvatnet, western Norway. The Holocene 12, 211–222.

Seppä, H., Birks, H.J.B., 2001. July mean temperature and annual precipitation trends during the Holocene in the Fennoscandian tree-line area: pollen-based climate reconstructions. The Holocene 11, 527–539.

Seppä, H., Birks, H.J.B., 2002. Holocene climate reconstructions from the Fennoscandian tree-line area based on pollen data from Toskaljarvi. Quaternary Research 57, 191–199.

Seppä, H., Birks, H.H., Birks, H.J.B., 2002. Rapid climatic changes during the Greenland stadial 1 (Younger Dryas) to early Holocene transition on the Norwegian Barents Sea coast. Boreas 31, 215–225.

Shemesh, A., Charles, C.D., Fairbanks, R.G., 1992. Oxygen isotopes in biogenic silica: global changes in ocean temperature and isotopic composition. Science 256, 1434–1436.

Shemesh, A., Rosqvist, G., Rietti-Shati, M., Rubensdotter, L., Bigler, C., Yam, R., Karlén, W., 2001. Holocene climate change in Swedish Lapland inferred from an oxygen isotope record of lacustrine biogenic silica. The Holocene 11, 447–454.

Sigmond, E.M.O., Gustavson, M., Roberts, D., 1984. Berggrunnskart over Norge (Bedrock map of Norway) M. 1:1 million. Norges geologiske undersøkelse.

Sletten, K., Blikra, L.H., Ballantyne, C.K., Nesje, A., Dahl, S.O., 2003. Holocene debris flows recognized in a lacustrine sedimentary succession: sedimentolgy, chronostratigraphy and cause of triggering. The Holocene 13, 907–920.

Snowball, I., 1993. Mineral magnetic properties of Holocene lake sediments and soils from the Kårsa valley, Lappland, Sweden, and their relevance to palaeomagnetic reconstruction. Terra Nova 5, 258–270.

Snowball, I., Sandgren, P., 1996. Lake sediment studies of Holocene glacial activity in the Kårsa valley, northern Sweden: contrast in interpretation. The Holocene 6, 367–372.

Snowball, I., Thompson, R., 1990. A mineral magnetic study of Holocene sedimentation in Lough Catherine, Northern Ireland. Boreas 19, 127–146.

Snowball, I., Sandgren, P., Petterson, G., 1999. The mineral magnetic properties of an annually laminated Holocene lake sediment sequence in northern Sweden. The Holocene 9, 353–362.

Snyder, J.A., Werner, A., Miller, G.H., 2000. Holocene cirque glacier activity in western Spitsbergen, Svalbard: sediment redords from proglacial Linnévatnet. The Holocene 10, 555–563.

Sohlenius, G., 1996. Mineral magnetic properties of late Weichselian-Holocene sediments from the northwestern Baltic Proper. Boreas 25, 79–88.

Solignac, S., de Vernal, A., Hillaire-Marcel, C., 2004. Holocene sea-surface conditions in the North Atlantic—contrasted trends and regimes in the western and eastern sectors (Labrador Sea vs. Iceland Basin). Quaternary Science Reviews 23, 319–334.

Staffelbach, T., Stauffer, B., Sigg, A., Oeschger, H., 1991. CO_2 measurements from polar ice cores—more data from different sites. Tellus Series B—Chemical and Physical Meteorology 43, 91–96.

Stockhausen, H., Zolitschka, B., 1999. Environmental changes since 13,000 cal BP reflected in magnetic and sedimentological properties of sediments from Lake Holzmaar (Germany). Quaternary Science Reviews 18, 913–925.

Stuiver, M., Braziunas, T.F., Grootes, P.M., Zielinski, G.A., 1997. Is there evidence for solar forcing of climate in the GISP2 oxygen isotope record? Quaternary Research 48, 259–266.

Stuiver, M., Grootes, P.M., Braziunas, T.F., 1995. The GISP2 $\partial^{18}O$ climate record of the past 16,500 years and the role of the sun, ocean, and volcanoes. Quaternary Research 44, 341–354.

Stuiver, M., Reimer, P.J., Bard, E., Beck, J.W., Burr, G.S., Hughen, K.A., Kromer, B., McCormac, G., VanderPlicht, J., Spurk, M., 1998. INTCAL98 radiocarbon age calibration, 24,000-0 cal. BP. Radiocarbon 40, 1041–1083.

Sutherland, R.A., 1998. Loss on ignition estimates of organic matter and relationships to organic carbon in fluvial bed sediments. Hydrobiologia 389, 153–167.

Svendsen, J.-I., Mangerud, J., 1997. Holocene glacial and climatic variations on Spitsbergen, Svalbard. The Holocene 7, 45–57.

ter Braak, C.J.F., Juggins, S., 1993. Weighted averaging partial least squares regression (WA-PLS): an improved method for reconstructing environmental variables from species assemblages. Hydrobiologia 269/270, 485–502.

Thompson, R., Oldfield, F., 1986. Environmental Magnetism. George Allen and Unwin, London.

Thompson, R., Batterbee, R.W., O'Sullivan, P.E., Oldfield, F., 1975. Magnetic susceptibility of lake sediments. Limnology and Oceanography 20, 687–698.

Thun, T., 2002. Dendrochronological constructions of Norwegian conifer chronologies providing dating of historical material. Doctoral Philosophical Thesis, Department of Biology, Faculty of Natural Sciences and Technology, Norwegian University of Science and technology (NTNU), 336pp.

Weliky, K., Suess, E., Ungerer, C.A., Müller, P.J., Fischer, K., 1983. Problems with accurate carbon measurements in marine sediments and particulate organic matter in sea water: a new apprach. Limnology and Oceanography 28, 1252–1259.

Wetzel, R.G., 2001. Limnology: Lake and River Ecosystems, third ed. Academic Press, San Diego, CA.

Willemse, N.W., Törnquist, T.E., 1999. Holocene century-scale temperature variability from West Greenland lake records. Geology 27, 580–584.

Winkler, S., Matthews, J.A., Shakesby, R.A., Dresser, P.Q., 2003. Glacier variations in Breheimen, southern Norway: dating Little Ice Age moraine sequences at seven low-altitude glaciers. Journal of Quaternary Science 18, 395–413.

Zielinski, G.A., Mayewski, P.A., Meeker, L.D., et al., 1994. Record of explosive volcanism since 7000 B.C. from the GISP2 Greenland ice core and implications for the volcano-climate system. Science 264, 948–952.

Zielinski, G.A., Mershon, G.R., 1997. Paleoenvironmental implications of the insoluble microparticle record in the GISP2 (Greenland) ice core during the rapidly changing climate of the Pleistocene–Holocene transition. Geological Society of America Bulletin 109, 547–559.

Quaternary Science Reviews 23 (2004) 2207–2218

Synchronous Holocene sea surface temperature and rainfall variations in the Asian monsoon system

S.J.A. Jung*, G.R. Davies, G.M. Ganssen, D. Kroon

Vrije Universiteit Amsterdam, de Boelelaan 1085, 1081 HV Amsterdam, The Netherlands

Abstract

An increasing number of high-resolution paleoclimate records show substantial natural variation during the Holocene. In order to improve climate projections on human lifetime, the processes that potentially control teleconnections between different parts of the climate system need to be understood. A highly suitable area to study these processes is the Asian monsoon system, as it is one of the most dynamic climate systems on Earth and largely controls climate in Asia and the Indo-W-Pacific realm. Here, we present a Holocene stable O-isotope record from the summer-dwelling planktic foraminifer *G. bulloides* in Core 905 off Somalia. Initially dated by the radiocarbon method the record was tuned to atmospheric ^{14}C-variations without violating the radiocarbon dates. The O-isotope variations in Core 905 imply monsoon-controlled Holocene average summer sea surface temperature variations of up to 2–2.5 °C within decades to centuries. Other proxy records from marine and continental settings representing key locations in the Asian monsoon system were compared with the record of Core 905. Within the resolution of the individual age models, Holocene monsoonal records vary in concert for the Arabian Sea, continental China and the South China Sea and imply simultaneous warm/wet and cool/dry alternations in the Asian monsoon within centuries to decades. Two potential scenarios are discussed to explain the coherent monsoonal records. Scenario one involves trade wind-induced temperature variations in the W-Pacific warm pool as the driving force. The second scenario advocates the controlling process to synchronous variations in trade wind strength simultaneously occurring over the Pacific and Indian Ocean.
© 2004 Elsevier Ltd. All rights reserved.

1. Introduction

In recent years progress has been made in documenting a number of multi-annual oscillatory climate systems, e.g. the ENSO (El Nino – Southern oscillation) (Cobb et al., 2003; Philander and Fedorov, 2003; Trenberth and Hoar, 1997). On slightly longer decadal time scales other oscillatory systems, e.g. the North Atlantic Oscillation (NAO) or the Pacific Decadal scale Oscillation (PDO) (Biondi et al., 2001; Hurrel, 1995 and references therein), contribute significantly to the observed climate variations on Earth. A similarly dynamic and large-scale system is the Asian monsoon, reaching from the Arabian Sea into the W-Pacific

Ocean. It directly affects the socio-economic development of Asia through the monsoonal rainfall.

Numerous studies have documented that the long-term (10^5 years) climate variations of the Asian monsoon on orbital time scales are largely controlled by variations in the incoming solar radiation, due to modulations of the Earth's orbital parameters, e.g. (Anderson and Prell, 1993; Clemens et al., 1991; Morley and Heusser, 1997). On the decadal-century scale, climate variations in the Asian monsoon system also have been ascribed to variations in solar insolation (probably resulting from variations in the output of solar insolation or related to sun spot activity) (Gupta et al., 2003; Jung et al., 2002a; Neff et al., 2001—see also Sirocko, 1994; and Overpeck et al., 1996). Little is known, however, about the teleconnections involved in linking different climate compartments in the monsoon area on a decadal-century time scale. Here, we

*Corresponding author. Tel.: +31-20-444-7424; fax: +31-20-646-6257.

E-mail address: jung@geo.vu.nl (S.J.A. Jung).

0277-3791/$ - see front matter © 2004 Elsevier Ltd. All rights reserved.
doi:10.1016/j.quascirev.2004.08.009

reconstruct a regional Holocene monsoon record at a decadal resolution off Somalia and compare this record with available high-resolution Holocene records representing key locations of the Asian monsoon system. This comparison shows that within the resolution of the individual age models, the Asian monsoon system witnessed coherent alternations of warm/wet and cool/dry conditions at a decadal-century time scale.

2. Methodological approach

Piston Core 905 was retrieved from a water depth of 1580 m offshore Somalia during the Netherlands Indian Ocean Program (NIOP) in 1993 (Fig. 1). Based on an initial age model (Jung et al., 2001) and results of the

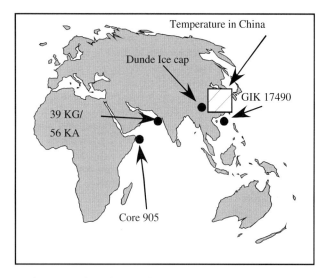

Fig. 1. Location of new and published sites used in this study.

lower Holocene section of Core 905 (Jung et al., 2002a) the upper 150 cm were sampled at 0.5 cm steps in order to complete the Holocene section. Due to the very high Holocene sedimentation rate of roughly 30 cm/ka this sampling strategy resulted in an age difference between neighbouring samples of roughly 20–30 years. Subsequent to standard sample preparation (washing, sieving), 5 specimens of the planktic foraminifer species *G. bulloides* were picked for stable O-isotope analysis. *G. bulloides* dwells during the summer period in the upwelling area form the western Arabian Sea and records sea surface temperatures late in the upwelling cycle (Kroon and Ganssen, 1989; Peeters et al., 2002). Two to five replicate analyses (average ~3) per sample were performed. The replicate analysis strategy was designed to minimize any noise introduced by natural variation within a sample due to short-term oceanographic variations and the effect of bioturbation.

The age model of the Holocene section of Core 905 is based on 27 AMS ^{14}C-dates (Jung et al., 2001, 2002a, b). Initially, we corrected the dates for the marine reservoir effect (Table 1) of 600 years (von Rad et al., 1999; Staubwasser et al., 2002). We felt, however, that the age model could be further improved by correlating the stable O-isotope record of *G. bulloides* to the δ^{18}O-record of a nearby stalagmite that has been absolutely dated using U-series (Neff et al., 2001). In a previous study it was clear that the Hoti Cave δ^{18}O-record revealed striking similarities with the early Holocene δ^{18}O-record of Core 905 and, within the tolerance of the AMS ^{14}C-dates, allowed a near-perfect match (Fig. 2; (Jung et al., 2002a)). Neff et al. (2001) demonstrated that there is a close relation between the monsoon-controlled rainfall in Oman and the ^{14}C-concentration in the atmosphere. Consequently, this strategy indirectly

Table 1a
Age control Late Holocene[a]

| Depth (cm) | AMS ^{14}C-dates | | | | Main tie points used for tuning the Late Holocene section | | |
	AMS ^{14}C-dates	Calendar age (2 sigma) min (avg) max	Cal. age in tuned age model		Tie point in Fig. 3a	Interpolated Cal. age Based on AMS ^{14}C-dates	Tuned age of tie points
0.25	327	(assumed) −43	−43				
10	935	272 (398) 485	390				
22.75					1	990	1030
26.25	1761	961 (1122) 1254	1230				
30.25					2	1350	1390
40.25	2420	1654 (1818) 1978	1865				
44.25					3	2140	2170
53.75					4	2670	2660
80.25	4050	3622 (3815) 3979	3805				
91.25					5	4390	4380
111.25					6	5270	5260
120.25	5410	5440 (5578) 5708	5610				
126.25					7	5910	5870

[a]Summary of the radiocarbon dates for the late Holocene and the tie points used to graphically tune the O-isotope record of *G. bulloides* from Core 905 to the atmospheric Δ^{14}C-record (Intcal 98 data set (Stuiver et al., 1998)).

Table 1b
Age control Early Holocene (see Jung et al., 2002)[b]

Depth (cm)	AMS ^{14}C-dates		
	AMS ^{14}C-dates	Calendar age (2 sigma) min (avg) max	Cal. age in tuned age model
167.25	6653 (based on 6 dates)	6745 (6931) 7147	7078
183.75	7125 (based on 6 dates)	7229 (7422) 7557	7391
250.25	8620 (based on 3 dates)	8800 (8919) 9278	8986
291.25	9748 (based on 6 dates)	9865 (10287) 11090	10287

[b]Summary of radiocarbon dated intervals for the Early Holocene (details see Jung et al., 2002).

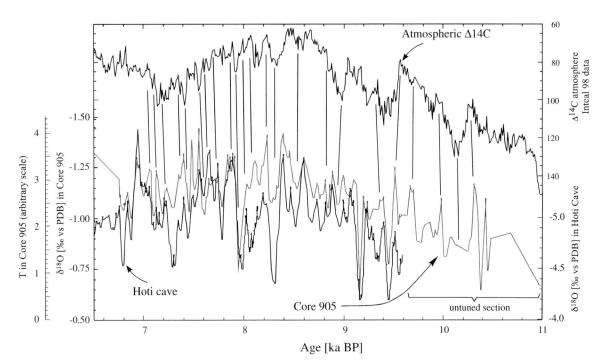

Fig. 2. The stable O-isotope record of *G. bulloides* from Core 905 and from a stalagmite from Hoti Cave (Neff et al., 2001) based on the age model used for the Hoti Cave record. A 50-year moving averaging filter was applied to both records in order to improve visual comparison. The tree-ring derived atmospheric Δ^{14}C-record (Intcal 98 data set (Stuiver et al., 1998)) illustrates the close correlation between variations in solar radiation and the terrestrial and marine monsoon records. The potential temperature variation implied by the δ^{18}O-record of Core 905 was calculated assuming a ratio of 0.23‰/ °C (Erez and Luz, 1983).

links the O-isotope variations in Core 905 and the decadal-scale variations in solar insolation (details see Jung et al., 2002a). In the present study, we extend the same strategy to the upper Holocene section of Core 905 and cross correlate the δ^{18}O-record to the variation of the ^{14}C-concentration in the atmosphere (Stuiver et al., 1998). Fig. 3a illustrates the main tie points used to tune both records and shows that the cross-tuned (revised) age estimates for Core 905 fall within the tolerance of the AMS ^{14}C-based chronology. Fig. 3b shows a near-perfect match between variations in the atmospheric Δ^{14}C and the δ^{18}O-changes in Core 905. This kind of tuning exercise is useful because marine ^{14}C-ages suffer from variable marine reservoir effects through time (Sarnthein et al., 2002 and references

therein) and hampers accurate chronology at a decadal resolution. Tuning the δ^{18}O-record of *G. bulloides* to the atmospheric Δ^{14}C curve reveals that deviations from the original age model based on AMS ^{14}C-dates (Table 1a, b) are in the order of decades and only in one case more than 100 years.

Before placing the results of Core 905 into the context of large-scale variations in the Asian monsoon system, we must assess the effect of bioturbation on the paleo-records. Bioturbation of sediments resembles a low-pass filter (e.g. Trauth et al., 1997). In general, the higher sedimentation rate will improve the potential of high-frequency climate change to be preserved, particularly if bioturbation depth is constant. The absolute effect of bioturbation on the preserved record of Core 905 is

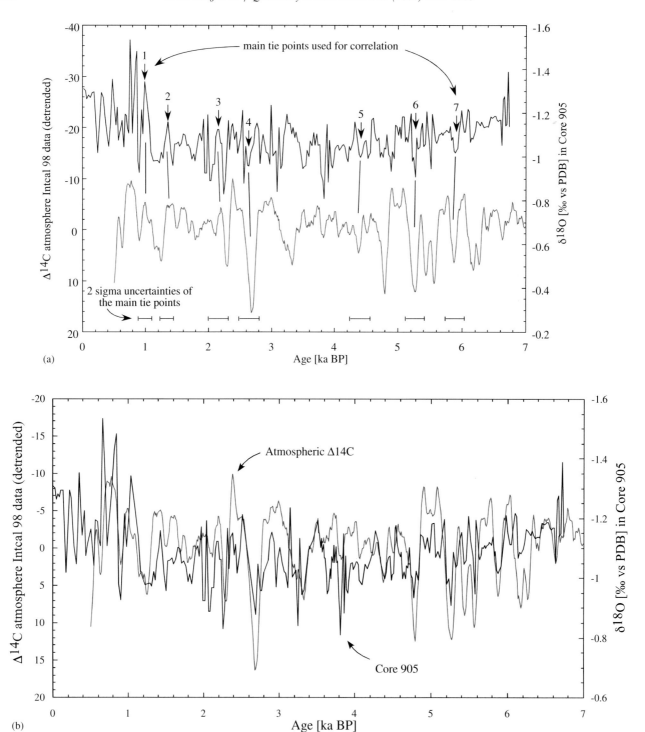

Fig. 3. Correlation of the late Holocene O-isotope record of *G. bulloides* to $\Delta^{14}C$ in the atmosphere (Stuiver et al., 1998). In order to improve visual comparability, both records were smoothed by a 50-year running mean filter. The $\Delta^{14}C$-record was accordingly smoothed and detrended. Black lines in (a) illustrate the main correlation tie points used. (b) shows the resulting correlation. We have shown previously that the uncertainties of the AMS ^{14}C-dates for the Holocene vary. In order to illustrate the age tolerance of the tie points in the O-isotope record the theoretical 2-sigma error bars were calculated based on a linear interpolation between the neighbouring AMS-^{14}C-dated samples (compare Table 1). These 2-sigma age error bars include a 200-year uncertainty in the marine reservoir age (see Jung et al. (2002a) for details of the argument).

difficult to assess. Given the high Holocene sedimentation rate of ~30 cm/kyr we believe it is conservative to assume that its influence is minor. In order to assess the variability in $\delta^{18}O$ recorded by each sample, the

standard deviation was calculated. Fig. 4 shows that for the Holocene there is significant variation, in particular on centennial time scales, although some of the $\delta^{18}O$-variation is within error. This finding suggests

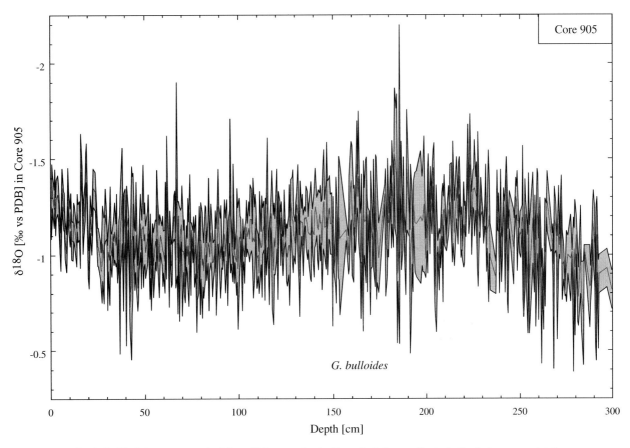

Fig. 4. Stable O-isotope record of Core 905 versus depth. Grey shaded areas illustrate the 1-sigma uncertainty.

that a substantial part of the variability shown reflects the natural variability within the time period covered by each sample. Accordingly, we assume that the small δ^{18}O-variations, although clearly close to the detection limit of the method, reflect the natural variability off Somalia.

To evaluate whether the variation documented off Somalia represents a local or Asian monsoon-wide signature we compared the record of Core 905 with results from sites that encompass regional climate variations within the Asian monsoon system on a decadal to century-scale. In order to do so, we compared the record of Core 905 with a sea surface temperature record from the Pakistani margin (von Rad et al., 1999; Doose-Rolinski-H et al., 2001), a δ^{18}O-record from the Dunde ice core (Tibetan Plateau) that predominantly reflects temperature variations (Thompson et al., 1989; Thompson, 2000), a 2000-year temperature reconstruction from China (Yang et al., 2002) and a sea surface salinity record from the South China Sea as a monitor of continental run-off from China (Wang et al., 1999). In order to improve the graphical comparability, all records were smoothed by a 50-year running mean filter. Visual inspection establishes that major peaks and troughs of the different records can be correlated with

those of Core 905 within the precision of the individual dating methods. The average difference between the tuned and the original age model of the individual records varies between 14 and 250 years. The maximum age offset in all records is <80–90 years for the last 2.5 ka BP (only in a few cases slightly higher offsets of 120–140 years were found). Slightly higher differences of up to 350 years occurred between 2.5 and 10 ka BP (Fig. 5). A general problem in high-resolution studies is the lack of sufficient age control down to a decadal time scale. All dating techniques used in the different studies involve a 5–10% error. Therefore, the revised/tuned age estimates for the different age models do not violate the original dates, although final proof of the proposed correlations requires better dating techniques. Larger age offsets of up to 900–1000 years occur solely in the lowest Holocene (>7.5 ka BP) of the Dunde ice core where the original age model is based on ice flow modelling. This particular section is close to the bottom of the ice core and hence potentially suffers from additional errors. These factors may limit the accuracy of the age model for this section. Hence, we expect that the tuned ages for the lowest Holocene section are still within the two-sigma error bars of the original age model.

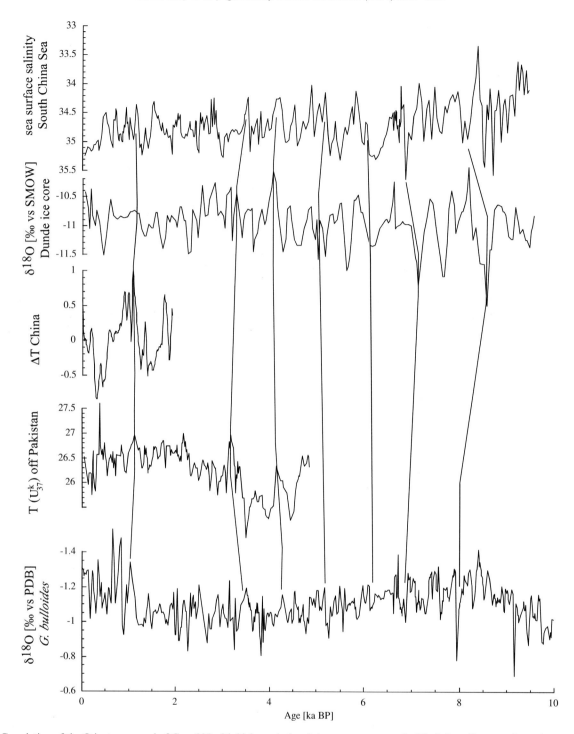

Fig. 5. Correlation of the O-isotope record of Core 905 with high-resolution Asian monsoon records. Black lines illustrate the main correlation tie points used.

3. Decadal scale monsoon variations: Core 905 and an Asian perspective

The modern Asian monsoon system consists of the W-Asian-, the Tibetan plateau- and the E-Asian monsoon system. The seasonal climate in these regions is controlled by the annual variation in solar insolation. During boreal summer the differential continent–ocean heating results in an overall airflow from the tropical Indian/W-Pacific Ocean to continental Asia. During winter a generally oceanward airflow is established. The resulting monsoon winds largely control the seasonal precipitation and temperature change.

The present study aims to assess the Holocene history of the subsystems of the Asian monsoon at a decadal-century time scale. Our comparison evolves sequentially from a local scale to a cross-continental assessment of potential teleconnections in the Asian monsoon.

The nucleus for the present work is a high-resolution study on Core 905 from the Arabian Sea off Somalia. Figs. 3 and 4 show the Holocene O-isotope record of the planktic foraminifer *G. bulloides*. It depicts rapid decadal scale O-isotope variations of 0.3–0.6‰ with the highest amplitudes during the early and the late Holocene. The δ^{18}O-record of *G. bulloides* mainly reflects the temperature and the stable O-isotope composition of the ambient seawater (δ^{18}O$_{water}$). Present day variations in δ^{18}O$_{water}$ offshore Somalia, however, are small because there are no major rivers flowing into the area and there is no significant correlation between δ^{18}O$_{water}$ and salinity (Delaygue et al., 2001; Jung et al., 2001). The potential effect of variations in salinity on the O-isotope record of *G. bulloides* can be assessed based on a maximum-change-scenario, i.e. assuming a Glacial to Holocene shift in salinity of 1.3 PSU (Rostek et al., 1997) and taking into account a δ^{18}O$_{water}$ value in (modern) NE-African rain of ~-3‰ [GNIP data set] and a δ^{18}O$_{water}$ in surface water from the region of 0.3‰ (Jung et al., 2001). Assuming that the entire change in salinity would result from fluvial run-off, such a scenario would only explain a change of –0.12‰ (for the full argument see Jung et al., 2002a). Given that the change in the δ^{18}O-record of *G. bulloides* frequently is in the order of 0.6‰ (unsmoothed record), we conclude that the observed δ^{18}O-record predominantly reflects temperature variations of up to $\sim2\,^{\circ}$C during summer (see Jung et al. (2002a)) for details of the argument). A comparison of the time series of Core 905 with a recently published high-resolution temperature record from a sediment core off Pakistan (Doose-Rolinski-H et al., 2001) supports this conclusion. The alkenone (U_{37}^{k})-based temperature estimates for Core 39 KG/56

K17 off Pakistan shown in Fig. 6 imply temperature variations (annual temperature variations with a strong summer component) of ~1.5–$2\,^{\circ}$C. These estimates are, although some minor differences occur, overall strikingly similar to the temperature variations deduced from the stable O-isotope record of Core 905 (Jung et al., 2002a; this study). Accordingly, the correlation between the two data sets suggests that coherent basin wide decadal-century scale temperature variations occurred over the entire Arabian Sea during the Holocene. The temperature variations probably result from variations in the trade wind-influenced Somali current system as it flows NE-ward across the Arabian Sea and affects both the upwelling area off Somalia and the Arabian Sea off Pakistan.

In order to assess the variations in other parts of the Asian monsoon system, a comparison is made with a 2000-year monsoon record from continental China (Yang et al., 2002) and a Holocene ice core record from the Dunde ice cap (Thompson, 2000). These data potentially link marine and continental monsoon records. Both continental time series reflect monsoon-controlled temperature variations in the Tibetan plateau/E-Asian monsoon system. They strongly resemble the record of Core 905 on a peak-to-peak basis as well as in the overall change in amplitude for the entire Holocene (Fig. 7a and b). Given the large distance between the records from the Arabian Sea and continental China, their strong resemblance may imply that coherent monsoonal change occurred in these areas during the Holocene. A recently published record from the South China Sea (Wang et al., 1999) represents climate variations in the easternmost Asian monsoon system and completes the coverage of the full W-E-extent of the Asian monsoon system. Fig. 8 correlates the surface salinity/run-off record from Core GIK 17940

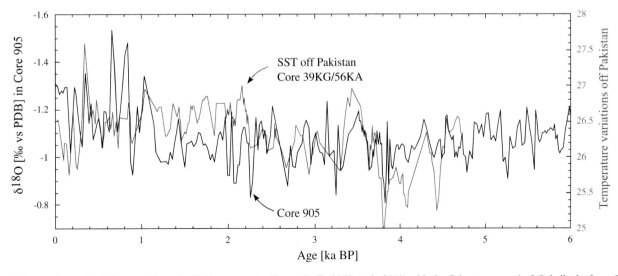

Fig. 6. Comparison of a SST record from the Pakistan margin (Doose-Rolinski-H et al., 2001) with the O-isotope record of *G. bulloides* from Core 905.

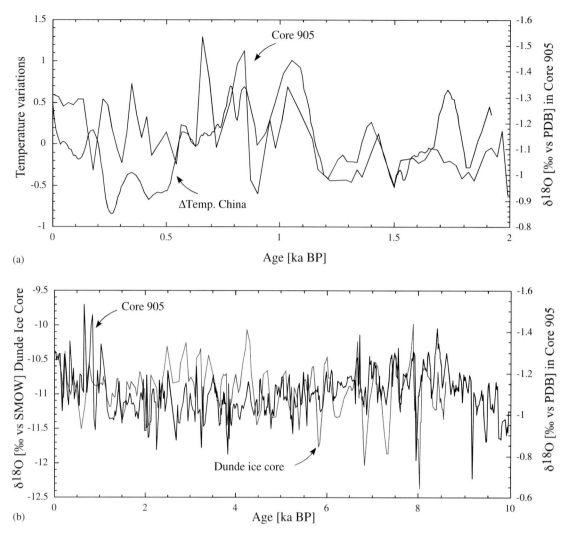

Fig. 7. (a) Comparison of a temperature record from China (Yang et al., 2002) with the record of Core 905; (b) comparison of an O-isotope record from the Dunde ice core (Tibetan Plateau) (Thompson et al., 1989) with the O-isotope record of *G. bulloides* from Core 905.

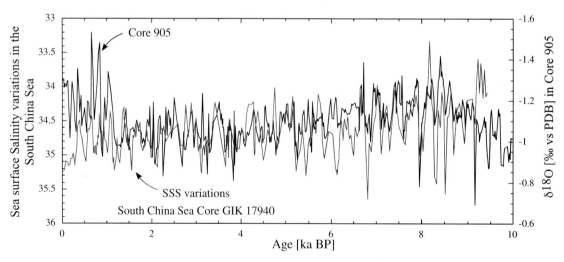

Fig. 8. Comparison of a SSS record from the South China Sea (Wang et al., 1999) with the O-isotope record of *G. bulloides* from Core 905.

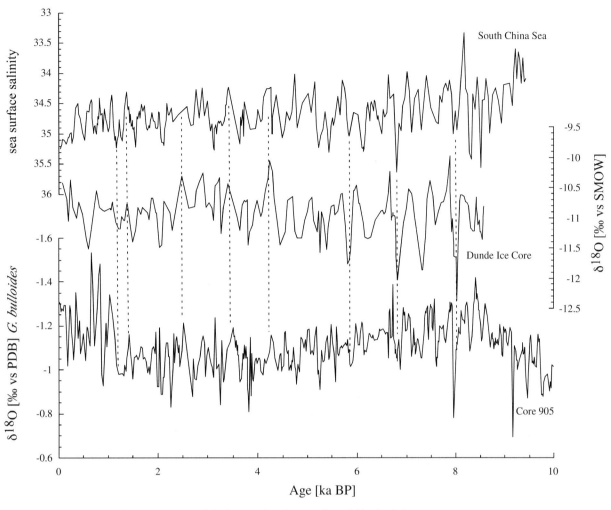

Fig. 9. Summary of the large-scale teleconnection within the Asian monsoon system.

(Wang et al., 1999) with the δ^{18}O-variations off Somalia. The variation in surface water salinity recorded in Core GIK 17490 reflects summer monsoon-controlled changes in run-off from the Pearl River (Wang et al., 1999). Similar to the other records, the comparison of the GIK 17490 record with that of Core 905 shows (despite some minor deviations) that the overall pattern is surprisingly similar. Fig. 9 illustrates that within the resolution of the individual age models, the available high-resolution paleorecords imply a coherent climate change pattern in the Holocene for the entire monsoonal system with alternations between warm/moist periods and cooler/drier periods.

4. Trade winds: pacemaker of the Asian monsoon?

The potential implications of the coherent paleo-monsoon records can be deduced based on an assessment of submodern analogue climate data. These data sets were taken from the TYN 1.1 data set at the Tyndale Center for Climate Change Research in

Norwich and provide monthly temperature and rainfall data for the period 1901–2000 (Mitchell et al., 2003, submitted). These climate records were used to summarise the entire Asian monsoon system by selecting key observational records from throughout the region. To assess variations in the W-Asian monsoon a rainfall record from India was used. Temperature and rainfall records from China were selected to cover monsoonal change over the continent as well as the induced changes in runoff into the South China Sea. In order to include the "source area" of E-Asian precipitation, a rainfall record from Indonesia was chosen. Given that most paleorecords either reflect the summer monsoon season (Jung et al., 2002a) or variations that predominantly occurred during summer (Thompson, 2000; Doose-Rolinski-H et al., 2001; Yang et al., 2002), the observed monsoonal records were adapted by only using temperature and rainfall data from the summer month (for details see Fig. 10). In order to assess the observed monsoonal variation on the time scales resolved in the paleorecords, the temporal resolution of the submodern analogue data was adjusted to a decadal-scale (Fig. 10a).

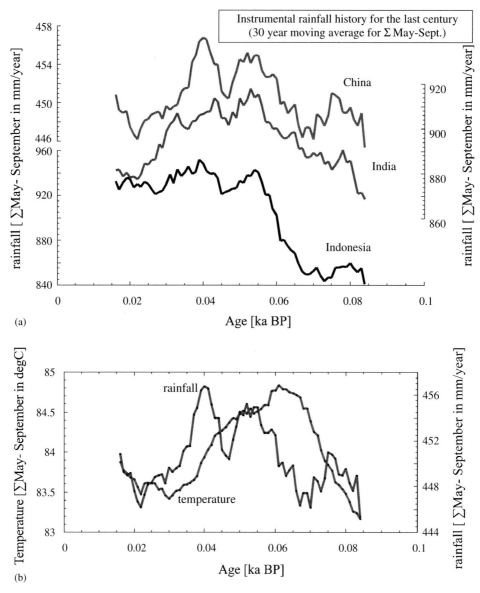

Fig. 10. Instrumental climate records from key locations within the Asian monsoon system. (a) rainfall data for China, Indonesia and India. (b) rainfall and temperature data for China. All data were taken from the TYN 1.1 data set at the Tyndale Center for Climate Change Research in Norwich and provide monthly temperature and rainfall data for the period 1901–2000 (Mitchell et al., 2003 submitted). In order to highlight the decadal scale variability a 40-year running mean filter was applied (0 = year 2000).

Although minor differences exist between the local rainfall records, the long-term changes over India and China are strikingly similar (Fig. 10a). Interestingly, these variations are also coherent with the long-term rainfall variations in Indonesia suggesting that climate in the subtropical-temperate and tropical regions of the Asian monsoon varied in phase on a decadal time scale. Despite some minor differences, there is a generally positive correlation at a decadal resolution between the long-term precipitation and the temperature variation over China (Fig. 10b). The fact that temperature and rainfall covary for the past 100 years strongly supports a conclusion based on the paleorecord comparison in

Fig. 9, i.e. that a joint process probably controls the coherent climate variation in the entire region affected by the monsoon system.

The processes that potentially control a coherent climatological change across the Asian monsoon system may be analogous to those involved in oscillatory systems such as ENSO. The modern ENSO involves piling up of warm waters in the W-Pacific Ocean due to the predominant easterly trade winds. Every few years (for a recent summary, see (Cobb et al., 2003; Philander and Fedorov, 2003; Trenberth and Hoar, 1997)) these trade winds weaken resulting in a flow of warm surface water from the W-Pacific to the E-Pacific Ocean causing unusually high surface water temperatures off S-America. In recent years, a number of studies have shown that the tropical Indian Ocean witnessed similar repeated temperature anomalies, although the nature of these variations is still intensely debated (Saji et al., 1999; Webster et al., 1999 and references therein). Based on these modern examples two potential scenarios can be developed to explain the coherency in the Asian monsoon system. Scenario 1 explains the coherent variation in the Asian monsoon by trade wind-induced temperature changes in the west Pacific warm pool. The warm pool region directly impacts on the rainfall in the E-Asian monsoon region as the moisture originates from the W-Pacific Ocean. In addition, the warm pool waters affect the throughflow of surface water from the Pacific to the Indian Ocean as part of the global thermohaline circulation. A significant portion of these surface waters flows across the tropical Indian Ocean and is eventually entrained into the Somali Current (Schott and McCreary Jr, 2001) that controls the surface water hydrography in the Arabian Sea causing the increased SST off Somalia recorded in Core 905. A second scenario would involve synchronous trade wind and summer monsoon variations in the Indian and the Pacific Ocean possibly occurring in tune with ITCZ-shifts. In this scenario coherent variations in trade wind strength in both ocean basins would result in synchronous "warm pool"/temperature changes. This scenario would be in line with recently published hypotheses on Holocene shifts of the ITCZ for the Atlantic Ocean (Chiang et al., 2003), the Caribbean Sea (Haug et al., 2001), for the Pacific warm pool (Stott et al., 2002) and the Arabian Sea (Jung et al., 2004). If true a shift would affect the atmospheric pressure gradient between the ITCZ and the subtropical high-pressure cells. Accordingly, variations in the pressure gradient would lead to changes in the trade wind strength, as in our scenario 2.

In both scenarios higher temperatures in the western part of the tropical Pacific and Indian Ocean would result in higher evaporation and enhanced continent-ward moisture transport by the respective monsoonal winds. The stalagmite record from Oman (Neff et al., 2001) and the surface water salinity record from the South China Sea (Wang et al., 1999) document such variations. Similarly, the monsoonal records from continental Asia (Thompson, 2000; Yang et al., 2002) reflect temperature variations in surface-ocean-warmed summer monsoon winds. The induced higher temperatures in the Somali current would account for the coherent positive temperature anomalies as recorded in the Arabian Sea off Somalia (Jung et al., 2002a) and off Pakistan (Doose-Rolinski-H et al., 2001).

Studies on the long-term observational climate records and short-term proxy records are broadly in line with our findings. They imply that a substantial amount of the observed climate variation occurs on a decadal time scale (Bond et al., 2001; Haug et al., 2001; Neff et al., 2001; Jung et al., 2002a; Fleitmann et al., 2003). Interestingly, there are indications that these temperature anomalies are similar to the ENSO-phenomenon and occur with the same sign in all tropical oceans (Latif et al., 1997). Hence, on a decadal time scale a more pronounced warm-pool co-exist with warm water piling up in the W-Indian Ocean. If true, these results would favour our scenario 2 although a definite resolution of the oceanographic relationships requires additional research.

The increasing number of records resolving Holocene climate change at a decadal time scale consistently demonstrates an unexpectedly large natural climate variation that is in phase with variations in the ^{14}C-concentration in the atmosphere. In particular, recently published studies from the Arabian Sea region (Neff et al., 2001; Jung et al., 2002a; Fleitmann et al., 2003) and the Atlantic region (Bond et al., 2001; Haug et al., 2001) imply a strong sensitivity to insolation changes on decadal time scales. The advocated variations in the trade wind strength probably are also controlled by variations in solar insolation. All these proxy records appear to emphasize the important role of decadal-scale variations in solar insolation in controlling Earth climate and may point to a particular climatological sensitivity to external forcing on these time scales. In order to prove this conclusion, however, better dating techniques with accuracy down to a decadal time scale are required.

Acknowledgements

We thank Stefan Mulitza and one anonymous reviewer for their constructive criticisms that have helped to improve the manuscript. This study is part of EU-TMR network ERBFMRXCT960046 "The marine record of continental tectonics and erosion", NWO Grant 854.00.002 and is NSG publication number 2004-0903.

References

Anderson, D.M., Prell, W.L., 1993. A 300 kyr record of upwelling off Oman during the late Quaternary; evidence of the Asian southwest monsoon. Paleoceanography 8, 193–208.

Biondi, F., Gershunovk, A., Cayan, D., 2001. North Pacific decadal scale variability since 1661. Journal of Climate 14, 5–10.

Bond, G., Kromer, B., Beer, J., Muscheler, R.N.E.M., Showers, W., Hoffmann, S., Lotti-Bond, R., Hajdas, I., Bonani, G., 2001. Persistent solar influence on North Atlantic climate during the Holocene. Science 294, 2130–2136.

Chiang, J.C., Biasutti, M., Battisti, D.S., 2003. Sensitivity of the Atlantic Intertropical Convergence Zone to Last Glacial maximum boundary conditions. Paleoceanography 18, Doi: 10.1029/2003PA000916,2003.

Clemens, S., Prell, W.L., Murray, D., Shimmield, G.B., Weedon, G.P., 1991. Forcing mechanisms of the Indian Ocean monsoon. Nature 353, 720–725.

Cobb, K.M., Charles Christopher, D., Cheng, H., Edwards, R.L., 2003. El Niño/Southern Oscillation and tropical Pacific climate during the last millennium. Nature 424, 271–276.

Delaygue, G., Bard, E., Rollion, C., Jouzel, J., Stievenard, M., Duplessy, J.-C., Ganssen, G., 2001. Oxygen isotope/salinity relationship in the northern Indian Ocean. Journal of Geophysical Research 106, 4565–4574.

Doose-Rolinski-H, Rogalla, U., Scheeder, G., Lueckge, A., von Rad, U., 2001. High-resolution temperature and evaporation changes during the late Holocene in the eastern Arabian Sea. Paleoceanography 16, 358–367.

Erez, J., Luz, B., 1983. Experimental paleotemperature equation for planktonic foraminifera. Geochimica et Cosmochimica Acta 47, 1025–1031.

Fleitmann, D., Burns, S.J., Mudelsee, M., Neff, U., Kramers, J., Mangini, A., Matter, A., 2003. Holocene forcing of the Indian monsoon recorded in a stalagmite from southern Oman. Science 300, 1737–1739.

Gupta, A.K., Anderson, D., Overpeck, J.T., 2003. Abrupt changes in the Asian southwest monsoon during the Holocene and their links to the North Atlantic Ocean. Nature 421, 354–357.

Haug, G., Hughen, K.A., Sigman, D.M., Peterson, L.C., Röhl, U., 2001. Southward migration of the Intertropical Convergence Zone through the Holocene. Science 293, 1304–1308.

Hurrel, J.W., 1995. Decadal trends in the North Atlantic Oscillation: regional temperatures and Precipitation. Science 269, 676–679.

Jung, S., Davies, G.R., Ganssen, G.M., Kroon, D., 2002a. Decadal-Centennial scale monsoon variations in the Arabian Sea during the Early Holocene. Geochemistry, Geophysics, Geosystems 3, doi 10.10.1029.

Jung, S.J.A., Ganssen, G., Davies, G.R., 2001. Multi-decadal variations in the Early Holocene outflow of Red Sea Water into the Arabian Sea. Paleoceanography 16, 658–668.

Jung, S.J.A., Davies, G.R., Ganssen, G.M., Kroon, D., 2002b. Centennial-millennial scale monsoon variations off Somalia over the last 35 kyr. In: Clift, P., Kroon, D. (Eds.), Tectonic and Climatic Evolution of the Arabian Sea Region. Special publication Journal of the Geological Society London, London, pp. 341–352.

Jung, S.J.A., Davies, G., Ganssen, G., Kroon, D., 2004. Stepwise Holocene aridification in NE-Africa deduced from dust borne radiogenic isotope records. Earth and Planetary Science Letters 221, 27–37.

Kroon, D., Ganssen, G., 1989. Northern Indian Ocean upwelling cells and the stable isotope composition of living planktonic foraminifers. Deep-Sea Research 36, 1219–1236.

Latif, M., Kleeman, R., Eckert, C., 1997. Greenhouse warming, decadal variability, or El Niño? An attempt to understand the anomalous 1990s. Journal of Climate 10, 2221–2239.

Mitchell, T.D., Carter, T.R., Jones, P.D., Hulme, M., New, M., 2003. A comprehensive set of high-resolution grids of monthly climate for Europe and the globe: the observed record (1901–2000) and 16 scenarios (2001–2100). Journal of Climate, Submitted for publication.

Morley, J.J., Heusser, L.E., 1997. Role of orbital forcing in East Asian monsoon climates during the last 350 kyr: evidence from terrestrial and marine climate proxies from core RC14-99. Paleoceanography 12, 483–493.

Neff, U., Burns, S.J., Mangini, A., Mudelsee, M., Fleitmann, D., Matter, A., 2001. Strong coherence between solar variability and the monsoon in Oman between 9 and 6 kyr ago. Nature 411, 290–293.

Overpeck, J., Anderson, D., Trumbore, S., Prell, W., 1996. The southwest Indian monsoon over the last 18000 years. Climate Dynamics 12, 213–225.

Peeters, F.J.C., Brummer, G.-J.A., Ganssen, G., 2002. The effect of upwelling on the distribution and stable isotope composition of Globigerina bulloides and Globigerinoides ruber (planktic foraminifera) in modern surface waters of the NW Arabian Sea. Global and Planetary Change 34, 269–291.

Philander, S.G., Fedorov, A., 2003. Is El Niño sporadic or cyclic. Annual Reviews in Earth Planetary Sciences 31, 579–594.

Rostek, F., Bard, E., Beaufort, L., Sonzogni, C., Ganssen, G., 1997. Sea surface temperature and productivity records for the past 240 kyr in the Arabian Sea. Deep Sea Research (ii) 44, 1461–1480.

Saji, N.H., Goswami, B.N., Vinayachandran, P.N., Yamagata, T., 1999. A dipole mode in the tropical Indian Ocean. Nature 401, 360–363.

Sarnthein, M., Kennett James, P., Allen, J.R.M., Beer, J., Grootes, P., Laj Carlo, E., McManus, J., Ramesh, R., 2002. Decadal-to-scale climate variability—chronology and mechanisms: summary and recommendations. Quaternary Science Reviews 21, 1121–1128.

Schott, F.A., McCreary Jr., J.P., 2001. The monsoon circulation of the Indian Ocean. Progress in Oceanography 51, 1–123.

Sirocko, F., 1994. High-frequency periodicities in the Asian monsoonal climate. Terra Nostra.

Staubwasser, M., Sirocko, F., Grootes, P., Erlenkeuser, H., 2002. South Asian monsoon climate change and radiocarbon in the Arabian Sea during early and middle Holocene. Paleoceanography 17 (4), doi:10.1029/2000PA000608.

Stott, L.D., Poulsen-Chris, J., Lund, S., Thunell, R.C., 2002. Super ENSO and global climate oscillations at millennial time scales. Science 297, 222–226.

Stuiver, M., Reimer, P.J., Bard, E., Beck, J.W., Burr, K.A., Hughen, K.A., Kromer, B., McCormac, J., van der Plicht, J., Spurk, M., 1998. INTCAL98 Radiocarbon age calibration. Radiocarbon 40, 1041–1083.

Thompson, L.G., 2000. Ice core evidence for climate change in the tropics; implications for our future. Quaternary Science Reviews 19, 19–35.

Thompson, L.G., Mosley-Thompson, E., Davis, M.E., Bolzan, F., Dai, J., Tao, T., Gundestrup, N., Wu, X., Klein, L., Xie, Z., 1989. Holocene-late Pleistocene climatic ice core records from Qinghai-Tibetan Plateau. Science 246, 474–477.

Trauth, M.H., Sarnthein, M., Arnold, M., 1997. Bioturbational mixing depth and carbon flux at the sea floor. Paleoceanography 12, 517–526.

Trenberth, K.E., Hoar, T.J., 1997. El Niño and climate change. Geophysical Research Letters 24, 3057–3060.

von Rad, U., Schaaf, M., Michels, K.H., Schulz, H., Berger, W.H., Sirocko, F., 1999. A 5000-yr record of climate change in varved sediments from the oxygen minimum zone off Pakistan, northeastern Arabian Sea. Quaternary Research (New York) 51, 39–53.

Wang, L., Sarnthein, M., Erlenkeuser, H., Grimalt Joan, O., Grootes, P., Heilig, S., Ivanova, E., Kienast, M., Pelejero, C., Pflaumann, U., 1999. East Asian monsoon climate during the late Pleistocene; high-resolution sediment records from the South China Sea. Marine Geology 156, 245–284.

Webster, P.J., Moore, A.M., Loschnigg, J.P., Leben, R.R., 1999. Coupled ocean-atmosphere dynamics in the Indian ocean during 1997–98. Nature 401, 356–360.

Yang, B., Braeuning, A., Johnson, K.R., Shi, Y., 2002. General characteristics of temperature variation in China during the last two millennia. Geophysical Research Letters 29, doi 10.1029/2001GL014485.

Quaternary Science Reviews 23 (2004) 2219–2230

Glacial–interglacial contrast in climate variability at centennial-to-millennial timescales: observations and conceptual model

Michael Schulz[a,*], André Paul[a], Axel Timmermann[b,c]

[a]*Fachbereich Geowissenschaften and DFG-Forschungszentrum "Ozeanränder", Universität Bremen, Postfach 330440, Bremen 28334, Germany*
[b]*Leibniz Institut für Meereswissenschaften, Universität Kiel, Germany*
[c]*IPRC, SOEST, University of Hawaii, 2525 Correa Road, Honolulu, HI 96822, USA*

Abstract

A set of published paleoclimate proxy records from the northern hemisphere, capturing different climate processes, is used to study glacial–interglacial differences in climate variability at centennial-to-millennial timescales during the past fifty thousand years. These proxy records reveal the existence of distinct oscillatory modes of the climate system. Glacial climate variability is dominated by a single mode, the Dansgaard–Oeschger cycles, composed of stadial and interstadial states. This glacial mode results in well-expressed covariations of the proxies, which are paced by a *fundamental* 1470-year signal. In contrast, there is no compelling evidence for a dominant and persistent centennial-to-millennial climate cycle during the Holocene. Interglacial climate variations seem to covary less pronounced than those of the last glacial period, suggesting the simultaneous activity of independent climate modes, each characterized by its own natural periods, between approximately 400–3000 years. A conceptual model is introduced to interpret this contrast in covariation at glacial–interglacial timescales. It is assumed that different climate modes can be represented by relaxation oscillators with different natural periods in the centennial-to-millennial band. Interactions among such oscillators may lead to a phase-synchronization and the development of a new climate mode with a joint frequency. We suggest that the coupled state with its synchronized dynamics resembles a glacial whereas the decoupled state represents an interglacial with its reduced covariations of climate fluctuations. The synchronization greatly enhances the frequency stability of the coupled system, and has the potential to reconcile the stability of the glacial 1470-year pacing cycle with an origin within the Earth's climate system.
© 2004 Elsevier Ltd. All rights reserved.

1. Introduction

Late Pleistocene climate variability at centennial-to-millennial timescales differs significantly between glacials and interglacials. Whereas the former are dominated by rapid transitions between cold stadials and warm Dansgaard–Oeschger (DO) interstadials, the latter show only little variability around the mean state (McManus et al., 1999; Helmke et al., 2002; Pahnke et al., 2003). Despite intensive research efforts, the origin of the glacial DO-type climate variability remains controversial. Suggested hypotheses range from internal

oscillations of the ocean–atmosphere system (Broecker et al., 1990; Winton, 1993; Sakai and Peltier, 1997; Schulz et al., 2002; Timmermann et al., 2003) over periodic calving of the Greenland ice sheet (van Kreveld et al., 2000), to external forcing mechanisms (Ganopolski and Rahmstorf, 2001). In addition, it remains also unresolved whether Holocene climate variability merely corresponds to extremely damped DO-type variations or represents a truly different state of the climate system.

Another complication in understanding late Pleistocene climate variations arises from the fact that the onset of DO-interstadials appears to be paced by a fundamental period of 1470 years, that is, they are separated by multiples of 1470 years (Schulz, 2002; Rahmstorf, 2003). Neither the origin of the pacing nor

*Corresponding author. Tel.: +49-421-218-7136; fax: +49-421-218-7040.

E-mail address: mschulz@palmod.uni-bremen.de (M. Schulz).

0277-3791/$ - see front matter © 2004 Elsevier Ltd. All rights reserved.
doi:10.1016/j.quascirev.2004.08.014

its apparent frequency stability, which has been estimated to range from $\pm 12\%$ to $\pm 20\%$ (Schulz, 2002; Rahmstorf, 2003), have been determined as yet. It has been suggested that this inferred stability of the pacing period makes an origin within the climate system rather unlikely (Wunsch, 2000) and that the regularity of the DO cycles would be easier to reconcile with an extraterrestrial cause (Berger and von Rad, 2002; Rahmstorf, 2003).

Using existing paleoclimate records we show that Holocene paleoclimate proxy records are incompatible with the derived glacial 1470-year pacing cycle. We introduce a conceptual model in which different interactions of oscillatory modes of the climate system are utilized to explain the apparent contrast between covarying glacial and largely independent interglacial climate fluctuations. Moreover, our conceptual model offers a means to reconcile the high frequency stability of the glacial 1470-year pacing with an Earth-bound origin.

2. Glacial–interglacial contrast in climate variability

It is our main interest to understand climate variability at centennial-to-millennial timescales at large spatial scales (i.e., regional to global) and across different components of the climate system. To investigate differences in climate variability between glacials and interglacials we choose two paleoclimate proxy records from the same archive covering a significant portion of a full glacial–interglacial cycle at multi-decadal resolution. This approach minimizes any bias due to stratigraphic uncertainties as well as to possible offsets between different archives across glacial-to-interglacial transitions.

Our analysis is based on the $\delta^{18}O$ and potassium records from the Greenland GISP2 ice core (Mayewski et al., 1997; Stuiver and Grootes, 2000). To a first order $\delta^{18}O$ composition of the ice scales linearly with the air temperature above Greenland (Stuiver and Grootes, 2000), whereas the potassium concentration (Mayewski et al., 1997) is a proxy for the strength of the Siberian High (Meeker and Mayewski, 2002). (Since dry deposition at the GISP2 ice-core location appears to be negligible (Meeker et al., 1997), it follows that variations in ice-core ion concentration are proportional to changes in atmospheric ion concentration (cf. Alley et al., 1995). Accordingly, we follow Mayewski et al. (1997) and use potassium concentration instead of the accumulation rate to infer changes in atmospheric potassium loading.)

During the last glacial period both proxies recorded the sequence of DO events in a very similar manner (Fig. 1a), resulting in a high correlation between the two time series ($R^2 = 0.73$). Specifically, the proxy data

suggest a strengthening of the Siberian High during stadials and a weakening during interstadials. In contrast, during the Holocene the records show no discernable similarity (Fig. 1b) and their correlation is greatly reduced ($R^2 = 0.08$) compared to the glacial interval. Even the largest event recorded in the $\delta^{18}O$ series during the Holocene, at $\sim 8.2\,ky\,BP$ (thousand years before present), has no corresponding counterpart in the potassium record. Nevertheless, both proxy records show clearly discernable variability at centennial-to-millennial timescales during the entire Holocene, which, however, seems to be unrelated between the records. Taken together, these observations suggest that the strength of the Siberian High and temperature above Greenland covary during the course of DO events, but fluctuate independent of each other in the absence of DO events.

The pronounced covariation of glacial DO-type climate fluctuations at a global scale is further supported by a recent compilation of glacial paleoclimate records (Voelker et al., 2002). According to this study, DO-type climate-change signals can be found in proxies characterizing the state of the atmosphere (e.g., large-scale circulation patterns; Mayewski et al. (1997)), the ocean (e.g., thermohaline circulation; Sarnthein et al. (2001)), and the carbon cycle (e.g., atmospheric methane concentration; Brook et al. (1999)). The notion of an ubiquitous DO-signal in the climate system does neither imply that all proxies record identical "patterns" of climate histories nor that all time series vary in phase which each other. As an example, consider the modified "see-saw" concept (Stocker and Johnsen, 2003), in which temperature in the North and South Atlantic Ocean *change* simultaneously, but in opposite direction. This leads to the generation of two signals with identical "patterns" which are out of phase by 180°. The delayed response of temperature in Antarctica results in a modified pattern of the temperature evolution in this region compared to the South Atlantic (Stocker and Johnsen, 2003). Despite these regional differences in the actual course of climate change, the entire system considered in this conceptual model is pervaded by the DO events.

Based on the chronology of the GISP2 $\delta^{18}O$ record (Stuiver and Grootes, 2000), the recurrence times of the onset of DO-interstadials occur as multiples of 1470 years (Schulz, 2002; Rahmstorf, 2003). Taken together with the well-expressed covariations of glacial climate fluctuations this finding implies commensurate recurrence times of climate events, which are linked to DO-type climate variability, irrespective of the paleoclimate proxy being considered. The estimated pacing period of the DO interstadials and its stability depend critically on the chronology of the GISP2 ice-core (Meese et al., 1997). Any inaccuracy of the ice-core chronology will likely affect both estimates. Unfortunately, available

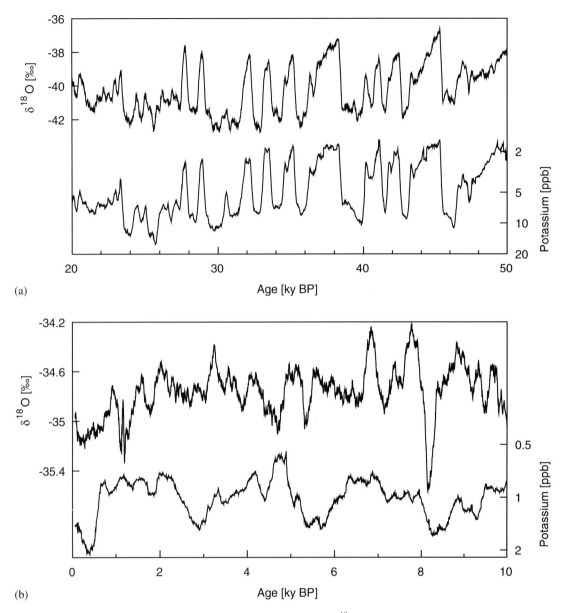

Fig. 1. Glacial–interglacial contrast in covariations of climate fluctuations. (a) GISP2 $\delta^{18}O$ (Grootes and Stuiver, 1997) and potassium concentration (Mayewski et al., 1997) during the last glacial period. Since our focus is on variations at centennial-to-millennial timescales, both series were smoothed with a 200-year running mean filter. Both records depict the same course of events. (b) as (a) but for the Holocene. In contrast to the glacial interval (a) the time series indicate no obvious covariations. Note that the length of the ages axes differs between (a) and (b).

radiometric datings (Wang et al., 2001; Spötl and Mangini, 2002; Burns et al., 2003; Genty et al., 2003) are inconclusive in this respect due to their uncertainties. With respect to possible modifications of the GISP2 chronology, we feel confident that the inferred high degree of covariations of glacial DO-type climate fluctuations is a robust feature.

To contrast the pronounced covariations of glacial climate fluctuations with the more independent interglacial temporal patterns further, we turn to a selection of Holocene paleoclimate records (Fig. 2). These records are thought to reflect various climate processes in the Atlantic region and encompass atmospheric and

marine proxies. It is obvious from the inspection of Fig. 2 that most time series exhibit no obvious common denominator in terms of variability at centennial-to-millennial timescales. At first glance this lack of low-frequency climate variability is surprising, since some of the proxies are considered to be closely coupled due to their link to changes in North Atlantic deep-water production: drift-ice record (Bond et al., 2001); sea-surface temperature off the western Barents Sea (Sarnthein et al., 2003); $\delta^{13}C$ of overflow water (Oppo et al., 2003). The overall lack of a common temporal pattern is especially obvious, when focusing on different climate-proxy records from the same archive

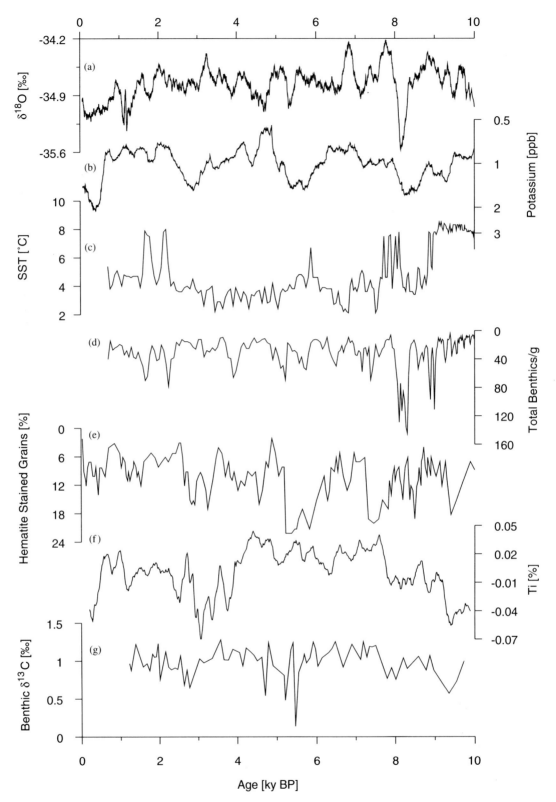

Fig. 2. Selection of Holocene paleoclimate records from the Atlantic region, indicating largely independent interglacial climate variations. (a) GISP2 $\delta^{18}O$ (Grootes and Stuiver, 1997), a proxy for air temperature above Greenland. (b) GISP2 potassium concentration (Mayewski et al., 1997), indicating the strength of the Siberian High (Meeker and Mayewski, 2002). (c) Reconstructed sea-surface temperature (SST) off the western Barents shelf (Sarnthein et al., 2003). (d) Total number of benthic foraminifera in a core off the western Barents shelf (Sarnthein et al., 2003), documenting winter storm activity. (e) Hematite stained grains in a sediment core off Ireland (Bond et al., 2001), measuring the amount of drift ice. (f) Titanium concentration (linearly detrended) in a Cariaco–Basin record, indicating changes in the hydrological cycle and river runoff (Haug et al., 2001). (g) Benthic $\delta^{13}C$ from south of Iceland (Oppo et al., 2003), interpreted to reflect changes in North-Atlantic deep-water formation. Time series in (a), (b) and (f) were smoothed with a 200-year running mean filter. Orientation of ordinate scales is such that inferred cold intervals point downward.

(Fig. 2a/b and c/d), that is, for situations in which any stratigraphic bias is rather small. Low signal-to-noise ratios and stratigraphic uncertainties could potentially hamper the comparison of low-frequency climate variability between the Holocene records. Whenever these obstacles are minimal, individual records documenting coupled climate processes may indeed reveal similar variations (e.g., the log-transformed potassium concentration (Mayewski et al., 1997; Fig. 2b) and titanium-concentration record (Haug et al., 2001; Fig. 2f) exhibit significant [$\alpha = 0.01$] cross-spectral coherency at a period of approximately 2500 years; not shown).

This picture of largely independent climate variability at centennial-to-millennial timescales during the Holocene is further supported by a large range of recurrence times between climate events that have been reported for different proxy records from the North-Atlantic region (e.g., O'Brien et al., 1995; Bond et al., 1997; Bianchi and McCave, 1999; Chapman and Shackleton, 2000; Schulz and Paul, 2002; Risebrobakken et al., 2003; Sarnthein et al., 2003; Hall et al., 2004). While the recurrence times range from approximately 400–3000 years, they appear to be clustered around 400–500 years and 900–1100 years, but not at approximately 1500 years or multiples thereof (Table 1). It should be kept in mind that the notion of recurrence time only reflects the fact that a record contains a distinct temporal pattern which is repeated after some time. Neither the exact repetition of such a pattern nor an exact timing of its recurrence is implied. Furthermore, reported recurrence times during the Holocene vary by as much as approximately $\pm 50\%$ within the same record (Bond et al., 1997; Sarnthein et al., 2003). In summary, analyses of many Holocene paleoclimate proxy records failed to confirm an ubiquitous glacial-type "1500-year cycle" (or

multiples thereof) and display variations at centennial-to-millennial timescale that seem to be largely unrelated to each other.

3. Glacial-type pacing of Holocene climate variations?

A record of hematite-stained grains from the north-eastern Atlantic Ocean (Bond et al., 1997), which has been interpreted to reflect the amount of drift ice reaching this area, is of great importance for understanding climate variability at centennial-to-millennial timescales. The glacial variability in this record is tightly linked to the occurrence of DO events and, therefore, also reflects the fundamental 1470-year pacing associated with these events. In contrast to other climate proxy data sets, the millennial-scale climate variability documented by this record is unaffected by the last glacial–interglacial transition giving rise to a Holocene recurrence time of inferred cold events of 1470 ± 500 years (Bond et al., 1997). Moreover, it has been postulated that the Holocene variations seen in this data set are controlled by variations in solar output (Bond et al., 2001).

If the Holocene drift ice-events are indeed manifestations of muted glacial-type DO stadials (Bond et al., 1997), one would expect the Holocene recurrence time of drift-ice anomalies to be consistent with the fundamental 1470 year pacing of the glacial DO events. To conclude for such a consistency two aspects have to be fulfilled: (i) a match of the period of the fundamental pacing cycle across the glacial termination and (ii) a consistent phase of a pacing cycle between glacial and interglacial.

To test if the drift-ice anomalies are consistent with the estimated glacial 1470 year pacing cycle, we extrapolate the 1470 year template for the onset of glacial DO interstadials (Schulz, 2002) into the Holocene (Fig. 3). If the timing of events is unaffected by the last deglaciation, the predicted times from the template should coincide with transitions from drift-ice maxima to minima, i.e., inferred warmings in the North Atlantic. For this comparison we take into account the $\pm 12\%$ uncertainty estimated for the glacial pacing period (Rahmstorf, 2003). During the last ~10 ky, the drift-ice record shows nine clearly discernable "warmings" (Fig. 3). However, a match between the actual timing of the "warmings" and those predicted by the template occurs only in three cases. Based on a binomial distribution (cf. Schulz, 2002) the probability for three matches out of nine "trials" by chance is 22%. Hence, the occurrence of Holocene climate events, as recorded by the drift-ice proxy, is consistent with a random origin and does not require pacing by the glacial 1470 year cycle. This suggests that the Holocene climate events recorded in this proxy record may not

Table 1
Selected Holocene (quasi)-periods from paleoclimate proxy records from the North Atlantic region (NATL = North Atlantic Ocean)

Proxy record	Period(s) (years)	Reference
NATL, drift ice	400–500 and 900–1100	Bond et al. (2001)
NATL, benthic foraminifera abundance	~400–1300	Sarnthein et al. (2003)
NATL, sediment color	550, 1000, ~1600 (?)	Chapman and Shackleton (2000)
NATL, planktonic foraminifera δ^{18}O	550 and 1150	Risebrobakken et al. (2003)
NATL, mean size of sortable silt	400 and ~1000	Hall et al. (2004)
GISP2, δ^{18}O	900	Schulz and Paul (2002)
GISP2, potassium	2600	O'Brien et al. (1995)

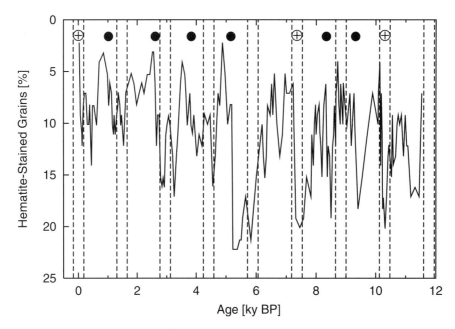

Fig. 3. Drift-ice record from site VM 29-191 from the northeastern North Atlantic (Bond et al., 2001, solid line) and 1470-year pacing template (dashed lines) extrapolated from the last glacial period. Adjacent dashed lines indicate the ±12% uncertainty associated with the pacing (Rahmstorf, 2003). Orientation of the drift ice-curve and the template are such that warm events appear at the top of the graph. Encircled plus-signs indicate where the onset of warming events in the drift-ice record coincides with the onset predicted by the template, whereas filled circles indicate mismatches. The number of matches is too small to exclude a random origin. The pacing template was constructed by fitting a trapezoidal time-series model to Dansgaard–Oeschger events 5–7 in the GISP2 $\delta^{18}O$ series (see Schulz, 2002 for details).

be a simple (muted) continuation of glacial Dansgaard–Oeschger events. (This finding is robust with respect to the technique used to estimate the glacial pacing of the DO events (Schulz, 2002; Rahmstorf, 2003). While both studies suggested identical periods of the pacing cycle, the actual templates are offset by 166 years.)

Bond et al. (2001) reconstructed variations in solar output by means of ^{10}Be and ^{14}C measurements and argued that the Holocene climate variations are controlled by fluctuations in solar forcing. We use a combined ^{10}Be record (Finkel and Nishiizumi, 1997; Yiou et al., 1997; converted to fluxes as in Bond et al. (2001)) to test if the reconstructed Holocene variations in solar forcing are compatible with the glacial-type 1470 year pacing, following the same approach as before. Between 3 and 11 ky BP the ^{10}Be accumulation rate shows four transitions between high and low values, which are thought to correspond to climate "warmings" (Bond et al., 2001). Only one out of four of these transitions is consistent which the glacial 1470 year pacing cycle (Fig. 4). This finding strongly suggests that the glacial 1470 year pacing cycle is not controlled by variations in solar forcing, unless one accepts the unlikely scenario that variations in solar-controlled climate pacing change across glacial–interglacial transitions. Hence, we have to seek for an alternative origin of the glacial 1470 year pacing cycle and its perplexing frequency stability.

4. Modeling the synchronization of interacting oscillators

The question arises whether the difference between covarying glacial climate variations and more independently fluctuating interglacial climate variations at centennial-to-millennial timescales holds the key for explaining the regularity of the glacial pacing cycle. To address this question, we consider a collection of oscillators within the climate system, which are thought to represent various oscillatory modes at centennial-to-millennial timescales. Examples for such oscillators include ice caps (e.g., MacAyeal, 1993) or the ocean–atmosphere system (e.g., Winton, 1993). Our working hypothesis is that during glacials a physical process is present which couples these oscillators, whereas the coupling is absent or insignificant during interglacials. Such a mechanism may then explain the lack of covariance in low-frequency climate fluctuations among climate proxies during the Holocene, which is indicative of the simultaneous presence of independent climate modes (see Sections 2 and 3).

To elucidate this contrast in synchronization, we explore the potential of the so-called mean-field coupling of oscillators (e.g., Pikovsky et al., 2001). We illustrate this concept by considering three different oscillatory modes that are represented by relaxations oscillators with different periods of approximately 400, 800 and 2400 years (Fig. 5; see Appendix A for details). These periods were chosen to represent the periods of

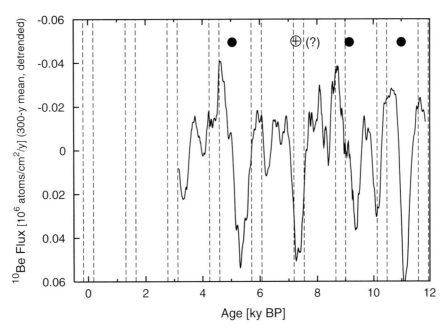

Fig. 4. Reconstructed solar forcing (Bond et al., 2001, solid line) versus 1470-year pacing template (as in Fig. 3). Changes in solar forcing are inferred from the beryllium-10 flux derived from Greenland ice-core data (Bond et al., 2001 and refs. therein). Encircled plus-sign indicates a match between the onset of a "warming event" in the [10]Be record with the onset predicted by the template, whereas filled circles indicate mismatches. The overall mismatch makes a solar origin of the 1470-year pacing cycle unlikely. The [10]Be data were smoothed with a 300-year running-mean filter and subsequently detrended. Orientation of [10]Be series and template is such that warm events appear at the top of the graph.

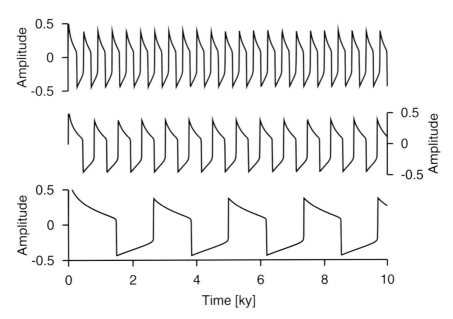

Fig. 5. Relaxation oscillations of three uncoupled oscillators with periods of approximately, 400, 800 and 2400 years (see appendix A for details).

the identified clusters of low-frequency climate variability during the Holocene (Table 1). It should also be noted that the actual number of oscillators is not critical for the following discussion, as long as at least two oscillators are considered. Initially, the oscillators are uncoupled, that is, they are not "aware" of each other and can thus oscillate at their "own" frequencies (Fig. 6). After 10 ky each oscillator is allowed to influence

the remaining oscillators (see Appendix A for details). As a result, the three oscillations become synchronized within approximately 1000 years. Even more important is that the period of the synchronized oscillations of 1300 years differs from the periods of the uncoupled oscillators. In other words: a new timescale, corresponding to the natural period of a new oscillatory mode, arises from the synchronization of oscillators. (This behavior is not

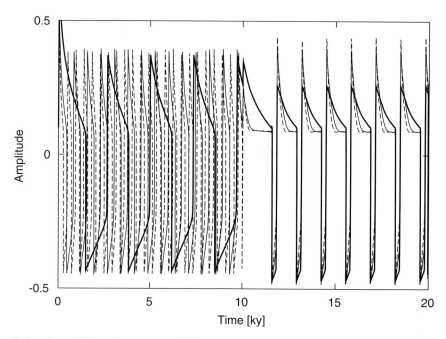

Fig. 6. Synchronization of relaxation oscillators. Between 0 and 10 ky the three oscillators are uncoupled ($\alpha = 0$), leading to independent oscillations (as in Fig. 5). At 10 ky the coupling strength is increased instantaneously to $\alpha = 0.3$, resulting in a synchronization of the oscillators within ~1 ky. The period of the synchronized oscillations is 1300 years.

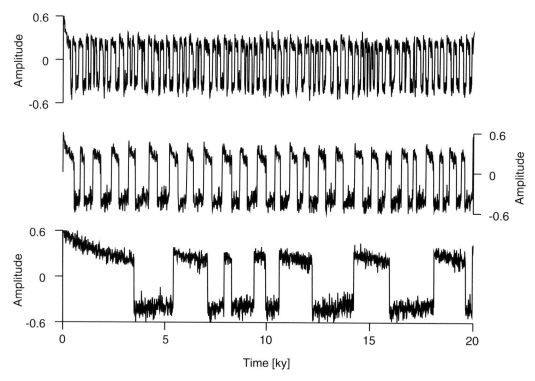

Fig. 7. Relaxation oscillations of three uncoupled oscillators with randomly perturbed periods of approximately 300 ($\pm 20\%$), 700 ($\pm 20\%$) and 2500 ($\pm 50\%$) years (see appendix A for details).

limited to three oscillators and can easily be achieved with a larger number of oscillators; not shown.)

In a second set of experiments, we explore whether the synchronization of the three oscillators works also in the presence of randomly driven frequency fluctuations. For

this experiment the periods of the three oscillators are set to 300 ($\pm 20\%$), 700 ($\pm 20\%$) and 2500 ($\pm 50\%$) years (Fig. 7). Shortly after the oscillators have been coupled (at 20 ky), the three oscillations become again synchronized (Fig. 8). Thus, despite the presence of

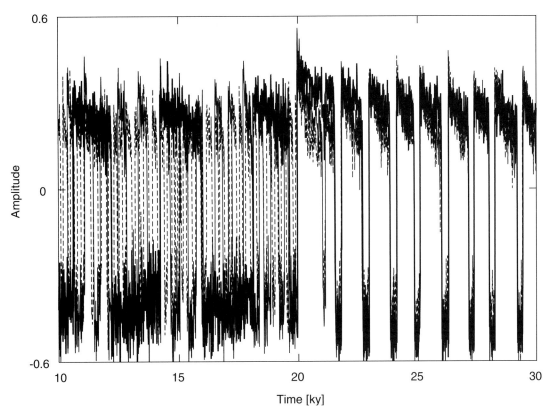

Fig. 8. Synchronization of relaxation oscillators in the presence of noise. Until 20 ky the three oscillators are uncoupled ($\alpha = 0$), leading to independent oscillations (as in Fig. 6; for clarity, only the portion after 10 ky is shown). At 20 ky the coupling strength is increased instantaneously to $\alpha = 0.6$, resulting in a synchronization of the oscillators within ~1 ky. The period of the synchronized oscillations is 1100 years with a variation of $\pm 10\%$. The frequency stability of the synchronized oscillations is thus higher than that of the uncoupled oscillators.

noise, the mean-field coupling mechanism is still capable of synchronizing the oscillators. As in the previous experiment, the mean period of the synchronized oscillations (1100 years; Fig. 8) has no direct counterpart in the periods of the non-synchronized oscillations. Even more striking is the fact that the period of the synchronized oscillation varies by only $\pm 10\%$, that is, the frequency stability of the synchronized oscillations is better than that of any of its constituents. Hence, this experiment strongly suggests that the synchronization of oscillators with time-varying periods can result in remarkable stable oscillations once the oscillators become synchronized.

5. Making sense of the glacial–interglacial contrast in climate variability—a concept

Based on the synchronization experiments we suggest that the inferred contrast in glacial–interglacial climate variability can be understood in terms of the presence or absence of interactions among different low-frequency climate modes. Accordingly, interglacials correspond to an uncoupled (or weakly coupled) state, whereas glacials provide a mechanism to couple low-frequency modes,

thereby synchronizing their dynamics and giving rise to a new mode that dominates the entire climate system.

Our heuristic approach involving different interactions of oscillatory modes of the climate system between glacials and interglacials leads immediately to the following questions: (i) Which components of the climate system give rise to these modes and what are their "natural" periods? (ii) What controls the coupling strength between independent modes at glacial–interglacial timescales? (iii) What are the physical mechanisms that provide the actual coupling (communication) between the modes?

Based on our evaluation of paleoclimate proxy data and our conceptual model we can give some tentative answers to these questions. We set out by assuming that the identified variance peaks at timescales of approximately, 400–500, 900–1100 and maybe 2600 years during the Holocene (Table 1), are the result of oscillatory climate modes. We note that neither the period of the oscillation is assumed to be stable during the Holocene nor that the oscillations are present throughout the entire Holocene. So far, our understanding of natural climate variability at centennial-to-millennial timescales seems to be too limited (cf. Folland et al., 2001) to pin down the actual sources of the variability. However, it is reasonable to assume that components of the climate system are involved that have a sufficiently large "inertia" for generating low-frequency climate oscillations. Such components include the deep ocean, ice caps and glaciers, groundwater storage, as well as vegetation (e.g., Saltzman, 2002). Using a continuation method, Weijer and Dijkstra (2003) demonstrated that the large-scale ocean circulation can exhibit low-frequency modes. Further modeling experiments are required to test the robustness of this result as well as to identify other potential mechanisms for low-frequency oscillations within the global climate system.

The most obvious factor controlling the coupling strength during a glacial–interglacial cycle are northern-hemisphere ice sheets. While this control of the coupling strength is a rather abstract concept, the physically more challenging questions surround the mechanisms which provide the potential coupling between the independent oscillatory modes of the climate system. Without detailed knowledge of how these oscillatory modes are generated we can offer only some general ideas along these lines. Atmospheric circulation and the associated heat and moisture transports offer an efficient and fast link to couple the climate modes mentioned above. Interhemispheric linkages could be accommodated by the oceanic thermohaline circulation, whereas an effective coupling between low and mid latitudes is provided by the ventilation of the thermocline. Finally, sea-level could communicate between climate modes at a global scale.

From the previous discussion it should be clear that oscillators and communicators may not always be clearly differentiated. For example, the oceanic thermohaline circulation may be an integral part of an oscillatory mode and at the same time provide the coupling between different modes. Moreover, one can anticipate that the existence of at least some oscillatory modes may depend on the general state of the climate system (e.g., DO-type events triggered by massive meltwater events in a cold climate (Paul and Schulz, 2002; Timmermann et al., 2003)).

In summary, the invoked mean-field coupling of interacting oscillators is a potential mechanism to explain the glacial–interglacial contrast in covariance of climate variations and to synchronize oscillatory climate modes over a wide range of natural frequencies. Moreover, this mechanism can account for a high degree of frequency stability in the coupled system and may hold the key to reconcile the stability of the glacial 1470 year pacing cycle with an origin within the Earth's climate system. It is tempting to speculate whether a combination of (i) the synchronization between Heinrich Events and DO-cycles (Schulz et al., 2002) and (ii) the triggering of a sequence DO-events by a Heinrich Event could lead to a self-synchronization, capable of generating the observed sequence of glacial climate events and a high degree of frequency stability of a fundamental pacing cycle of the DO events (this will be the subject of a forthcoming study).

6. Conclusions

Our analysis of climate variations at centennial-to-millennial timescales revealed that glacial climate changes covary to a high degree and are locked to a common pacing cycle. In contrast, we found no compelling evidence for a dominant interglacial climate cycle. Specifically, Holocene millennial-scale climate variability does not appear to be a (muted) continuation of glacial Dansgaard–Oeschger events. Moreover, a solar origin of the estimated glacial 1470 year pacing cycle is unlikely (keeping in mind that this inference is based on a rather limited set of observations).

The notion of oscillatory climate modes offers a framework to explain the glacial–interglacial contrast in climate covariations at multicentennial-to-millennial timescales by means of different coupling strength between these modes. While strong coupling during glacials leads to synchronized variations corresponding to a single dominating climate mode, weak or absent coupling during interglacials allows the climate modes to exist independently of each other and to result in rather independent climate variations.

Mean-field coupling is capable of synchronizing oscillators over a wide frequency range, even in the presence of noise. The synchronization greatly enhances the frequency stability of the coupled system, and has the potential to reconcile the stability of the glacial 1470 year pacing cycle with an origin within the Earth's climate system.

Acknowledgments

We greatly appreciate the careful reviews by Trond Dokken and an anonymous referee. This work was supported by the *Deutsche Forschungsgemeinschaft* through DFG Research Center "Ocean Margins" (MS and AP) and Collaborative Research Project SFB 460 (AT). This is RCOM publication 0167.

Appendix A

The Morris–Lecar system (after Somers and Kopell, 1995) was used to generate relaxation oscillations. The equations for the ith, $i = 1, \ldots, N$ oscillator read

$$dv_i/dt = -a_1 m_\infty(v_i)(v_i - b_1) - a_2 w_i(v_i - b_2)$$
$$-a_3(v_i - b_3) + \Omega_i, \qquad (A.1)$$

$$\mathrm{d}w_i/\mathrm{d}t = \varepsilon_i[w_\infty(v_i) - w_i]/\tau_w(v_i), \qquad (A.2)$$

where

$$m_\infty(v_i) = 0.5[1 + \tanh\{(v_i - c_1)/c_2\}],$$

$$w_\infty(v_i) = 0.5[1 + \tanh\{(v_i - c_3)/c_4\}],$$

$$\tau_w(v_i) = 1/\cosh\{(v_i - c_3)/c_5\}$$

and t denotes time. The following parameter values were employed for all simulations: $a_1 = 1.0$, $a_2 = 2.0$, $a_3 = 0.5$, $b_1 = 1.0$, $b_2 = -0.7$, $b_3 = -0.4$, $c_1 = -0.01$, $c_2 = 0.15$, $c_3 = 0.1$, $c_4 = 0.145$, $c_5 = 0.29$. Furthermore $\Omega_i = 0.1 \ \forall i$ was applied in the standard case, whereas random perturbations were achieved by setting $\Omega_i = \zeta_i(t)$ where ζ_i is an evenly distributed random number in $[-1,1]$. The value of ε_i in Eq. (A.2) controls the frequency of the oscillation. For the standard case we used $\varepsilon_i = \{0.0004, 0.0014, 0.0024\}$ for the $N = 3$ oscillators, whereas in case of random perturbations the following values were adopted: $\varepsilon_i = \{0.00008, 0.00058, 0.00108\}$. Global mean-field coupling of the oscillators was achieved by adding the term

$$-\alpha \kappa a_1(v_i - b_1)$$

to Eq. (A.1), where

$$\kappa = \frac{1}{N} \sum_{i=1}^{N} 0.5[1 + \tanh\{(v_i - c_6)/c_7\}].$$

Here α determines the coupling strength between the oscillators. Values of α ranged from 0.0 to 0.6 as stated in the figure captions. Additional parameter values are $c_6 = 0.05$ and $c_7 = 0.15$. A fourth-order Runge–Kutta scheme was used for integrating the system of equations.

References

Alley, R.B., et al., 1995. Changes in continental and sea-salt atmospheric loadings in central Greenland during the most recent deglaciation. Journal of Glaciology 41, 503–514.

Berger, W.H., von Rad, U., 2002. Decadal to millennial cyclicity in varves and turbidites from the Arabian Sea: hypothesis of tidal origin. Global and Planetary Change 34, 313–325.

Bianchi, G.G., McCave, I.N., 1999. Holocene periodicity in North Atlantic climate and deep-ocean flow south of Iceland. Nature 397, 515–517.

Bond, G., et al., 1997. A pervasive millennial-scale cycle in North Atlantic Holocene and glacial climates. Science 278 (5341), 1257–1266.

Bond, G., et al., 2001. Persistent solar influence on North Atlantic climate during the Holocene. Science 294, 2130–2136.

Broecker, W.S., Bond, G., Klas, M., Bonani, G., Wolfli, W., 1990. A salt oscillator in the glacial Atlantic? 1. The concept. Paleoceanography 5, 469–477.

Brook, E.J., Harder, S., Severinghaus, J., Bender, M., 1999. Atmospheric methane and millennial-scale climate change. In: Clark, P.U., Webb, R.S., Keigwin, L.D. (Eds.), Mechanisms of Global Climate Change at Millennial Time Scales. Geophysical Monograph. American Geophysical Union, Washington, D.C., pp. 165–175.

Burns, S.J., Fleitmann, D., Matter, A., Kramers, J., Al-Subbary, A.A., 2003. Indian Ocean climate and an absolute chronology over Dansgaard/Oeschger events 9 to 13. Science 301 (5638), 1365–1367.

Chapman, M.R., Shackleton, N.J., 2000. Evidence of 550-year and 1000-year cyclicities in North Atlantic circulation patterns during the Holocene. The Holocene 10, 287–291.

Finkel, R.C., Nishiizumi, K., 1997. Beryllium 10 concentrations in the Greenland Ice Sheet Project 2 ice core from 3–40 ka. Journal of Geophysical Research—Oceans 102 (C12), 26699–26706.

Folland, C.K., et al., 2001. Observed climate variability and change. In: Houghton, J.T., et al. (Eds.), Climate Change 2001: The Scientific Basis. Cambridge University Press, New York, pp. 99–181.

Ganopolski, A., Rahmstorf, S., 2001. Rapid changes of glacial climate simulated in a coupled climate model. Nature 409, 153–158.

Genty, D., et al., 2003. Precise dating of Dansgaard–Oeschger climate oscillations in western Europe from stalagmite data. Nature 421, 833–837.

Grootes, P.M., Stuiver, M., 1997. Oxygen 18/16 variability in Greenland snow and ice with 10^{-3}- to 10^5-year time resolution. Journal of Geophysical Research C 102 (C12), 26455–26470.

Hall, I.R., Bianchi, G.G., Evans, J.R., 2004. Centennial to millennial scale Holocene climate-deep water linkage in the North Atlantic. Quaternary Science Reviews 23, 1529–1536.

Haug, G.H., Hughen, K.A., Sigman, D.M., Peterson, L.C., Röhl, U., 2001. Southward migration of the intertropical convergence zone through the Holocene. Science 293, 1304–1308.

Helmke, J.P., Schulz, M., Bauch, H.A., 2002. Sediment-color record from the northeast Atlantic reveals patterns of millennial-scale climate variability during the past 500,000 years. Quaternary Research 57, 49–57.

MacAyeal, D.R., 1993. A low-order model of the Heinrich event cycle. Paleoceanography 8, 767–773.

Mayewski, P.A., et al., 1997. Major features and forcing of high-latitude northern hemisphere atmospheric circulation using a 110,000-year-long glaciochemical series. Journal of Geophysical Research C 102 (C12), 26345–26366.

McManus, J.F., Oppo, D.W., Cullen, J.L., 1999. A 0.5-million-year record of millennial-scale climate variability in the North Atlantic. Science 283, 971–975.

Meeker, L.D., Mayewski, P.A., 2002. A 1400 year long record of atmospheric circulation over the North Atlantic and Asia. The Holocene 12 (3), 257–266.

Meeker, L.D., Mayewski, P.A., Twickler, M.S., Whitlow, S.I., Meese, D., 1997. A 110,000-year history of change in continental biogenic emissions and related atmospheric circulation inferred from the Greenland Ice Sheet Project Ice Core. Journal of Geophysical Research—Oceans 102 (C12), 26489–26504.

Meese, D.A., et al., 1997. The Greenland Ice Sheet Project 2 depth-age scale: methods and results. Journal of Geophysical Research C 102 (C12), 26411–26423.

O'Brien, S.R., et al., 1995. Complexity of Holocene climate as reconstructed from a Greenland ice core. Science 270, 1962–1964.

Oppo, D.W., McManus, J.F., Cullen, J.L., 2003. Deepwater variability in the Holocene epoch. Nature 422, 277–278.

Pahnke, K., Zahn, R., Elderfield, H., Schulz, M., 2003. 340,000 Year centennial-scale marine record of southern hemisphere climatic oscillation. Science 301, 948–952.

Paul, A., Schulz, M., 2002. Holocene climate variability on centennial-to-millennial time scales: 2. Internal feedbacks and external forcings as possible causes. In: Wefer, G., Berger, W.H., Behre, K.-E., Jansen, E. (Eds.), Climate Development and History of the North Atlantic Realm. Springer, Berlin, pp. 55–73.

Pikovsky, A., Rosenblum, M., Kurths, J., 2001. Synchronization: A Universal Concept in Nonlinear Sciences. Cambridge University Press, Cambridge 411pp.

Rahmstorf, S., 2003. Timing of abrupt climate change: a precise clock. Geophysical Research Letters 30 (10) 10.1029/2003GL017115.

Risebrobakken, B., Jansen, E., Andersson, C., Mjelde, E., Hevrøy, K., 2003. A high resolution study of holocene paleoclimatic and paleoceanographic changes in the Nordic Seas. Paleoceanography 18 (1), 1017 10.1029/2002PA000764.

Sakai, K., Peltier, W.R., 1997. Dansgaard–Oeschger oscillations in a coupled atmosphere–ocean climate model. Journal of Climate 10 (5), 949–970.

Saltzman, B., 2002. Dynamical Paleoclimatology. Academic Press, San Diego, 350pp.

Sarnthein, M., et al., 2001. Fundamental modes and abrupt changes in North Atlantic circulation and climate over the last 60 ky—Concepts, reconstructions and numerical modeling. In: Schäfer, P., Ritzrau, W., Schlüter, M., Thiede, J. (Eds.), The Northern North Atlantic: A Changing Environment. Springer, Berlin, pp. 365–410.

Sarnthein, M., et al., 2003. Centennial-to-millennial-scale periodicities of Holocene climate and sediment injections off the western Barents shelf 75°N. Boreas 32, 447–461.

Schulz, M., 2002. On the 1470-year pacing of Dansgaard–Oeschger warm events. Paleoceanography 17 (3) 10.1029/2000PA000571.

Schulz, M., Paul, A., 2002. Holocene climate variability on centennial-to-millennial time scales: 1. Climate records from the North-Atlantic realm. In: Wefer, G., Berger, W.H., Behre, K.-E., Jansen, E. (Eds.), Climate Development and History of the North Atlantic Realm. Springer, Berlin, pp. 41–54.

Schulz, M., Paul, A., Timmermann, A., 2002. Relaxation oscillators in concert: a frame-work for climate change at millennial timescales during the late Pleistocene. Geophysical Research Letters 29 (24) 10.1029/2002GL016144.

Somers, D., Kopell, N., 1995. Waves and synchrony in networks of oscillators of relaxation and non-relaxation type. Physica D 89, 169–183.

Spötl, C., Mangini, A., 2002. Stalagmite from Austrian Alps reveals Dansgaard–Oeschger events during isotope stage 3: Implications for the absolute chronology of Greenland ice cores. Earth and Planetary Science Letters 203, 507–518.

Stocker, T.F., Johnsen, S.J., 2003. A minimum thermodynamic model for the bipolar see-saw. Paleoceanography 18 (4), 1087, 10.1029/2003PA000920.

Stuiver, M., Grootes, P.M., 2000. GISP2 oxygen isotope ratios. Quaternary Research 53, 277–284.

Timmermann, A., Gildor, H., Schulz, M., Tziperman, E., 2003. Coherent resonant millennial-scale climate oscillations triggered by massive meltwater pulses. Journal of Climate 16, 2569–2585.

van Kreveld, S.A., et al., 2000. Potential links between surging ice sheets, circulation changes and the Dansgaard–Oeschger cycles in the Irminger Sea, 60–18 kyr. Paleoceanography 15, 425–442.

Voelker, A.H.L., et al., 2002. Global distribution of centennial-scale records for Marine Isotope Stage (MIS) 3: a database. Quaternary Science Reviews 21, 1185–1212.

Wang, Y.J., et al., 2001. A high-resolution absolute-dated late Pleistocene monsoon record from Hulu Cave, China. Science 294, 2345–2348.

Weijer, W., Dijkstra, H.A., 2003. Multiple oscillatory modes of the Global Ocean circulation. Journal of Physical Oceanography 33 (11), 2197–2213.

Winton, M., 1993. Deep decoupling oscillations of the oceanic thermohaline circulation. In: Peltier, W.R. (Ed.), Ice in the Climate System. Springer, Berlin, pp. 417–432.

Wunsch, C., 2000. On sharp spectral lines in the climate record and the millennial peak. Paleoceanography 15, 417–424.

Yiou, F., et al., 1997. Beryllium 10 in the Greenland Ice Core Project ice core at Summit, Greenland. Journal of Geophysical Research—Oceans 102 (C12), 26783–26794.

Quaternary Science Reviews 23 (2004) 2231–2246

Palaeoceanographic changes off North Iceland through the last 1200 years: foraminifera, stable isotopes, diatoms and ice rafted debris

Karen Luise Knudsen[a,*], Jón Eiríksson[b], Eystein Jansen[c], Hui Jiang[d], Frank Rytter[a], Esther Ruth Gudmundsdóttir[b]

[a]*Department of Earth Sciences, University of Aarhus, DK-8000 Århus C, Denmark*
[b]*Science Institute, University of Iceland, IS-101 Reykjavík, Iceland*
[c]*Bjerknes Centre for Climate Research and Department of Earth Science, University of Bergen, N-5007 Bergen, Norway*
[d]*Laboratory of Geographic Information Science, East China Normal University, Shanghai 200062, PR China*

Abstract

Palaeoclimatic changes through the last 1200 calibrated years have been documented by high-resolution multi-proxy studies of three cores from about 400 m water depth on the North Icelandic shelf. Benthic and planktonic foraminiferal assemblages and stable isotope values, as well as ice rafted debris (IRD) concentrations, are compared with diatom-based sea-surface water temperatures and the reconstructed mean temperature for the Northern Hemisphere. Changes in surface and bottom water characteristics are mainly due to variations in the strength of the relatively warm, high-salinity Irminger Current and the cold East Icelandic Current. The time period between 1200 and around 7–800 cal. (years) BP, including the Medieval Warm Period, was characterized by relatively high bottom and surface water temperatures due to the inflow of Atlantic water masses. After that, a general temperature decrease in the area marks the transition to a period with increased influence of the East Icelandic Current and, at the sea floor, the Norwegian Sea Deep Water. This corresponds to the transition to the Little Ice Age. After about 3–400 cal. BP, the inflow of cold East Icelandic Current was further enhanced. In particular, this had a strong influence on the surface waters, while the sea floor was under some influence of Atlantic water masses, resulting in stratification of the water masses. There is no clear indication of any warming in the area during the last decades.
© 2004 Elsevier Ltd. All rights reserved.

1. Introduction

The North Icelandic shelf is located in an area of sharp oceanographic and atmospheric gradients. The northern part of the shelf is close to the marine Polar Front (Fig. 1) separating the cold, low-salinity East Icelandic Current with an Arctic component in the north and the relatively warm, high-salinity Irminger Current, an offshot of the North Atlantic Current, that flows clockwise around Iceland (e.g. Stefánsson, 1962; Swift, 1986; Hansen and Østerhus, 2000). The area is also within the realm of the Arctic Front, that delineates the

maximum extent of winter sea ice, which periodically extends from Greenland to Iceland (i.a. Hurdle, 1986).

The strength of the Irminger Current is generally related to deep water formation in the Nordic Seas (e.g. Malmberg and Jónsson, 1997), and it is expected to be strong during active deep water formation and weak during periods of freshening of the surface waters of the Nordic Seas, i.e. during periods with strong input of Polar water from the East Greenland and the East Icelandic Current. Lateral and vertical mixing as a result of the parallel flow of the Irminger Current and the East Icelandic Current and winter cooling leads to the formation of an intermediate water mass in the area which disintegrates by spring and summer. At present, the Norwegian Sea Deep Water replaces the mixed surface water masses at a depth of 3–400 m off North

*Corresponding author. Tel.: +45-8942-3557; fax:+45-8618-3936.

E-mail address: karenluise.knudsen@geo.au.dk (K.L. Knudsen).

0277-3791/$ - see front matter © 2004 Elsevier Ltd. All rights reserved.
doi:10.1016/j.quascirev.2004.08.012

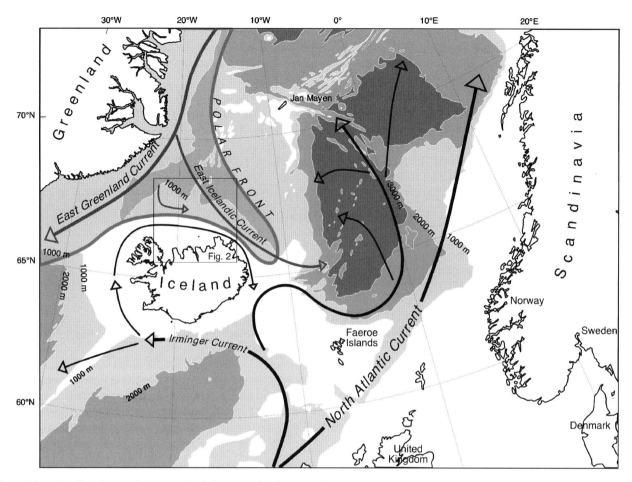

Fig. 1. The regional modern surface water circulation around Iceland and the position of the Polar Front. Depth contour intervals 1000 m. Modified after Hurdle (1986), Map 8, and Knudsen and Eiríksson (2002).

Iceland (Stefánsson, 1962), and the modern salinity and temperature data from the area (Figs. 2 and 3) show that the cold deep water masses may be expected to encroach into topographic lows and basins on the shelf during periods of active deep water formation in the Nordic Seas. Recently, Jónsson and Valdimarsson (2004) detected a new path for Denmark Strait overflow water along the North Icelandic slope. A strong westward current was found to have its velocity core as high up as 500 m depth on the slope. The cold deep waters, that crawl high up over the Icelandic slope, would presumably also bring Norwegian Sea Deep Water (below 0 °C) to the deeper parts of the shelf.

Instrumental records show that the northern North Atlantic has experienced oceanographic anomalies on a decadal time-scale in recent times, the most prominent ones being the so-called salinity anomalies (Dickson et al., 1988; Belkin et al., 1998). These periods were generally initiated by increased advection of Arctic sea water into the the North Atlantic either via the Fram Strait or the Canadian Archipelago (Belkin et al., 1998). A strengthened East Greenland Current and associated increase in freshwater and sea-ice input along the east coast of Greenland would not necessarily cause a

freshening of the surface waters on the North Icelandic shelf. However, there would be an enhanced effect of atmospheric conditions (local northerly winds) on the advection of sea ice and freshwater along the East Icelandic Current (see also Ólafsson, 1999; Knudsen and Eiríksson, 2002).

Changes in sea-surface temperature and salinity during the time period 1948–2002, including the Great Salinity Anomaly (Dickson et al., 1988) in the 1960s and early 1970s, are shown in Fig. 4. The measurements are from the Siglunes section off North Iceland (Figs. 2 and 3) at a site very close to cores HM107-01 and HM107-03. Since the available data sets from the spring and summer months are much more complete than those from the winter months, spring (mean values of April–June) has been chosen as a representative for the cold part of the year. It appeared from the data that the mean spring temperatures are not much higher than those from the winter months (mean January–March), and that the salinities are almost the same during winter and spring. The warm summer season is shown as mean values for July–September. The time series (Fig. 4) clearly show a period of lowered temperatures and salinities both at 20 and 50 m depth during the Great

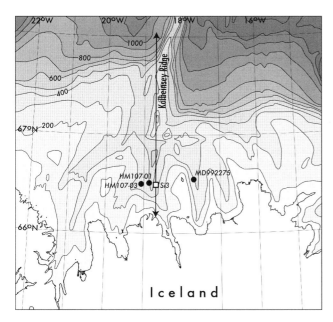

Fig. 2. Location of cores off North Iceland (contour intervals 100 m). The arrow indicates the location of the oceanographic section of Fig. 3. Si3 = the oceanographic site of Fig. 4.

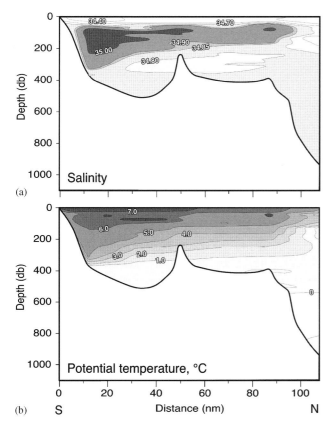

Fig. 3. Oceanographic section from the north coast of Iceland towards north across the shelf to the slope (the Siglunes hydrographic profile, Fig. 2). Summer salinity (a), and temperature (b), data (August 1998) are presented as examples of the distribution of water masses (data from the Marine Research Institute, Reykjavík, Iceland, Hedinn Valdimarsson, pers. comm.). db = decibar ≈ 1 m; nm = nautical mile.

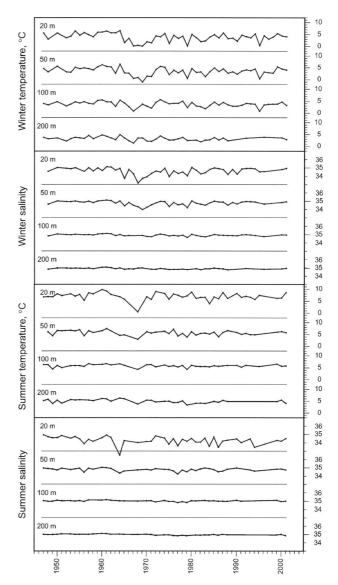

Fig. 4. Seasonal temperature and salinity measurements at one specific site (Si3) in the Siglunes profile, close to cores HM107-01 and HM107-03. Salinity and temperature were measured at 20, 50, 100 and 200 m depths. Spring measurements (mean of April–June) represent the cold season (winter, see text), and summer measurements (mean of July–September) show the warm season (data from the Marine Research Institute, Reykjavík, Iceland, Hedinn Valdimarsson, pers. comm.).

Salinity Anomaly. In addition, the measurements at a water depth of 100 and 200 m show that only small temperature variations occurred and that the salinity has been fairly constant through the period of the Great Salinity Anomaly at these depths.

Historical and instrumental data as well as palaeo-data show significant fluctuations in the position of the frontal systems north of Iceland through the Holocene (Ogilvie, 1996; Eiríksson et al., 2000; Knudsen and Eiríksson, 2002; Knudsen et al., 2004). Even relatively small changes in the current pattern in the northern North Atlantic and in the atmospheric circulation above

Greenland are likely to become archived in the sedimentary record in this sensitive boundary region. Ultra high sedimentation rates in the shelf sediments (up to 200 cm/ka) yield the possibility of a maximum time resolution of 5–10 years per sample (1-cm sediment slices). In addition, the presence of historical and terrestrially dated air-fall tephra markers from Icelandic volcanoes makes it possible to construct reliable age models for the marine cores, minimising the problem of changes in the marine reservoir ages related to incursions of Polar and Arctic water masses with a reservoir age, which is higher than the 400 years that is generally obtained for Atlantic water masses (Knudsen and Eiríksson, 2002; Larsen et al., 2002; Eiríksson et al., 2004).

This paper presents a detailed reconstruction of climatic changes through the last 1200 years based on high-resolution proxy data from three North Icelandic shelf cores. An extensive data set of modern variables, including foraminifera and diatoms, has previously been compiled for the area (Jiang et al., 2001; Rytter et al., 2002) as a basis for the reconstruction of the history of marine environmental changes in the region. The study of benthic as well as planktonic foraminiferal distribution and their oxygen isotope values allows us to discuss temporal variations in the stratification of the water masses as well. The present results are discussed in relation to previously described climatic indications from both marine and atmospheric data in the region, including the Medieval Warm Period and the Little Ice Age.

2. Material and methods

Core locations for material discussed in the present study are shown in Fig. 2. Two gravity cores, HM107-01 (66°30′29″N; 18°51′54″W, 464 m water depth) and HM107-03 (66°30′09″N; 19°04′20″W, 400 m water depth), were obtained on the *R/V Haakon Mosby* BIOICE cruise in 1995. Two supplementary boxcores, 2730 and 2732, were taken at the same locations. The MD992275 core (66°33′06″N; 17°41′59″W, 410 m water depth) is a piston core retrieved on the *R/V Marion Dufresne* IMAGES V cruise in 1999.

Due to temporal variations in marine reservoir ages through the Late Holocene off North Iceland (Eiríksson et al., 2000; Knudsen and Eiríksson, 2002; Larsen et al., 2002), the chronology for the three cores in this work is based on tephrocronological age models only (for age zones, see Eiríksson et al., 2004). The age models for cores HM107-01 and HM107-03 are constructed on the basis of three tephra marker horizons, Hekla 3, Hekla AD 1104 and KOL AD 1372. The age of Hekla 3 (2879 ± 34 ^{14}C BP; Dugmore et al., 1995; Eiríksson et al., 2000) is determined by radiocarbon datings on terrestial

material on Iceland, while the Hekla AD 1104 and the KOL AD 1372 are historically dated (see also Eiríksson et al., 2004). The top part of each of the gravity cores HM107-01 and HM107-03 is supplemeted with samples from boxcores obtained at the same sites. The age of the combined records thus corresponds to the time of sampling, i.e. AD 1995. The boxcores have been linked to the gravity cores by comparison of sedimentological, micropalaeontological and isotopic data.

The age model for core MD992275 is based on seven tephra marker horizons, The Snæfellsjökull 1, the Settlement layer, Hekla AD 1104, Hekla AD 1300 and Veidivötn AD 1410, 1477 and 1717 (see also Larsen et al., 2002; Eiríksson et al., 2004). The age of the Snæfellsjökull 1 tephra (1855 ± 55 ^{14}C BP; Larsen et al., 2002) is based on radiocarbon dates of terrestrial material, the Settlement layer, 1080 cal. BP (AD 871 and 877; Grönvold et al., 1995; Zielinski et al., 1997), on correlation with the GRIP and GISP2 ice cores, while the upper five tephra layers are historically dated. The age of the top of core MD992275 is estimated at AD 1950 on the basis of ^{210}Pb determinations of the top part of the piston core.

The samples for ice rafted debris (IRD) analyses and for foraminiferal and diatom analyses of the three cores, each represent a 1-cm thick sediment slice, and the sediment records were analysed at 1–5 cm intervals.

IRD is defined as sediment released from melting ice in the ocean and deposited on the sea floor. As a proxy for this process, specific minerals and grains, i.e. crystals (mainly quartz), rock fragments, and altered volcanics showing signs of abrasion were counted in the size fraction 125–1000 μm. In order to separate IRD from primary air-fall tephra from the frequent volcanic events in Iceland, it is necessary to check the mineralogy of sand grains in the cores. At least 300 grains were counted in each sample, and the IRD concentration is reported as flux (number of grains/cm²/year), using bulk sediment density (g/cm³) and sedimentation rate (cm/year) for each of the cores. This grain size has also been used effectively in the Nordic Seas to identify IRD variability (Henrich et al., 1995). The provenance of the sand grains in North Icelandic shelf sediments is discussed in Eiríksson et al. (2000), Knudsen and Eiríksson (2002) and Knudsen et al. (2004).

Since all the sediments reported here from the North Icelandic shelf were deposited at water depths of around 400 m, reworking by storm events can be ruled out (see e.g. Leeder, 1999). The coring sites are all located in basins dominated by mud deposition from suspension and weak bottom currents transporting sortable silt. The cores were checked for possible turbidity current deposits on X-ray films, and no signs of graded beds or sorting events were observed.

Foraminiferal contents were analysed in the 125–1000 μm size fraction by counting at least 300 specimens of benthic as well as planktonic foraminifera, when possible. The benthic foraminiferal percentages are calculated on the basis of calcareous species only, because the fossil preservation of arenaceous species is poor. The benthic foraminiferal flux (number of specimens/cm^2/year) is also reported for each of the cores. The planktonic assemblages are dominated by sinistrally coiled *Neogloboquadrina pachyderma* (Ehrenberg) (NPS). The percentages of this species are calculated relative to the total planktonic content.

Oxygen isotopes were measuered on the planktonic sinistrally coiled *N. pachyderma* as well as on the benthic species *Islandiella norcrossi* (Cushman) from the 125–1000 μm fraction of each foraminiferal sample, when possible. The $\delta^{18}O$ values were measured on a Finnigan MAT 251 mass spectrometer at the Stable Isotope Laboratory, University of Bergen, Norway. Results are given with respect to PeeDee Belemnite, after calibration to the National Institute of Standard and Technology NPS 19 standard.

Based on the oxygen isotope composition, palaeo-temperatures were calculated, using the Shackleton (Shackleton and Opdyke, 1973) equation. In this calculation, the oxygen isotopic composition of sea water was assumed to be zero. For the benthic temperature estimates, this is a fair assumption, given that the salinity of the water masses is on average 34.9 at this location with less than 0.1 variability, seasonally or interannually, over the past 30 years available instrumental time series from the nearby Siglunes section and the 55-year time series available for 200 m water depth (Fig. 4). This salinity variability would account for less than 0.05 per million change in the oxygen isotopic composition of sea water, regardless of the salinity vs. oxygen isotope relation applied, and would thus impose an uncertainty in the temperature estimate of not more than 0.35 °C. Due to the uncertainties in terms of possible deviations from oxygen isotopic thermodynamic equilibrium, we did not impose any vital effect correction on the results. The salinity variability is somewhat larger for the upper 100–50 m depth (Fig. 4), which is the likely appropriate depth for the calcification of *N. pachyderma* sinistral. The mean salinity is close to that of the bottom water, characteristic of Arctic water masses, hence the zero value assumption for the oxygen isotopic composition of sea water, which was applied in the calculation is well founded. The interannual salinity change is of the order of 0.5 over the past 55 years. If this is also true for the longer time scales, such a variation would account for approximately 1–1.5 °C uncertainty in the oxygen isotope-based temperature estimates based on planktonic foraminifera. Since colder waters are generally less saline, the effect is such that salinity changes would lead to an underestimation of the cooling when colder, less saline Polar waters affect the area.

A modern diatom data set with nine environmental variables from around Iceland was established. It covers areas with adequate environmental gradients for the deduction of quantitative relationships between diatom species and environmental variables (Jiang et al., 2001; Jiang et al., 2002). Non-linear ordination method of canonical correspondence analysis (CCA) was used to elucidate the relationships between diatom assemblages and environmental variables in the area. A Monte Carlo permutation test (99 permutations) with forward selection shows that summer sea-surface temperature (SSTs) captures 30% of the total variance, suggesting that the SSTs largely controls the distribution of diatoms from around Iceland and is therefore a potential environmental variable for quantitative reconstruction.

The SSTs record for core MD992275 is reconstructed using the CALIBRATE program (Juggins and ter Braak, 1992). Six numerical reconstruction methods (Juggins and ter Braak, 1992; Jiang et al., 2002) were tested for the extended modern diatom-SST data set. The results show that the best performances are weighted averaging partial least squares (WA-PLS) using six components for the SSTs as inferred from apparent root mean squared error (RMSE) and root mean squared error of prediction based on the leave-one-out jackknifing (RMSEP$_{(Jack)}$) test. The RMSE for the SSTs is 0.49 and RMSEP$_{(Jack)}$ is 0.98 °C.

3. Benthic foraminiferal assemblages

The percentage distributions of eight selected species of benthic foraminifera in the interval 0–1200 cal. BP of the three cores MD992275, HM107-03 and HM107-01 are shown in Figs. 5–7. The displayed species are among the most common benthic foraminifera. Due to different environmental preferences for each of these species, the assemblage composition indicates specific bottom water-mass properties. The benthic foraminiferal flux (no/cm^2/year), which is a measure of the productivity in the area, is shown for each of the cores as well. It should be noted that the flux values are generally much higher in core MD992275 than in cores HM107-03 and HM107-01. Since the abundance distributions and the percentage distributions for each species generally show similar patterns, only the relative distributions are shown.

3.1. Core MD992275

The percentage distribution of benthic species varies considerably through the last 1200 years in core MD992275 (Fig. 5). The Atlantic water indicator *Cassidulina neoteretis* (Seidenkrantz) (see i.a. Seidenkrantz, 1995; Hald and Steinsund, 1996; Rytter et al., 2002) is most common in the lower and again in the uppermost part of the record. A marked decrease

MD992275

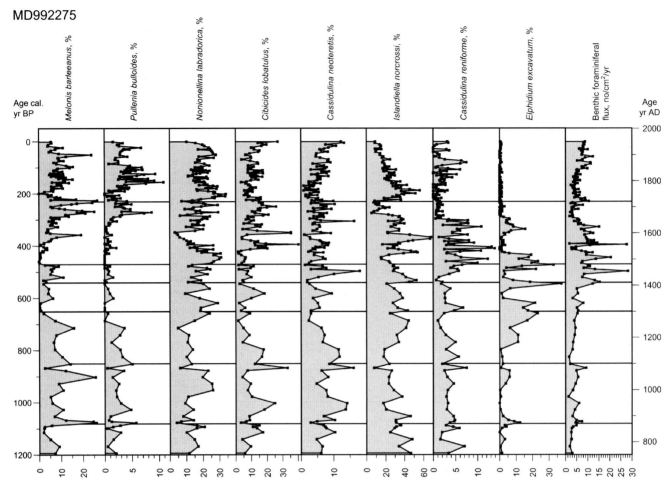

Fig. 5. Benthic foraminifera in core MD992275: percentage distribution of eight selected species and total benthic foraminiferal flux (no/cm²/year) in the time interval 0–1200 cal. BP. Tephra layers are marked with horizontal lines (see also Fig. 8).

occurred at around 700 cal. BP, and a subsequent gradual increase is seen through the last ca 200 years. There is a distinct inverse relationship between this species and the cold water indicator *I. norcrossi* (i.a. Nørvang, 1945; Rytter et al., 2002). *Nonionellina labradorica* (Dawson), which is an indicator for the position of the Polar Front (e.g. Hald and Steinsund, 1996; Rytter et al., 2002), became especially abundant after ca 700 cal. BP. This is also the level of a marked change to very low frequencies of the two species *Melonis barleeanus* (Williamson) and *Pullenia bulloides* (d'Orbigny), which again became an important part of the assemblages from about 400 and 300 cal. BP, respectively. These two species, which show similar distribution patterns in the core, are found to have highly resembling ecological preferences (Rytter et al., 2002). They are indicators of a specific nutrient supply, i.a. preference for slightly decomposed organic matter (Mackensen et al., 1985; Caralp, 1989). It is remarkable that the period with minimum amounts of these two species (700 to 3–400 cal. BP) closely corresponds to the interval of low productivity (low flux values) in the area

(Fig. 5). Low percentages of the Atlantic water indicator *C. neoteretis* and maximum amounts of the Polar Front indicator *N. labradorica* were found in the same time interval.

The opportunistic cold-water species *Elphidium excavatum* (Terquem), which occurs as the arctic forma *clavata* Cushman (see Feyling-Hanssen, 1972; Hald et al., 1994), was particularly common, although variable, between 700 and 300 cal. BP, after which it almost disappeared. Maximum percentages of the arctic species *Cassidulina reniforme* Nørvang (see i.a. Nørvang, 1945; Rytter et al., 2002) were found between 500–300 cal. BP. This corresponds in time to the period of maximum flux values. After that it displayed low frequencies.

In summary, there is strong faunal evidence of a period with relatively cold bottom-water conditions between 700 and 300 cal. BP. In addition, decreased frequencies of the high energy indicator *Cibicides lobatulus* (Walker and Jacob) point to low bottom current velocity during the same period of time, probably due to low influence of Irminger Current waters at the sea floor.

HM107-03

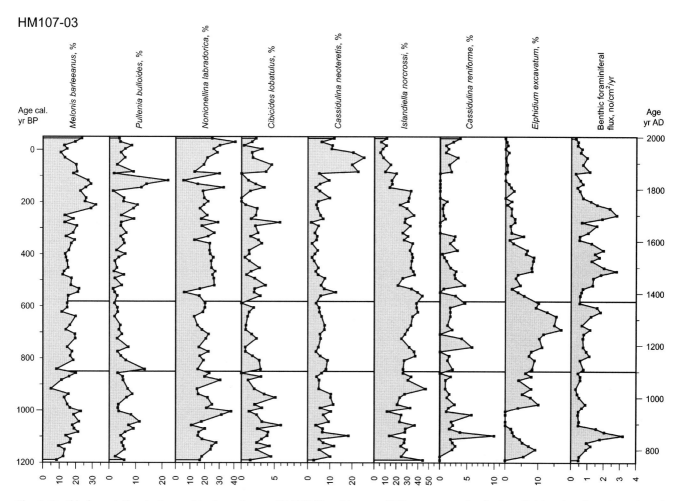

Fig. 6. Benthic foraminifera in the combined gravity core HM107-03 and boxcore 2732: percentage distribution of eight selected species and total benthic foraminiferal flux (no/cm²/year) in the time interval 0–1200 cal. BP. Tephra layers are marked with horizontal lines (see also Fig. 9).

3.2. Core HM107-03

The general distribution pattern for most of the benthic species in core HM107-03 (Fig. 6) is similar to that found in core MD992275, even though the variations are less distinct. Relatively low percentage values of *C. neoteretis*, *P. bulloides* and *C. lobatulus* occur between 8–700 and 300 cal. BP, and an increase in *M. barleeanus* is also seen through the last 300 years (since 250 cal. BP). The arctic species *C. reniforme* and *E. excavatum* decreased in frequency after 300 cal. BP. As recorded in core MD992275, an inverse relationship between the Atlantic water indicator *C. neoteretis* and the arctic *I. norcrossi* is clearly seen.

Some clear trends are also seen in the faunal distributions of the uppermost part of the record. The Atlantic water indicator *C. neoteretis* and the high-energy indicator *C. lobatulus* increase, while several cold-water species, including *I. norcrossi* and *E. excavatum,* decrease in frequency. This suggests a gradual increase in the inflow of the Irminger Current at the sea floor during the last ca 200 years. In the topmost part

there is, however, also an increase in the Polar Front indicator species *N. labradorica*.

3.3. Core HM107-01

In some respects, the benthic foraminiferal distribution exhibits a slightly different pattern at the deeper core site HM107-01 (Fig. 7) than in the other two cores. The total amount of arctic species is higher at this core site, and the amount of the Atlantic water indicator *C. neoteretis* is lower than in the other cores. As seen in cores MD992275 and HM107-03, there is also a marked decrease in arctic species *E. excavatum*, *C. reniforme* and *I. norcrossi* through the last 300 at this core site, but in contrast to the others, there is also a decrease in Atlantic water species *C. neoteretis*. The percentages of the Polar Front indicator *N. labradorica*, however, increase considerably through the last ca 150 years, while the percentages of all other benthic species decrease. During this interval, the abundances of all species decrease. In general, the temperature at the sea floor was lower at site HM107-01 than at sites MD992275 and HM107-03, and

HM107-01

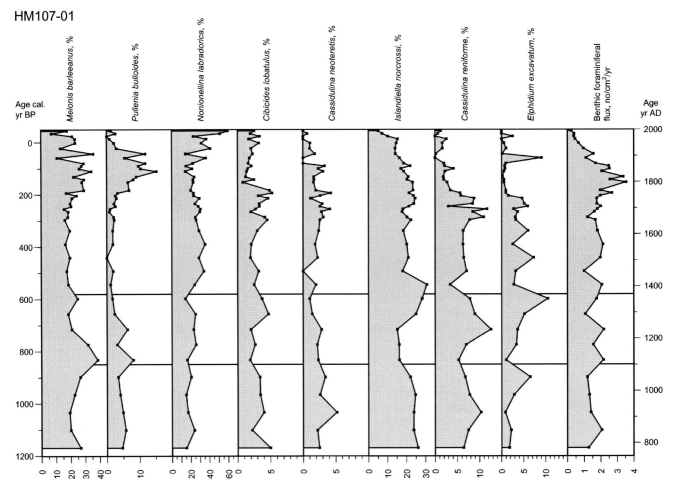

Fig. 7. Benthic foraminifera in the combined gravity core HM107-01 and boxcore 2730: percentage distribution of eight selected species and total benthic foraminiferal flux (no/cm^2/year) in the time interval 0–1200 cal. BP. Tephra layers are marked with horizontal lines (see also Fig. 10).

in contrast to the other cores, there is faunal evidence of a temperature decrease through the last few hundred years.

As seen in cores MD992275 and HM107-03, the two species *M. barleeanus* and *P. bulloides* show distinct minimum values between about 700 and 300 cal. BP. This would imply that the overall changes in the supply of organic matter was similar throughout the area. Maximum flux values, however, occur later at core HM107-01 than at the other core sites, i.e. around 200–100 cal. BP.

4. Palaeoclimatic proxies and results

Selected palaeoclimatic proxies for surface waters at each of the three cores sites are shown in Figs. 8–10. In order to obtain an overview of possible changes in the stratification of the water masses, benthic oxygen isotope values are also presented. Surface and bottom-water temperature records, reconstructed based on the oxygen isotopes (Figs. 11 and 12), and the isotopic

results are evaluated and discussed in relation to the faunal distribution patterns described above.

4.1. Core MD9922275

The percentages of NPS in core MD992275 are generally very high, particularly after 650 cal. BP (Fig. 8). This indicates that cold surface waters prevailed at the core site, and that the temperatures have been especially low since 650 cal. BP. There is an indication of periodically higher temperatures around 500–400 cal. BP, and a distinct cold peak occurred at around 350 cal. BP. After that, low temperature conditions prevailed until a slight rise occurred at about 100 cal. BP, followed by generally higher, but fluctuating values.

The planktonic (NPS) oxygen isotope values are relatively low in the interval between 1200 and 400 cal. BP, though with a small change to heavier values from about 800 cal. BP. After that, there is a distinct drop to heavy values at around 350 cal. BP, which is in accordance with the planktonic foraminiferal indication (high NPS percentage) of especially cold surface waters at that time.

MD992275

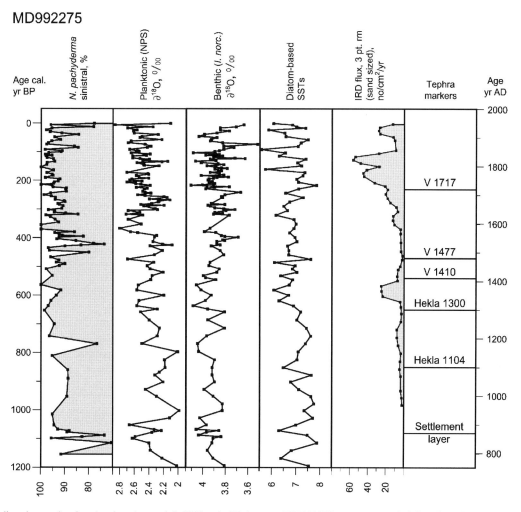

Fig. 8. Palaeoclimatic proxies for the time interval 0–1200 cal. BP in core MD992275: percentage of sinistrally coiled *N. pachyderma* (NPS), planktonic $\delta^{18}O$ (NPS), benthic $\delta^{18}O$ (*I. norcrossi*), diatom-based reconstruction of the SSTs and IRD flux (no/cm²/year, 3-point running mean). Tephra layers are marked on the right-hand side. V = Veidivötn (see also Eiríksson et al., 2004).

In general, the diatom-based reconstructed SSTs record (Fig. 8) appears to be in phase with the planktonic isotope fluctuations through the lower part of the record, i.e. relatively heavy isotope values correspond to periods of low sea-surface temperatures. The first marked drop in SSTs is seen at around 650 cal. BP and a second one occurs at 350 cal. BP. The latter temperature drop is especially distinct for the winter sea-surface temperature (SSTw) (unpublished data) and corresponds in time to the indication of a considerable drop in temperature by planktonic assemblages as well as heavier isotope values. After 150 cal. BP, the diatom-based SSTs values fluctuate greatly, with occasionally very low temperatures. The planktonic oxygen isotope values show a similar pattern. Due to the diatom indication of generally low, but fluctuating, sea-surface temperatures in this time interval, these relatively light isotope values and fluctuations are assumed to be a result of the influence of freshwater rather than an increase in temperature.

The benthic and planktonic oxygen isotope curves generally display a similar pattern between 1200 and about 300 cal. BP. The relatively high temperatures between 1200 and 700 cal. BP, as indicated by benthic foraminiferal assemblages, are, however, not revealed by the oxygen isotopes. Instead these show heavier values, which is suggested to be due to high-salinity Atlantic water masses at the sea floor. After 300 cal. BP, the benthic and planktonic oxygen isotope curves are out of phase. In contrast to the planktonic values, the benthic values show a decreasing trend towards the top of the record. This corresponds to the benthic faunal indication of an increasing influence of Atlantic water masses. The relatively higher salinity of this water mass compared to the Arctic waters would suppress an increase in oxygen isotope values due to a temperature rise.

The first slight increase in IRD flux values from a low background level is seen at about 800 cal. BP in core MD992275. This is contemporaneous with a decrease in

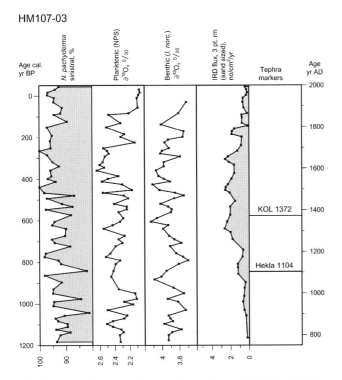

Fig. 9. Palaeoclimatic proxies for the time interval 0–1200 cal. BP in core HM107-03 combined with boxcore 2732: percentage of sinistrally coiled *N. pachyderma* (NPS), planktonic δ^{18}O (NPS), benthic δ^{18}O (*I. norcrossi*) and IRD flux (no/cm^2/year, 3-point running mean). Tephra layers are marked on the right-hand side (see also Eiríksson et al., 2004).

planktonic oxygen isotope values indicating increased influence of the East Icelandic Current water. A short-term IRD peak occurs between 600 and 550 cal. BP, corresponding in time to the first pronounced drop in diatom-based reconstructed sea-surface temperatures (Fig. 8). After about 350 cal. BP, there was a gradual increase in IRD, reaching maximum values for the record between 200 and 100 cal. BP (AD 1750–1850). A last minor peak is seen shortly after AD 1900.

4.2. Core HM107-03

The percentages of NPS fluctuate but are generally higher than 90 in core HM107-03 (Fig. 9). Lower percentages occur sporadically until ca 800 cal. BP, but an increase to maximum values occurs after ca 500 cal. BP. This indicates that Arctic water masses prevailed throughout the record, but that especially cold surface waters influenced the area during deposition of the later part. There is a general agreement between the fluctuations in NPS percentages and the planktonic and benthic oxygen isotope curves between 1200 and about 300 cal. BP. The planktonic isotope values indicate particularly low sea-surface temperatures at around 350 cal. BP, which is in agreement with observations from core MD992275. The relatively high bottom-water

temperatures towards the top of the record, as indicated by the benthic foraminifera after 300 cal. BP, would be suppressed in the benthic isotopic record due to the higher salinities of the Atlantic waters. This is also the case in core MD992275.

The IRD flux in core HM107-03 increases after ca 850 cal. BP. The period with high IRD fluxes corresponds to an interval of generally heavy planktonic oxygen isotope values, i.e. indication of a relatively high influence of the cold East Icelandic Current. The IRD flux values at this core site decreased again during the last 200 years to a level of almost zero. The general pattern of the IRD record is similar to that found in core MD992275, but the absolute values are much lower at this core site.

4.3. Core HM107-01

The percentages of NPS are generally over 90 in core HM107-01, indicating a sustained high influence of Arctic water masses from the East Icelandic Current (Fig. 10). Maximum values occur at about 500 cal. BP. The planktonic oxygen isotope values show an increasing trend through the lower part, reaching maximum values, i.e. minimum temperatures, at around 600 cal. BP. Minimum values occur at around 550 cal. BP, after an 0.8 per million reduction. This may either be due to a freshening of surface waters or a temperature increase. If calculated as salinity decrease, it would amount to approximately 1.3. This is probably a minimum value, since reduced temperatures would normally follow lower salinities in this area, hence reducing the oxygen isotopic change associated with reduced salinity. Such a high salinity reduction appears unlikely in view of the other data, as similar low salinities are typical only of the coldest and most coastal elements of the East Greenland Current. Thus, it seems more likely that the shift is a temperature effect. After this event, higher values prevailed between 400 and 200 cal. BP. This was followed by a decrease after 200 cal. BP. The mean value of the planktonic oxygen isotope values for the last 1200 years at site HM107-01 (2.44 per mil) is similar to the mean values of the record at the two other core sites (HM107-03: 2.38; MD992275: 3.43), suggesting that the sea-surface conditions in the area have experienced similar environmental development during this time interval.

The benthic oxygen isotope values are rather constant in core HM107-01, but decrease gradually to a slightly lower level and become more fluctuating after 300 cal. BP. The isotopic values suggest that the bottom-water temperatures were lower at site HM107-01 (mean value 4.07 per mil) than at the two other sites (HM107-03: 3.91; MD992275: 3.89), a result that fits the benthic foraminiferal distribution, which indicates consistently cold bottom-water conditions at the sea floor at HM107-01. This is presumably due to the location of

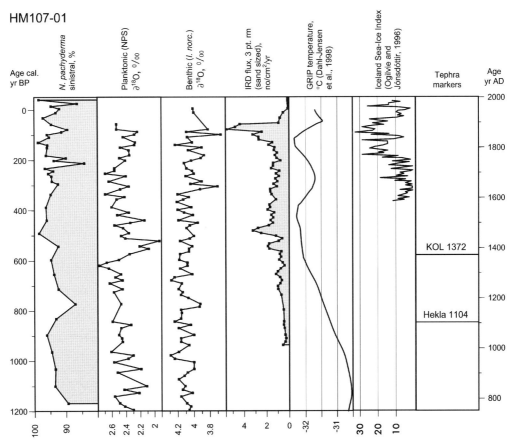

Fig. 10. Palaeoclimatic proxies for the time interval 0–1200 cal. BP in core HM107-01 combined with boxcore 2730: percentage of sinistrally coiled *N. pachyderma* (NPS), planktonic δ^{18}O (NPS), benthic δ^{18}O (*I. norcrossi*) and IRD flux (no/cm²/year, 3-point running mean). The proxies are compared with the reconstructed ice-sheet surface temperature changes at the GRIP ice-core (Dahl-Jensen et al., 1998) and with the Sea-Ice Index of Ogilvie and Jónsdóttir (1996). Tephra layers are marked on the right-hand side (see also Eiríksson et al., 2004).

this site in a deep trough, influenced by Norwegian Sea Deep Water enchroaching from the north. Also its location to the west of the Kolbeinsey Ridge may have an influence, since this ridge may change the local circulation of the East Icelandic Current and cause increased inflow of cold Polar surface waters to the site (see also Knudsen et al., 2004).

The pattern of the IRD flux record in core HM107-01 is remarkably similar to that in core MD992275, even though the absolute concentrations are considerably lower (Fig. 10). In both cores, the IRD flux values increase from a low background level at about 800 cal. BP in HM107-01. This occurs during a period of increasing planktonic oxygen isotope values, reflecting an increase in the influence of the East Icelandic Current. The first part of a double peak in IRD flux corresponds to a pronounced freshening reflected by low oxygen isotope values between 600 and 500 cal. BP, while the second part occurs shortly after the level of maximum in NPS percentages at 500 cal. BP. A later increase in the IRD flux from about 150 cal. BP (AD 1800) peaks shortly after AD 1850-1900, after which the flux decreases to a level of almost zero.

5. Comparison and discussion

A comparison between the benthic oxygen isotope-based reconstruction of bottom-water temperatures for cores HM107-03, HM107-01 and MD992275 is shown in Fig. 11, from the west to the east. Some general trends are similar, but in certain intervals the indications are out of phase. The above comparison of the isotopic results with benthic foraminiferal assemblages clearly shows that the salinity at the sea floor has varied through time in the area. In two of the cores, MD992275 and HM107-03, the influence of relatively warm, high-salinity Atlantic water masses was rather strong in the time period between 1200 and 700 cal. BP, and again after about 3–400 cal. BP. The sea floor at the deeper core site HM107-01, however, appears to have been influenced by cold Norwegian Sea Deep Water.

The difference between the oxygen isotope-based reconstruction of surface and bottom-water temperatures for core MD992275 is also shown in Fig. 11. Due to a relatively high bottom-water salinity in the time interval between 1200 and 7–800 cal. BP, the

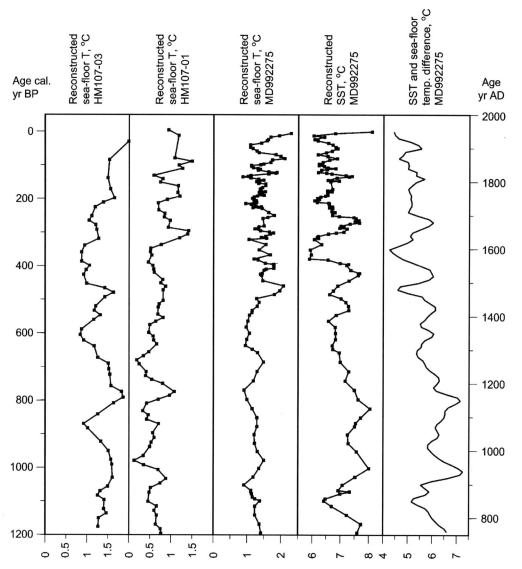

Fig. 11. Reconstructed bottom-water temperatures (3-point running mean) based on benthic oxygen isotopes (see text), from the west to the east on the North Icelandic shelf (cores HM107-03, HM107-01 and MD992275). On the right-hand side, the reconstructed sea-surface temperature record for core MD992275 (3-point running mean) is shown as well as the temperature difference between surface and bottom waters in core MD992275. This difference is calculated by interpolation of fixed 20 years distances between data points.

reconstructed temperature difference is presumably overstimated for that interval. For similar reasons, the temperature difference was actually also smaller than indicated by the reconstruction after 400 cal. BP. The highest temperature difference thus occurred between 700 and 400 cal. BP, corresponding to the early part of the Little Ice Age.

A comparison of sea-surface temperature reconstructions for cores HM107-03, HM107-01 and MD992275, from the west to the east, is shown in Fig. 12. The same general patterns are seen until about 300 cal. BP. An out-of-phase pattern in the curves after that is interpreted as a result of salinity fluctuations during a period of relatively high inflow of Arctic waters to the area. A higher salinity than 34.94 would imply that the calculated bottom temperatures are somwhat too low

for the intervals with raised salinities. Depending on the salinity/oxygen isotope slope, an 0.4 increase in salinity (which is probably a maximum value using modern hydrographical characteristics) would account for approximately 0.5 to 1 °C error towards the lower side in the calculated bottom temperature.

A comparison between the reconstructed sea-surface temperature curve based on isotope measurements in core MD992275 and the diatom-based sea-surface temperature curve (yearly mean, Fig. 12) show a high degree of correlation until about 300 cal. BP, when the freshwater influence in the area increases. The pattern of the diatom-based sea-surface temperature curve for the North Icelandic shelf is similar to that reconstructed from magnesium/calcium palaeothermometry from the Chesapeake Bay, Eastern Canada (Cronin et al., 2002)

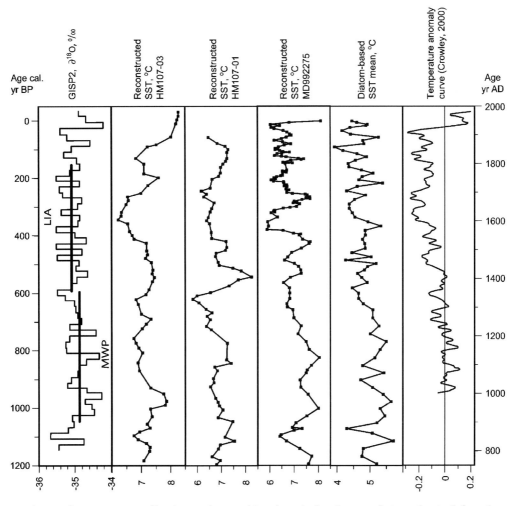

Fig. 12. Reconstructed sea-surface temperatures (3-point running mean) based on planktonic oxygen isotopes (see text), from the west to the east on the North Icelandic shelf (cores HM107-03, HM107-01 and MD992275), as well as the diatom-based sea-surface temperature reconstruction (annual mean, 3-point running mean) for core MD992275. The results are compared with the GISP2 oxygen isotope curve (20-year mean, Grove, 2001) and the reconstructed Northern Hemisphere temperature anomaly curve of Crowley (2000). MWP = Medieval Warm period, LIA = Little Ice Age.

and also to the Northern Hemisphere anomaly curve of Crowley (2000) (Fig. 12).

The time interval between 1200 and 7–800 cal. BP, when the Irminger Current had a marked influence on the North Icelandic shelf, particularly on the sea floor, corresponds in time to the Medieval Warm Period. The temperature drop, which is reflected in the data both for the bottom and surface waters after about 700 cal. BP (Fig. 12) can be correlated with the onset of the Little Ice Age. In a core from Reykjafjördur, further to the west on the North Icelandic shelf (Andrews et al., 2001a), the Little Ice Age is also marked by an increase in the cold water benthic foraminifera *E. excavatum* forma *clavata*. In other cores from that area, a pronounced decrease in carbonate content and flux (Andrews et al., 2001a, b) characterize the Little Ice Age record. This temperature decrease on the North Icelandic shelf at around 700 cal. BP (Fig. 12) corresponds in time to a decrease in the oxygen isotope values of the GISP2 ice core record (Grove, 2001) and to a decrease in the reconstructed

temperatures for the Northern Hemisphere (Crowley, 2000) (Fig. 12).

A comparison of the North Icelandic shelf record with diatom, foraminiferal and oxygen isotope-based sea-surface temperatures from the eastern Nordic Seas, shows that there is a general correspondence in terms of a warmer, but variable Medieval Warm Period, and a colder Little Ice Age period. However, on the eastern side the most marked cooling, signalling the onset of the Little Ice Age, happens at about 500 cal. year BP (Koç and Jansen, 2002; Andersson et al., 2003; Risebrobak-ken et al., 2003). This discrepancy can be due to phase differences in the climatic response on each side of the Nordic Seas, but chronological correlation problems cannot be ruled out at the present stage due to uncertainties in age control.

Minimum sea-surface temperatures were reached at around 350 cal. BP, when very cold conditions were indicated by several proxies. The cold East Iceland Current appears to have had an increasing influence on the sea-surface waters towards the top of all the cores, and the amplitude of fluctuations in the climate also increased through the last few hundred years. In cores HM107-01 and MD992275, the IRD flux values show a pronounced increase from around 200 to 150 cal. BP (AD 1750–1800), with maximum values in the 19th century (Figs. 8, 10). Oceanographic data from the southwestern Icelandic shelf (Jennings et al., 2001) also show that particularly cold conditions occurred during a period, corresponding in time to the interval of increased sea-ice influx around Iceland between AD 1780 and 1920 (Ogilvie, 1996) and to minimum reconstructed temperatures on the Greenland ice sheet (Dahl-Jensen et al., 1998).

On a decadal timescale, the response of the ocean north of Iceland to North Atlantic Oscillation (NAO) variability may be reflected in changes observed in the sedimentary record of the North Icelandic shelf (Hurrell et al., 2003; Visbeck et al., 2003). In a positive state of the NAO winter index sustained over a decade or longer, the wind directions across Iceland are characterised by a strong Iceland Low and predominantly southerly winds, carrying relatively warm air to the Greenland and Iceland Seas and pushing the sea-ice limit northward, probably by a combination of thermodynamic processes and wind stress. Conversely, a sustained negative NAO winter index leads to predominantly northerly winds east of Greenland caused by high pressure over Greenland. This enhances the flow of the East Greenland Current and the East Icelandic Current and leads to a southward movement of the sea-ice limit towards or onto the North Icelandic shelf. This scenario was most recently exemplified by the Great Salinity Anomaly of the 1960s (Dickson et al., 1988; Malmberg and Jónsson, 1997), which was a period of freshening and cooling of surface waters north of Iceland and prolonged sea-ice cover on the North Icelandic shelf (see also Fig. 4). The response of the oceanic conditions off North Iceland to NAO variability on a longer timescale is, however, not yet fully understood.

A modern warming of surface waters, as seen in the Crowley (2000) temperature anomaly curve for the Northern Hemisphere, is not registered in the proxy data from the North Icelandic cores, except perhaps in the topmost part of core MD992275, which is, however, not younger than AD 1950 (see Eiríksson et al., this issue). Depleted oxygen isotope values towards the top of core HM107-03 are suggested to be due to the increased inflow of freshwater from the East Icelandic Current at this site. This is supported by increased percentages of NPS.

6. Summary and conclusions

High-resolution multi-proxy studies of three cores from about 400 m water depth on the North Icelandic

shelf show that pronounced palaeoceanographic changes have occurred through the last 1200 calendar years. Benthic and planktonic foraminiferal assemblage distributions and stable isotope records show changes in bottom and surface water temperature and salinity as a result of variations in the relative strength of the Irminger Current, which brings relatively warm Atlantic water masses to the North Icelandic shelf on one hand, and the cold East Icelandic Current on the other. Also, the IRD flux and the diatom-based reconstruction of sea-surface temperatures reflect the fluctuations in the influence of Arctic surface waters. The highest IRD fluxes are observed in the eastern part of the area, where the surface waters are generally coldest.

The benthic productivity (number of benthic foraminifera/cm^2/year) is generally relatively high during cold periods and low during warmer periods. This is presumably determined by the position of the marine Polar Front close to the cores during cold spells. The productivity is considerably higher in the easternmost core MD992275 than in the others, perhaps due to an additional influence of the Jan Mayen Gyre in that part of the area.

The time period between 1200 and around 7–800 cal. BP, including the Medieval Warm Period, was characterized by relatively high bottom and surface water temperatures due to a rather strong influence of the Irminger Current. A general drop temperature in the area after 7–800 cal. BP marks the transition to a period with increased influence of the cold, low-salinity Arctic waters of the East Icelandic Current. As a result, the time interval between 7–800 and 3–400 cal. BP, corresponding to the early part of the Little Ice Age, experienced a pronounced temperature decrease, not only in surface waters but also at the sea floor. The deepest parts of the shelf sea floor may have have been influenced by Norwegian Sea Deep Water in this period of time.

The time interval after about 3–400 cal. BP is characterised by a further increase in the inflow of Arctic waters to the area. The surface waters were particularly influenced by highly variable inflow of cold and low-salinity waters, while the sea floor appears to have been under some influence of dense, high-salinity Atlantic waters. As a result, there was a pronounced stratification of the water masses during this later part of the Little Ice Age and into modern times. There is no clear indication of warming of water masses in the area during the last decades.

Acknowledgements

The present paper is a contribution to the HOLSM-EER and the PACLIVA projects funded by the European Commission's 5th Framework Programme

(Contracts no. CT-2000-00060 and EVK2-2002-00143), as well as to the PANIS project (Palaeoenvironments on the North Icelandic shelf) and the BIOICE project (Benthic Invertebrates of Icelandic Waters). Financial support by the Danish Natural Science Research Council, the Icelandic Research Council and the National Natural Science Foundation of China (Grant no. 40276013) is highly appreciated. We are grateful to the *Institut Paul-Emile Victor* (IPEV) for the IMAGES coring operations onboard the *R/V Marion Dufresne*.

References

Andersson, C., Risebrobakken, B., Jansen, E., Dahl, S.O., Meland, M., 2003. Late Holocene surface-ocean conditions of the Norwegian Sea (Vøring Plateau). Paleoceanography 18 (2), 1044, doi:10.1029/2001PA000654.

Andrews, J.T., Caseldine, C., Weiner, N., Hatton, J., 2001a. Late Holocene (ca 4 ka) marine and terrestrial environmental change in Reykjarfjördur, north Iceland: climate and/or settlement? Journal of Quaternary Science 16 (2), 133–143.

Andrews, J.T., Kristjánsdóttir, G.B., Geirsdóttir, Á., Hardardóttir, J., Helgadóttir, G., Sveinbjörnsdóttir, Á.E., Jennings, A.E., Smith, L.M., 2001b. Late Holocene (∼5 cal ka) trends and century-scale variability of N. Iceland marine records: measures of surface hydrography, productivity, and land/ocean interactions. In: Seidov, D., Maslin, M., Haupt, B. (Eds.), The Oceans and Rapid Climate Change: Past, Present and Future, Geophysical Monograph 126. American Geophysical Union, Washington, DC, pp. 69–81.

Belkin, I.M., Levitus, S., Antonov, J., Malmberg, S.-A., 1998. "Great salinity anomalies" in the North Atlantic. Progress in Oceanography 41, 1–68.

Caralp, M.H., 1989. Size and morphology of the benthic foraminifer *Melonis barleeanum*: relationships with marine organic matter. Journal of Foraminiferal Research 19, 235–246.

Cronin, T.M., Dwyer, G.S., Kamiya, T., Schwede, S., Willard, D.A., 2002. Medieval Warm Period, Little Ice Age and 20th century temperature variability from Chesapeake Bay. Global and Planetary Climate Change 36, 17–29.

Crowley, T.J., 2000. Causes of climate change over the past 1000 years. Science 289, 270–277.

Dahl-Jensen, D., Mosegaard, K., Gundestrup, N., Clow, G.D., Johnsen, S.J., Hansen, A.W., Balling, N., 1998. Past temperatures directly from the Greenland Ice Sheet. Science 282, 268–271.

Dickson, R.R., Meincke, J., Malmberg, S.A., Lee, A.J., 1988. The "Great Salinity Anomaly" in the northern North Atlantic 1968–1982. Progress in Oceanography 20, 103–151.

Dugmore, A.J., Cook, G.T., Shore, J.S., Newton, A.J., Edwards, K.J., Larsen, G., 1995. Radiocarbon dating tephra layers in Britain and Iceland. Radiocarbon 37, 379–388.

Eiríksson, J., Knudsen, K.L., Haflidason, H., Heinemeier, J., 2000. Chronology of late Holocene climatic events in the northern North Atlantic based on AMS ^{14}C dates and tephra markers from the volcano Hekla, Iceland. Journal of Quaternary Science 15 (6), 573–580.

Eiríksson, J., Larsen, G., Knudsen, K.L., Heinemeier, J., Símonarson, L.A., 2004. Marine reservoir age variability and water mass distribution in the Iceland Sea. Quaternary Science Reviews, this issue (doi:10.1016/j.quascirev.2004.08.002).

Feyling-Hanssen, R.W., 1972. The foraminifer *Elphidium excavatum* (Terquem) and its variant forms. Micropaleontology 18, 337–354.

Grönvold, K., Óskarsson, N., Johnsen, S.J., Clausen, H.B., Hammer, C.U., Bond, G., Bard, E., 1995. Ash layers from Iceland in the Greenland GRIP ice core correlated with oceanic and land sediments. Earth and Planetary Science Letters 135, 149–155.

Grove, J.M., 2001. The initiation of the 'Little Ice Age' in regions round the North Atlantic. Climatic Change 48 (1), 53–82.

Hald, M., Steinsund, P.I., 1996. Benthic foraminifera and carbonate dissolution in the surface sediments of the Barent and Kara Sea. In: Stein, R., Ivanov, G.I., Levitan, M.A., Fahl, K. (Eds.), Surface-sediment composition and sedimentary processes in the central Arctic Ocean and along the Eurasian Continental Margin, Berichte: Polarforschung 212, pp. 285–307.

Hald, M., Steinsund, P.I., Dokken, T., Korsun, S., Polyak, L., Aspeli, R., 1994. Recent and Late Quarternary distribution of *Elphidium excavatum* f. *clavatum* in Arctic seas. Cushman Foundation Special Publication 32, 141–153.

Hansen, B., Østerhus, S., 2000. North Atlantic–Nordic Seas exchanges. Progress in Oceanography 45, 109–208.

Henrich, H., Wagner, T., Goldschmidt, P., Michels, K., 1995. Depositional regimes in the Norwegian–Greenland Sea: the last two glacial to interglacial transitions. Geologische Rundschau 84, 28–48.

Hurdle, B.G., 1986. The Nordic Seas. Springer, New York, 777pp.

Hurrell, J.W., Kushnir, Y., Ottersen, G., Visbeck, M., 2003. An overview of the North Atlantic oscillation. In: Hurrell, J.W., Kushnir, Y., Ottersen, G., Visbeck, M. (Eds.), The North Atlantic Oscillation: Climatic Significance and Environmental Impact. Geophysical Monograph 134. American Geophysical Union, Washington, DC, pp. 1–35.

Jennings, A.E., Hagen, S., Hardardóttir, J., Stein, R., Ogilvie, A.E.J., Jónsdóttir, I., 2001. Oceanographic change and terrestrial human impacts in a post A.D. 1400 sediment record from the southwest Iceland shelf. Climatic Change 48, 83–100.

Jiang, H., Seidenkrantz, M.-S., Knudsen, K.L., Eiríksson, J., 2001. Diatom surface sediment assemblages around Iceland and their relationships to oceanic environmental variables. Marine Micropaleontology 41, 73–96.

Jiang, H., Seidenkrantz, M.-S., Knudsen, K.L., Eiríksson, J., 2002. Late Holocene summer sea-surface temperatures based on a diatom record from the north Icelandic shelf. The Holocene 12 (2), 137–147.

Jónsson, S., Valdimarsson, H., 2004. A new path for the Denmark Strait overflow water from the Iceland Sea to Denmark Strait. Geophysical Research Letters 31 (L03305), doi:10.1029/2003GL019214.

Juggins, S., ter Braak, C.F.J., 1992. CALIBRATE—A Program For Species-Environment Calibration By (Weighted-Averaging) Partial Least Squares Regression. Environmental Change Research Centre, University College, London.

Knudsen, K.L., Eiríksson, J., 2002. Application of tephrochronology to the timing and correlation of palaeoceanographic events recorded in Holocene and Late Glacial shelf sediments off North Iceland. Marine Geology 191, 165–188.

Knudsen, K.L., Jiang, J., Jansen, E., Eiríksson, J., Heinemeier, J., Seidenkrantz, M.-S., 2004. Environmental changes off North Iceland during the deglaciation and the Holocene: foraminifera, diatoms and stable isotopes. Marine Micropaleontology 50, 273–305.

Koç, N., Jansen, E., 2002. Holocene climate evolution of the North Atlantic Ocean and the Nordic Seas—a synthesis of new results. In: Wefer, G., Berger, W.H., Behre, K.E., Jansen, E. (Eds.), Climate and History in the North Atlantic Realm. Springer, Berlin, Heidelberg, pp. 163–165.

Larsen, G., Eiríksson, J., Knudsen, K.L., Heinemeier, J., 2002. Correlation of late Holocene terrestrial and marine tephra markers, north Iceland: implications for reservoir age changes. Polar Research 21 (2), 283–290.

Leeder, M., 1999. Sedimentology and Sedimentary Basins. From Turbulence to Tectonics. Blackwell Science Ltd, Oxford, 592pp.

Mackensen, A., Sejrup, H.P., Jansen, E., 1985. The distribution of living benthic foraminifera on the Continental Slope and Rise off Southwest Norway. Marine Micropaleontology 9, 275–306.

Malmberg, S.A., Jónsson, S., 1997. Timing of deep convection in the Greenland and Iceland Seas. ICES Journal of Marine Science 54, 300–309.

Nørvang, A., 1945. Foraminifera. The Zoology of Iceland II (2), 1–79.

Ogilvie, A.E.J., 1996. Sea-ice conditions off the coasts of Iceland A.D. 1601–1850 with special reference to part of the Maunder Minimum period (1675–1715). AmS-Varia 25, 9–12.

Ogilvie, A.E.J., Jónsdóttir, I., 1996. Sea-ice incidence off the coasts of Iceland A. D. 1601–1850: Evidence from historical data and early sea-ice maps. In: 26th International Arctic Workshop, Arctic and Alpine Environments, Past and Present. Program with Abstracts INSTAAR, 14—16 March, 1996. Boulder, Colorado, pp. 109–110.

Ólafsson, J., 1999. Connections between oceanic conditions off N-Iceland, Lake Mývatn temperature, regional wind direction variability and the North Atlantic Oscillation. Rit Fiskideildar 16, 41–57.

Risebrobakken, B., Jansen, E., Andersson, C., Mjelde, E., Hevrøy, K., 2003. A high-resolution study of Holocene paleoclimatic and paleoceanographic changes in the Nordic Seas. Paleoceanography 18, 764–777.

Rytter, F., Knudsen, K.L., Seidenkrantz, M.-S., Eiríksson, J., 2002. Modern distribution of benthic foraminifera on the north Icelandic shelf and slope. Journal of Foraminiferal Research 32, 217–244.

Seidenkrantz, M.-S., 1995. *Cassidulina teretis* Tappan and *Cassidulina neoteretis* new species (foraminifera): stratigraphic markers for deep sea and outer shelf areas. Journal of Micropalaeontology 14, 145–157.

Shackleton, N.J., Opdyke, N.D., 1973. Oxygen isotope and palaeomagnetic stratigraphy og equatorial core V28-238: oxygen isotope temperatures and ice volumes on a 105 and 106 year scale. Quaternary Research 3, 39–55.

Stefánsson, U., 1962. North Icelandic Waters. Rit Fiskideildar 3, 1–269.

Swift, J.H., 1986. The Arctic waters. In: Hurdle, B.G. (Ed.), The Nordic Seas. Springer New York Inc., New York, pp. 129–153.

Visbeck, M., Chassignet, E.P., Curry, R.G., Delworth, T.L., Dickson, R.R., Krahmann, G., 2003. The ocean's response to North Atlantic oscillation variability. In: Hurrell, J.W., Kushnir, Y., Ottersen, G., Visbeck, M. (Eds.), The North Atlantic Oscillation: Climatic Significance and Environmental Impact. American Geophysical Union, Washington DC, pp. 113–145.

Zielinski, G.A., Mayewski, P.A., Meeker, L.D., Grönvold, K., Germani, M.S., Whitlow, S., Twickler, M.S., Taylor, K., 1997. Volcanic aerosol records and tephrochronology of the Summit, Greenland, ice cores. Journal of Geophysical Research 102 (C12), 26625–26640.

ELSEVIER

Quaternary Science Reviews 23 (2004) 2247–2268

QSR

Marine reservoir age variability and water mass distribution in the Iceland Sea

Jón Eiríksson[a],*, Gudrún Larsen[a], Karen Luise Knudsen[b], Jan Heinemeier[c], Leifur A. Símonarson[a]

[a]*Science Institute, University of Iceland, IS-101 Reykjavík, Iceland*
[b]*Department of Earth Sciences, University of Aarhus, DK-8000 Århus C, Denmark*
[c]*AMS ^{14}C Dating Laboratory, Institute of Physics and Astronomy, University of Aarhus, DK-8000 Århus C, Denmark*

Abstract

Lateglacial and Holocene tephra markers from Icelandic source volcanoes have been identified in five sediment cores from the North Icelandic shelf and correlated with tephra layers in reference soil sections in North Iceland and the GRIP ice core. Land-sea correlation of tephra markers, that have been radiocarbon dated with terrestrial material or dated by documentary evidence, provides a tool for monitoring reservoir age variability in the region. Age models developed for the shelf sediments north of Iceland, based on offshore tephrochronology on one hand and on calibrated AMS ^{14}C datings of marine molluscs on the other, display major deviations during the last 4500 years. The inferred temporal variability in the reservoir age of the regional water masses exceeds by far the variability expected from the marine model calculations. The observed reservoir ages are generally considerably higher, by up to 450 years, than the standard model ocean. It is postulated that the intervals with increased and variable marine reservoir age reflect incursions of Arctic water masses derived from the East Greenland Current to the Iceland Sea and the North Icelandic shelf.

1. Introduction

Palaeoclimatic studies that extend beyond historical and instrumental records must rely on dating techniques which enable reconstruction of time series for climate-variability related proxies. Marine sediment cores from high resolution sedimentary basins on the North Icelandic shelf provide archives of Holocene palaeoceanographic changes in the vicinity of the oceanic Polar Front (Fig. 1). The purpose of this study is to construct reliable age models for palaeo-climatic data from the marine environment in the region and to interpret the discrepancies between the AMS ^{14}C datings and the tephrochronological age models in terms of changes in the water mass distribution.

The present position of the Polar Front separates Arctic surface water of the East Greenland Current and the East Icelandic Current from branches of the North Atlantic Current to the west and north of Iceland. The Arctic surface water of the East Icelandic Current is partly derived from the East Greenland Current (Polar water) and partly from westerly eddies of the Norwegian Atlantic Current (Atlantic water) (e.g. Stefánsson, 1962; Hansen and Østerhus, 2000). The strong gradients both in the ocean and in the atmosphere make this region extremely sensitive to climatic changes.

The chronology for recent studies of the North Icelandic shelf sediments has been based on radiocarbon dates (Andrews et al., 2000, 2001a, b; Andrews and Giraudeau, 2003) or on combined tephrochronology and AMS ^{14}C datings of either molluscs or benthic or planktonic foraminifera (Eiríksson et al., 2000a, b; Jiang et al., 2002; Knudsen and Eiríksson, 2002; Larsen et al., 2002). The application of tephra-based age models has

*Corresponding author. Tel.: +354-525-4475; fax: +354-552-1331.
E-mail address: jeir@rhi.hi.is (J. Eiríksson).

0277-3791/$ - see front matter © 2004 Elsevier Ltd. All rights reserved.
doi:10.1016/j.quascirev.2004.08.002

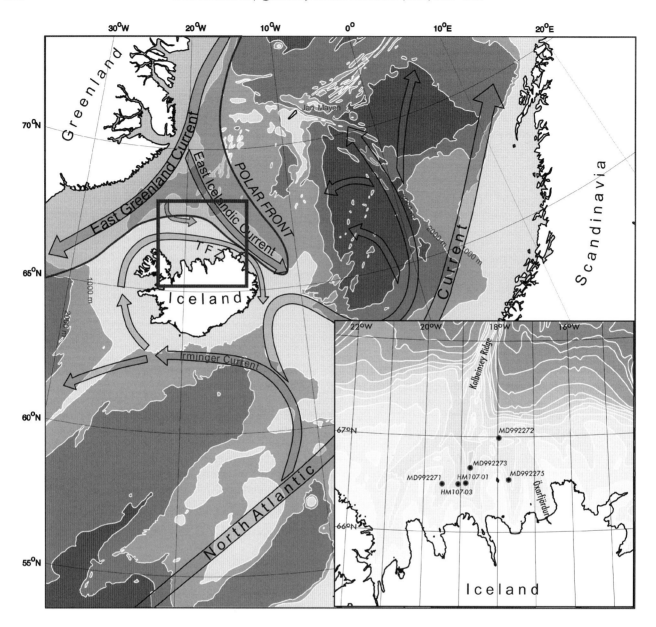

Fig. 1. The regional modern surface circulation around Iceland and the position of the marine Polar Front. Depth contour intervals 1000 m (modified after Knudsen and Eiríksson, 2002, and Hurdle, 1986, Map A8). TFZ = Tjörnes Fracture Zone. The inset location map shows the study area off North Iceland (depth contour intervals 100 m).

revealed that AMS ^{14}C dates from molluscs show variable deviation from tephrochronological age models, indicating that the ocean reservoir age at the coring sites has varied with time (Eiríksson et al., 2000a, b; Knudsen and Eiríksson, 2002; Larsen et al., 2002). The results of these studies demonstrate a real need for independent control on ^{14}C dating of marine sediment cores obtained from an oceanographic boundary region such as the Polar Front separating the Atlantic and Arctic water masses.

The study area on the North Icelandic shelf has the advantage of being close to numerous source volcanoes of Holocene tephras (Fig. 2). Tephra markers that can

be traced from volcanic source regions (see Fig. 2) into the marine depositional environment can provide control on radiocarbon dates from that environment. The ages of Icelandic tephra markers used in this study are based on historical records from Iceland for the last 900 years, on correlation to the Greenland ice-core chronology and on radiocarbon dates of terrestrial material. It was demonstrated by Larsen et al. (2002) that a high resolution tephra stratigraphy can be used to link chronologies in the terrestrial North Iceland and a high resolution marine sediment core (MD992275). The investigation presented a detailed land-sea correlation of the regional terrestrial tephrochronology with the

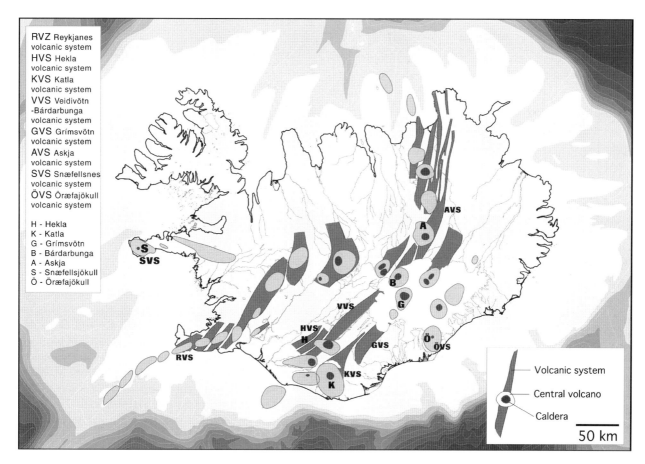

Fig. 2. The most relevant volcanic systems for the tephrochronology of northern Iceland and the North Icelandic shelf.

marine record. This comparison of a tephrochronological age model and a radiocarbon one makes it possible to study local reservoir age problems.

The pre-bomb reservoir age of the coastal waters of Iceland was determined by Håkansson (1983) to 365 ± 20 yr. This value is 35 years lower than the 400 yr reservoir age correction used by Andersen et al. (1989) for Icelandic coastal samples and the ca. 400 yr correction conventionally applied to marine ^{14}C ages (Stuiver et al., 1998a, b).

In discussing reservoir age variability in the region, Austin et al. (1994) noted that at present there is no surface ^{14}C gradient between 40° and 70 °N in the North Atlantic surface waters. The coastal and shelf surface waters around Iceland today are dominated by the Irminger Current (Fig. 1), which is derived from the North Atlantic Current and thus has the same affinity as the waters off western Norway, the Faroe Islands and the British Isles. In contrast, the reservoir age of the East Greenland Current north of the Polar Front, has been determined to 530 yr (Tauber and Funder, 1975) and to 515 ± 25 yr (Hjort, 1973; Håkansson, 1983). This means that the recent, pre-bomb apparent age difference across the Polar Front is about 100–150 yr.

It is suggested that discrepancies between the two age models are related to palaeoceanographic changes in the region and resulting changes in the reservoir age of the water masses on the North Icelandic shelf. The Holocene Climatic Optimum in this area is reflected by a pronounced influence of Atlantic waters, brought to the North Icelandic shelf by the Irminger Current already at about 10,300 cal. BP (Knudsen et al., 2004b). In that context it is notable that Eiríksson et al. (2000b) did show that a 400 year reservoir correction is applicable at the level of the Saksunarvatn ash (10,200 cal. BP). During Lateglacial conditions, however, Arctic water masses prevailed, and a reservoir correction of 750–800 years is required at the level of the Lateglacial Vedde Ash in this region (Eiríksson et al., 2000b; Haflidason et al., 2000). Results from the Late Holocene record of a series of shelf cores, analysed with a time resolution of up to 5 yr/cm, are included in this study. The investigations are organised jointly by the University of Iceland and the University of Aarhus, Denmark, as the umbrella project PANIS (Palaeoenvironments of the North Icelandic Shelf). Results from five of these cores are treated in this paper.

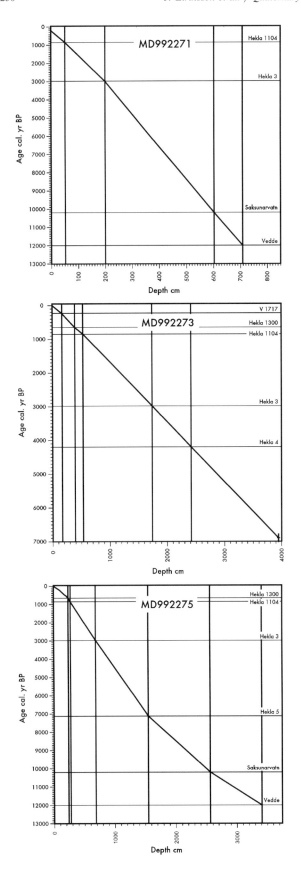

2. Materials and methods

Five sediment records from the Tjörnes Fracture Zone (TFZ) on the shelf north of Iceland provide data on palaeoceanographic variability in the region (Fig. 1). Samples for the present study were obtained from box cores, gravity cores and CALYPSO piston cores acquired on two research cruises (see inset map, Fig. 1). The gravity cores HM107-01 and HM107-03 and box cores at the same sites were obtained on the 1995 BIOICE cruise on *RV Haakan Mosby*. The piston cores MD992271, MD992273 and MD992275 were obtained on the 1999 GINNA cruise on *RV Marion Dufresne*, organised by the IMAGES programme. The Tjörnes Fracture Zone is topographically expressed as a series of north–south trending extensional troughs and ridges offsetting the Kolbeinsey Ridge spreading axis, which extends onto the northernmost part of the shelf—eastwards to Öxarfjördur, where the spreading axis continues southwards across north Iceland. Two of the cores, HM107–01 and MD992273 (464 and 665 m water depth) are located in Eyjafjardaráll, a deep extensional trough south of Kolbeinsey Ridge. Cores MD992271 and HM107–03 are located in shallower water (315 and 400 m depth) to the west of Eyjafjardaráll. The easternmost core, MD992275 (440 m depth), is located in the Skjálfandadjúp trough, which separates two of the TFZ volcanically active shallow submarine ridges.

Age models based on major Lateglacial and Holocene tephra markers indicate sustained high and relatively uniform sedimentation rates at the CALYPSO coring sites (Fig. 3), and the same is true for the gravity cores covering the upper Holocene. All cores were subsampled by extracting 1 cm thick slices of a split core. The various sedimentological, micropalaeontological and stable isotope parameters have generally been obtained at a time resolution of 20–50 years. Every cm of each core has been checked for available material for AMS ^{14}C dating. The sample levels for dating have mostly been chosen with respect to possible dating of important stratigraphic levels such as biozone boundaries and tephra layers.

Major tephra markers were identified by visual inspection of split cores, and on X-ray films. These were subsampled and sieved on a 63 μm sieve for the preparation of polished thin sections. Other tephra

◄

Fig. 3. Age-depth diagrams for cores MD992271, MD992273 and MD992275 (for details in the Late Holocene part, see Figs. 5, 8 and 9) showing tephra marker horizons back to the level of the Vedde Ash. The age models are based only on tephra marker horizons (tephrochronological age models). Some of the tephra layers are historically dated, others are dated on terrestrial material. The core depth is plotted on the *x*-axis. The position of each tephra layer in the cores is shown as a vertical line. The calibrated age is plotted on the *y*-axis and the age of each tephra layer is shown by a horizontal line. Cross indicates a ^{14}C age with one σ error bars.

layers were identified as grain size anomalies represented by peaks in the sand fraction. Major element geochemical analysis of volcanic glass shards (Table 1A–E) were carried out by standard wavelength dispersal technique on an ARL-SEMQ microprobe at the Geological Institute, University of Bergen, Norway, with an accelerating voltage of 15 kV, a beam current of 10 nA, and a defocused beam diameter of 6–12 μm. Natural and synthetic minerals and glasses were used as standards (see also Larsen et al., 2002).

Tephrochronological age models have been constructed for all the cores. The age models are constrained by first appearance depth of each marker, and the historical or terrestrial radiocarbon date of the marker. Between markers, linear interpolation has been used to estimate age of samples (straight line age model segments on Figs. 3 and 5–9). The calibrated radiocarbon dates are shown with 1 σ error bars on Figs. 5–9. The deviation between each calibrated radiocarbon age and the tephrochronological age of the same level, termed Δr, is also shown with 1 σ calibration error bars. The reservoir age offset (ΔR, Fig. 10) from the standard model ocean has been calculated for each radiocarbon date by means of the calibration curves. Although ΔR is the relevant physical parameter, directly related to the ocean ^{14}C concentration, we have chosen to show Δr because its value can be read directly off the graphs. As seen from Fig. 10, the two parameters vary almost identically.

Post-settlement tephra markers are dated to an exact or an approximate year AD. For pre-settlement tephra markers, there is an additional error associated with the terrestrial radiocarbon dating error. This affects the markers Snæfellsjökull 1 (1855±25 ^{14}C yr BP, Larsen et al., 2002), Hekla 3 (2879±34 ^{14}C yr BP, Dugmore et al., 1995), and Hekla 4 (3826±12 ^{14}C yr BP, Dugmore et al., 1995). Because these errors are relatively small compared to the deviations, they have not been incorporated in the graphs presented here.

Most of the tephra markers display a northerly dispersal pattern and were deposited as airfall tephra, both in the terrestrial and marine environment. Sea-ice cover may occasionally have delayed deposition on the sea-floor, but due to summer melting it is considered unlikely that this delay has ever exceeded one season in the upper Holocene records.

The foraminiferal samples, each representing a 1 cm sediment slice, were analysed at 1–5 cm intervals. They were washed through 1000, 125 and 63 μm sieves according to the methods described by Feyling-Hanssen et al. (1971) and Knudsen (1998). Total benthic foraminiferal assemblages were analysed in the 125 μm fraction by counting at least 300 specimens of benthic foraminifera, when possible. The palaeoecological interpretations are based on comparisons with modern faunal distributions in the North Atlantic region.

Modern foraminiferal faunal distribution on the North Icelandic shelf has been described by Rytter et al. (2002) and Weiner et al. (1999).

2.1. Marine dating material

The ^{14}C datings of marine samples (Table 2) were carried out at the AMS ^{14}C Dating Laboratory at the University of Aarhus, Denmark. The dates have been corrected for natural isotopic fractionation by normalisation to ^{13}C = −25‰ VPDB, and calibrated with CALIB 4.1 (Stuiver et al., 1998a), using the marine model calibration curve (Stuiver et al., 1998b). The marine model is a calculated smooth calibration curve for the mixed layer of a model ocean. To accommodate local effects in the actual oceans, the quantity ΔR is defined as the difference between the measured sample age and the global model curve. For a given region, ΔR is expected to be constant in time to a first approximation, to the extent that the regional reservoir, from which the sample is taken, parallels the global marine model. Since the reservoir age is the difference between the strongly fluctuating atmospheric calibration curve and the slowly responding smooth ocean curve, it is expected to vary (typically up to 100 yr) with time for a given location (water mass). A standard reservoir correction of about 400 ^{14}C years (ΔR = 0) is built into this model (see also Andersen et al., 1989). In this paper we simply refer to calibrated years BP (before 1950) as years.

The dating material selected for AMS ^{14}C dating consists of a range of carbonate shells from marine organisms. Available dating material did not permit the use of one single species throughout. Most of the dates are from molluscs, which have the advantage of being less prone to reworking by bottom currents than e.g. foraminifera (Heier-Nielsen et al., 1995a). AMS ^{14}C dating of benthic and planktonic foraminifera at the same level as molluscs in the research area, however, yielded no significant age difference (Eiríksson et al., 2000b). However, recent studies of shelf and shallow marine sediment dating in areas close to calcareous bedrock (Heier-Nielsen et al., 1995b; Dyke et al., 2002) and in sediments containing glacially eroded carbon (Forman and Polyak, 1997) have indicated anomalously high radiocarbon ages, specifically when dating deposit feeding molluscs. Consequently, the possible dating errors related to burrowing depths, feeding habits, and possible contamination by old carbon must be evaluated when reservoir problems are investigated.

If molluscs incorporate old carbon from pore waters that is not derived directly from the sea-water, this may affect the radiocarbon dates and lead to too high age values. Therefore, species which feed upon phytoplankton, smaller forms of zooplankton and detritus in suspension were selected whenever possible in the

Table 1
Representative microprobe glass analyses of tephra horizons relevant to this study from the five cores MD992271 (A), HM107-03 (B), HM107-01 (C), MD0992273 (D) and MD992275 (E)

	SiO_2	TiO_2	Al_2O_3	FeO	MnO	MgO	CaO	Na_2O	K_2O	P_2O_5	Sum
(A)											
MD992271/52											
Hekla 1104	71.70	0.24	13.79	3.41	n.d.	0.11	1.72	4.52	2.97	0.03	98.49
	71.55	0.25	14.43	3.32	n.d.	0.11	1.93	4.51	2.71	0.01	98.82
	71.17	0.24	14.24	3.20	n.d.	0.12	1.92	4.37	2.75	0.00	98.01
	70.95	0.19	14.18	3.37	n.d.	0.15	1.91	4.39	2.73	0.03	97.90
	70.61	0.25	14.29	3.16	n.d.	0.08	1.89	4.78	2.56	0.16	97.78
MD992271/200	70.80	0.16	14.21	3.28	n.d.	0.12	2.02	4.61	2.50	0.00	97.70
Hekla 3	69.83	0.19	14.44	3.88	n.d.	0.24	2.09	4.42	2.95	0.20	98.24
	64.80	0.43	14.75	6.81	n.d.	0.62	3.45	4.47	2.01	0.08	97.42
	63.87	0.57	15.42	7.09	n.d.	0.68	4.23	5.16	1.26	0.27	98.55
	61.78	1.02	15.01	8.17	n.d.	0.90	4.59	4.80	1.77	0.54	98.58
MD992271/602	50.40	2.93	13.09	14.49	0.26	5.42	9.77	2.64	0.46	0.35	99.81
Saksunarvatn	50.38	2.89	13.10	14.35	0.26	5.62	9.89	2.67	0.46	0.34	99.96
	49.76	2.80	13.23	13.21	0.24	6.10	10.58	2.43	0.39	0.38	99.12
	48.98	2.94	13.11	14.59	0.26	5.45	9.54	2.59	0.48	0.21	98.15
	48.55	2.77	13.32	14.67	0.29	5.30	9.69	2.56	0.51	0.45	98.11
MD992271/702	71.05	0.29	13.43	3.83	0.15	0.13	1.28	5.16	3.29	0.04	98.65
Vedde	70.95	0.32	13.58	4.01	0.12	0.21	1.33	5.33	3.53	0.06	99.44
	70.42	0.19	13.44	3.86	0.17	0.17	1.31	3.65	3.42	0.03	96.66
	70.31	0.32	13.49	3.92	0.19	0.24	1.27	5.26	3.31	0.00	98.31
	69.44	0.28	13.41	3.95	0.13	0.16	1.28	5.28	3.34	0.04	97.31
	51.38	3.27	13.19	12.65	0.25	3.56	8.05	3.84	1.23	0.43	97.85
	50.06	4.80	12.99	14.04	0.23	3.76	8.03	3.63	1.24	0.41	99.19
	49.01	4.23	12.94	14.21	0.19	4.42	9.34	2.50	0.92	0.57	98.33
	48.29	4.25	13.34	14.85	0.24	4.42	9.49	3.21	0.79	0.56	99.44
	47.31	4.29	13.03	14.25	0.35	4.56	9.31	3.29	0.89	0.50	97.78
(B)											
HM107-03/37											
KOL 1372	51.18	1.26	13.84	12.65	0.21	7.55	10.88	2.02	0.06	0.14	99.65
	51.11	1.18	13.63	12.62	0.18	7.33	10.97	2.01	0.05	0.11	99.08
	50.68	1.28	13.71	12.63	0.20	7.35	10.78	2.12	0.06	0.09	98.81
	50.64	1.32	13.96	12.69	0.25	7.25	11.17	2.19	0.12	0.06	99.59
	50.48	1.08	14.00	12.21	0.22	7.16	11.55	2.63	0.07	0.12	99.40
	50.36	1.16	13.82	12.08	0.20	7.20	11.48	1.98	0.07	0.08	98.35
	50.19	1.12	13.96	12.65	0.13	7.28	11.47	2.07	0.08	0.05	98.95
	49.96	1.13	14.03	12.62	0.20	7.33	11.47	2.38	0.10	0.03	99.22
	49.93	1.18	13.98	12.60	0.26	7.22	11.46	1.97	0.13	0.08	98.73
	49.25	1.11	13.89	12.55	0.21	7.33	11.12	2.06	0.06	0.09	97.58
(C)											
HM107-01/48											
KOL 1372	50.52	1.17	13.95	13.29	n.d.	7.19	11.51	2.03	0.06	0.14	99.72
	50.35	1.23	13.98	13.26	n.d.	7.09	11.48	2.10	0.05	0.09	99.54
	50.31	1.20	14.25	13.18	n.d.	7.13	11.46	2.11	0.14	0.07	99.78
	49.87	1.23	14.14	13.21	n.d.	7.54	11.50	2.03	0.13	0.04	99.65
	49.82	1.13	14.05	13.25	n.d.	7.26	11.62	2.17	0.08	0.13	99.38
HM107-01/74	70.40	0.18	14.14	3.25	n.d.	0.03	2.00	4.22	2.72	0.10	96.94
Hekla 1104	70.39	0.23	13.97	3.33	n.d.	0.07	1.96	4.36	2.64	0.18	96.95
	70.35	0.19	14.29	3.28	n.d.	0.17	1.96	4.53	2.88	0.00	97.65
	70.04	0.23	14.15	3.24	n.d.	0.12	1.91	4.48	2.93	0.02	97.10
	69.81	0.21	14.40	3.26	n.d.	0.05	1.96	4.35	2.67	0.00	96.71
(D)											
MD992273/168											
V 1717	50.75	1.77	14.19	12.65	n.d.	6.91	11.75	2.41	0.11	0.10	100.64
	50.73	1.76	14.08	12.74	n.d.	6.99	11.66	2.28	0.14	0.07	100.45
	50.30	1.70	14.17	12.56	n.d.	6.91	11.65	2.33	0.19	0.21	100.02
	49.58	1.60	14.17	12.17	n.d.	6.68	11.78	2.29	0.27	0.28	98.82

Table 1 (*continued*)

	SiO$_2$	TiO$_2$	Al$_2$O$_3$	FeO	MnO	MgO	CaO	Na$_2$O	K$_2$O	P$_2$O$_5$	Sum
	49.37	1.63	13.97	12.44	n.d.	6.47	11.42	2.30	0.23	0.19	98.02
MD992273/392	72.24	0.25	14.22	3.13	n.d.	0.12	1.97	4.95	2.83	0.06	99.77
Hekla 1300	70.40	0.24	14.18	3.06	n.d.	0.11	2.06	4.61	2.53	0.00	97.19
	63.08	0.91	14.96	8.35	n.d.	1.14	4.51	4.31	2.11	0.25	99.62
	62.11	1.48	14.99	8.78	n.d.	0.96	4.98	4.70	1.78	0.67	100.45
	61.09	1.24	14.93	10.07	n.d.	2.02	5.33	4.32	1.55	0.42	100.97
	59.78	1.21	14.64	10.42	n.d.	2.28	5.45	3.66	1.66	0.55	99.65
	56.61	2.93	13.47	11.25	n.d.	3.26	7.35	3.70	1.62	0.30	100.49
MD992273/534	73.63	0.21	13.62	3.22	n.d.	0.06	1.91	4.29	2.68	0.00	99.62
Hekla 1104	72.75	0.17	13.62	3.21	n.d.	0.09	1.89	4.32	2.75	0.03	98.83
	72.17	0.15	13.88	3.18	n.d.	0.04	1.93	4.45	2.72	0.03	98.55
	72.11	0.20	13.65	3.15	n.d.	0.09	1.92	4.52	2.79	0.03	98.46
	72.06	0.13	13.63	3.23	n.d.	0.04	1.95	4.31	2.74	0.03	98.12
MD992273/1745	74.86	0.14	13.29	1.89	0.06	0.05	1.37	4.74	2.89	n.d.	99.28
Hekla 3	71.75	0.16	14.33	3.06	0.11	0.11	2.13	4.57	2.52	n.d.	98.73
	67.88	0.40	14.89	5.12	0.17	0.30	3.09	4.57	2.19	n.d.	98.61
	66.45	0.45	15.66	5.57	0.24	0.48	3.60	4.83	1.85	n.d.	99.12
	64.24	0.69	15.24	7.36	0.22	0.73	3.97	4.51	1.81	n.d.	98.76
MD992273/2418	74.51	0.14	13.27	1.94	0.15	0.01	1.35	4.88	2.90	n.d.	99.15
Hekla 4	74.26	0.08	13.36	1.98	0.14	0.02	1.34	4.51	2.81	n.d.	98.49
	74.65	0.05	13.29	1.91	0.11	0.06	1.34	4.40	2.85	n.d.	98.67
	64.55	0.66	15.41	7.32	0.28	0.77	4.21	4.73	1.81	n.d.	99.72
	57.30	1.72	15.69	10.85	0.27	2.48	6.88	3.91	1.09	n.d.	100.19
(E)											
MD992275/101											
V 1717	50.14	1.73	14.32	12.91	n.d.	6.93	11.82	2.16	0.22	0.20	100.43
	50.12	1.62	14.13	12.77	n.d.	7.01	11.80	2.49	0.21	0.14	100.29
	49.60	1.71	14.07	12.49	n.d.	6.84	11.62	2.25	0.28	0.14	99.00
	49.45	1.87	14.11	12.43	n.d.	6.92	11.49	2.47	0.22	0.21	99.17
	49.28	1.72	14.14	12.39	n.d.	7.06	11.46	2.49	0.12	0.20	98.86
MD992275/1552											
Hekla 5	75.74	0.10	12.62	1.73	0.06	0.01	1.31	3.85	2.69	0.08	98.19
	75.42	0.08	12.76	1.71	0.03	0.00	1.30	4.08	2.82	0.00	98.20
	75.20	0.07	12.58	1.73	0.05	0.00	1.32	3.88	2.80	0.00	97.63
	75.05	0.06	12.34	1.80	0.09	0.11	1.23	3.71	2.86	0.04	97.29
	74.97	0.06	12.74	1.60	0.04	0.04	1.27	3.59	2.72	0.00	97.03
MD992275/2560											
Saksunarvatn	50.87	2.91	13.33	14.44	n.d.	5.50	10.06	2.50	0.48	0.37	100.46
	50.76	3.12	13.01	15.02	n.d.	5.56	9.75	2.44	0.44	0.38	100.48
	50.53	2.98	12.81	14.83	n.d.	5.50	9.72	2.50	0.47	0.32	99.66
	50.35	2.97	13.08	14.61	n.d.	5.75	10.04	2.56	0.37	0.36	100.09
	49.88	3.18	13.04	15.24	n.d.	5.60	9.84	2.79	0.40	0.27	100.24
MD992275/3405											
Vedde	72.79	0.28	13.48	3.77	n.d.	0.10	1.12	4.90	3.17	0.00	99.61
	72.74	0.26	12.89	3.72	n.d.	0.22	1.02	4.93	3.10	0.05	98.93
	72.68	0.27	12.92	3.68	n.d.	0.26	1.06	4.99	3.27	0.00	99.13
	71.79	0.35	13.64	3.79	n.d.	0.22	1.20	4.62	3.29	0.00	98.90
	71.65	0.28	13.26	3.68	n.d.	0.22	1.10	4.84	3.25	0.05	98.33
	50.94	4.70	12.50	14.20	n.d.	4.54	7.66	2.83	0.83	0.45	98.65
	50.88	3.93	13.27	13.06	n.d.	4.42	7.71	3.20	1.10	0.47	98.04
	50.58	4.30	13.24	13.84	n.d.	4.42	8.23	3.40	0.84	0.42	99.27
	50.16	4.20	13.22	13.29	n.d.	4.56	8.12	3.26	1.01	0.55	98.37
	48.22	4.85	12.97	14.76	n.d.	5.09	8.73	3.01	0.74	0.42	98.79

In most cases five analyses with sums between 97.5% and 100.5% are presented for each sample. The exceptions are Vedde tephra from cores MD992271 and MD992275 where both the rhyolitic and basaltic component is presented, KOL 1372 where 10 analyses from core HM107–03 are presented (locus typicus), and Hekla 1300 from core MD0992273 where 7 analyses demonstrate the range of composition. The analyses do not show the complete range of compostion of each tephra layer. Depth refers to core barrel depths.

Table 2
Tephra marker horizons and radiocarbon dates in cores MD992271, HM107-03 (including top 10 cm from box core), HM107-01 (including top 10 cm from box core), MD992273 and MD992275, calibrated with CALIB4 using the marine98 dataset (Stuiver et al., 1998a, b)

Core	Depth (cm)	Lab no.	Material	^{14}C age BP±1σ	Cal. yr BP±1σ	Cal. range (s), 1σ	∂^{13}C	Expected tephro-chronological age	Δr, cal. BP deviation	ΔR, ^{14}C deviation
MD992271	27.5	AAR-7697	*Siphonodentalium lobatum*	1157±42	686	659–728	+0.52	567	119	155
MD992271	52		Hekla 1104		850					
MD992271	53.5	AAR-7698	cf. *Lunatia pallida*, *Thyasira* cf. *equalis*	1422±35	953	926–985	+0.52	872	81	119
MD992271	93.5	AAR-7699	*Bathyarca glacialis*	1986±50	1532	1495–1596	+1.64	1447	85	83
MD992271	149.5	AAR-7800	*Yoldiella lenticula*	2762±42	2462	2361–2538	+0.32	2253	209	210
MD992271	158.5	AAR-7701	*Lunatia pallida*	2577±40	2286	2184–2308	+0.8	2383	–97	–149
MD992271	176	AAR-7768	*Thyasira equalis*, *Cuspidaria obesa*, *Thyasira* cf. *gouldi*	3203±46	2989	2933–3065	–8.04	2635	354	374
MD992271	197.5	AAR-7769	*Thyasira equalis*	3340±55	3202	3130–3264	–7.51	2944	258	173
MD992271	200		Hekla 3		2980					
MD992271	200	AAR-7770	*Yoldiella* sp., *Yoldiella intermedia*, *Thyasira equalis*	3356±35	3214	3167–3261	–2.23	2980	234	160
MD992271	207	AAR-7771	*Thyasira equalis*, cf. *Arctinula greenlandica*	3430±110	3318	3167–3417	–5.57	3106	212	155
MD992271	229.5	AAR-7772	*Bathyarca glacialis*	3664±47	3565	3487–3627	+1.69	3510	55	43
MD992271	263.5	AAR-7773	*Yoldiella* sp.	3915±43	3880	3827–3950	+0.36	4120	–240	–168
MD992271	602		Saksunarvatn ash		10200					
MD992271	708		Vedde Ash		12000					
HM107-03	41	AAR-3698	*Bathyarca glacialis*	840±55	480	440–510	+1.7	500	–20	–39
HM107-03	47		KOL 1372		580					
HM107-03	63		Hekla 1104		850					
HM107-03	64.5	AAR-5105	*Propebela* cf. *reticulata*, *Yoldiella* sp., *Oenopota* sp.	1200±35	728	689–766	+0.3	870	–142	–101
HM107-03	67.5	AAR-4029	*Siphonodentalium lobatum*	1515±45	1060	1000–1120	+1.2	910	150	160
HM107-03	82	AAR-5059	*Thyasira* sp., *Arctinula greenlandica*	1830±200	1360	1190–1590	0	1103	257	280
HM107-03	82	AAR-5747	*Foraminifera*, total benthic fauna	1750±60	1288	1253–1339	–1.7	1103	185	200
HM107-03	169.5	AAR-2930	*Yoldiella intermedia*, *Arctinula greenlandica*, *Yoldiella lenticula*, *Thyasira* sp.	2900±80	2700	2600–2740	+0.1	2268	432	343
HM107-03	186	AAR-5746	*Arctinula greenlandica*, *Yoldiella lenticula*, *Thyasira* sp.	2945±30	2728	2709–2742	+0.8	2487	241	163
HM107-03	217.5	AAR-3699	*Thyasira* cf. *gouldi*	3410±60	3300	3210–3350	0	2907	393	276
HM107-03	223		Hekla 3		2980					
HM107-03	224	AAR-6081	*Foraminifera*, total benthic fauna	3415±35	3306	3245–3334	–1.8	2992	314	210
HM107-03	266.5	AAR-3781	*Bathyarca* cf. *glacialis*, *Yoldiella* sp., *Bathyarca glacialis*	3710±40	3620	3570–3670	+1.7	3511	109	89

Core	Depth	Lab no.	Species	^{14}C age ±	Cal	Cal range	Δ			
HM107-03	295.5	AAR-3700	*Yoldiella* sp.	3865±55	3820	3730–3890	+0.7	3865	−45	−37
HM107-03	320.5	AAR-5744	*Yoldiella fraterna*	4180±50	4248	4173–4343	+1.3	4170	78	54
HM107-03	323		Hekla 4		4200					
HM107-03	323.5	AAR-5041	*Yoldiella lenticula*	4170±35	4240	4180–4290	+0.2	4206	34	32
HM107-03	325.5	AAR-5106	*Yoldiella fraterna*	4280±40	4400	4349–4425	0°	4231	169	117
HM107-03	334.5	AAR-2931	*Thyasira* sp., *Yoldiella* sp.	4320±90	4420	4340–4540	−2.9	4344	76	88
HM107-01	47.5	AAR-3779	*Siphonodentalium lobatum*	1045±45	627	564–649	+1.2	441	186	257
HM107-01	61		KOL 1372		580					
HM107-01	84		Hekla 1104		850					
HM107-01	94.5	KIA17226	*Siphonodentalium lobatum*	1595±25	1162	1131–1177	+0.6	992	170	130
HM107-01	138.5	AAR-4194	Foraminifera, total benthic fauna	2050±55	1607	1539–1689	−1	1589	18	18
HM107-01	228.5	AAR-4196	Foraminifera, total benthic fauna	3190±45	2968	2917–3058	−1.1	2810	158	131
HM107-01	234.5	AAR-5193	*Siphonodentalium lobatum*	3280±30	3114	3065–3159	0°	2892	222	149
HM107-01	241		Hekla 3		2980					
HM107-01	251.5	AAR-5104	*Thyasira equalis*	3525±45	3394	3353–3452	−8.2	3122	272	243
HM107-01	268.5	KIA17227	*Siphonodentalium lobatum*	3650±30	3552	3492–3586	+3.4	3353	199	174
HM107-01	271.5	AAR-3701	*Bathyarca glacialis*	3390±60	3258	3192–3333	+1.2	3393	−135	−134
HM107-01	289.2	AAR-2929	*Thyasira equalis*	4150±100	4220	4075–4368	−1	3633	587	428
MD992273	19.5	AAR-8426	*Thyasira equalis*	588±41	258	232–276	−8.82	27	231	134
MD992273	28.5	AAR-8337	*Thyasira equalis*	667±37	296	277–319	−7.45	39	257	217
MD992273	79.5	AAR-8428	*Thyasira equalis*	920±75	518	480–561	−8	109	409	429
MD992273	164.5	AAR-8427	*Thyasira equalis*	762±41	422	365–454	−7.5	225	197	220
MD992273	168		V1717		230					
MD992273	168.5	AAR-8330	*Thyasira equalis*	742±33	407	324–430	−7.45	231	176	197
MD992273	203.5	AAR-8331	*Thyasira equalis*	853±33	486	463–503	−8.19	297	189	184
MD992273	248.5	AAR-8412	*Thyasira equalis*	914±36	515	499–536	−14.7	381	134	187
MD992273	369	AAR-8333	*Thyasira equalis*	1214±35	739	706–781	−8.29	607	132	202
MD992273	392		Hekla 1300		650					
MD992273	419.5	AAR-8334	*Thyasira gouldi*	1313±27	881	828–903	−7.16	689	192	153
MD992273	435.5	AAR-8429	*Siphonodentalium lobatum*	1265±20	791	773–831	+0.39	711	80	84
MD992273	534		Hekla 1104		850					
MD992273	659	AAR-8335	*Thyasira equalis,* *Thyasira* sp.	1765±37	1297	1275–1332	−9.12	1070	227	233
MD992273	818.5	AAR-8562	*Thyasira equalis*	2005±40	1551	1517–1605	−9	1351	200	179
MD992273	902.5	AAR-8563	*Thyasira equalis*	2235±25	1832	1812–1866	−6.6	1499	333	298
MD992273	1060.5	AAR-8564	*Cuspidaria glacialis*	2470±25	2114	2088–2140	−7.77	1777	337	292
MD992273	1095.5	AAR-8565	*Thyasira equalis*	2470±30	2114	2080–2143	−8.89	1838	276	232
MD992273	1745		Hekla 3		2980					
MD992273	2418		Hekla 4		4200					
MD992275	34.5	AAR-7702	*Lunatia pallida*	475±36	73	0–132	+0.8	79	−6	−3
MD992275	63.5	AAR-7116	*Thyasira equalis*	696±45	313	287–406	−7°	145	168	164
MD992275	100.5	AAR-7117	*Thyasira* cf. *equalis*	785±40	438	410–468	−6.85	229	209	241
MD992275	101		V1717		230					
MD992275	122.5	AAR-6089	*Siphonodentalium lobatum*	895±45	507	483–530	+0.7	296	211	227
MD992275	134	AAR-7118	*Thyasira equalis*	815±45	461	427–491	−8.98	332	129	100

Table 2 (continued)

Core	Depth (cm)	Lab no.	Material	¹⁴C age BP ±1σ	Cal. yr BP ±1σ	Cal. range (s), 1σ	∂¹³C	Expected tephro-chronological age	Δr, cal. BP deviation	ΔR, ¹⁴C deviation
MD992275	162.5	AAR-7119	Nuculana sp.	945±35	532	513–552	+0.51	419	113	187
MD992275	179		V1477		470					
MD992275	209		V1410		540					
MD992275	222	AAR-7120	Thyasira equalis, Thyasira sp.	1265±45	791	745–879	-8.69	588	203	260
MD992275	239		Hekla 1300		650					
MD992275	259.5	AAR-7121	Thyasira equalis	1420±50	951	917–1003	-8	764	187	185
MD992275	275		Hekla 1104		850					
MD992275	289	AAR-6931	Thyasira equalis	1555±35	1114	1058–1160	-6.84	920	194	181
MD992275	321		Settlement tephra (V871/877)		1080					
MD992275	332.5	AAR-7122	Siphonodentalium lobatum	1710±45	1266	1233–1291	+0.44	1141	125	134
MD992275	373.5	AAR-6932	Siphonodentalium lobatum	1905±40	1451	1397–1507	+0.69	1359	92	74
MD992275	381.5	AAR-6933	Bathyarca glacialis	2020±40	1567	1526–1620	+1.65	1401	166	152
MD992275	439.5	AAR-7703	Thyasira equalis	2262±42	1866	1818–1912	-9.09	1709	157	128
MD992275	460		Snæfellsjökull-1		1818					
MD992275	472.5	AAR-6934	cf. Dentalium entalis	2245±40	1852	1809–1886	+0.72	1882	-30	-37
MD992275	512.5	AAR-6935	cf. Siphonodentalium lobatum	2530±45	2176	2125–2286	+0.91	2087	89	88
MD992275	537.5	AAR-7123	Thyasira equalis	2680±65	2341	2306–2438	-6.8	2215	126	131
MD992275	583	AAR-6936	Thyasira sp.	3110±70	2865	2779–2956	1°	2448	417	359
MD992275	632.5	AAR-7124	Thyasira cf. equalis	3345±45	3205	3147–3260	0.85	2701	504	456
MD992275	687		Hekla 3		2980					
MD992275	696.5	AAR-6937	Siphonodentalium lobatum	3265±50	3078	3006–3164	+0.96	3026	52	43
MD992275	796.5	AAR-6938	Bathyarca glacialis	3795±50	3714	3663–3812	+2.11	3505	209	175
MD992275	824.5	AAR-6088	Siphonodentalium lobatum	3980±55	3969	3890–4065	+0.7	3639	330	249
MD992275	934	AAR-7125	Siphonodentalium lobatum	4320±45	4424	4395–4502	+0.85	4164	260	195
MD992275	1552		Hekla 5		7125					
MD992275	2560		Saksunarvatn ash		10200					
MD992275	3407		Vedde Ash		12000					

A standard reservoir correction of about 400 years ($\Delta R = 0$) is built into the radiocarbon model. The three right-hand columns indicate the expected tephrochronological age of each dated sample, the deviation between each calibrated radiocarbon age and the tephrochronological age of the same level (Δr) and the calculated ΔR value for each radiocarbon date (additional reservoir age correction to the 400 years). ° = assumed standard $\partial^{13}C$ value.

present study. Previous research has shown that molluscs build their carbonate shells directly from CO_2 in the ocean or from pore water in the burrowed sediment (Mook, 1971; Eisma et al., 1976), indicating that in the case of deposit feeders, burrowing depth is more important than feeding habits. The dated species fall into the three following groups with regard to feeding habit and food: (1) scavengers (carnivores), (2) deposit feeders and (3) suspension feeders. The scaphopods *Dentalium entalis* and *Siphonodentalium lobatum* are among the scavengers, living in muddy or sandy sediment, only the small posterior aperture projects above the surface (Muus, 1959). They feed on microscopic organisms, especially foraminifera in the surrounding sediment, and are physically adapted to selecting and capturing food particles from the environment. There is a well-developed radula in the mouth and even small jaws. The gastropod species *Lunatia pallida*, *Propebela* cf. *reticulata* and *Oenopota* sp. also have a well-developed radula in the mouth and feed upon other animals, living or carrion (Thorson, 1944). In East Greenland and around Jan Mayen and Iceland the two bivalve species *Cuspidaria obesa* and *C. glacialis* feed by swallowing smaller animals (Ockelmann, 1958).

Five bivalve species used for AMS ^{14}C dating in the present project belong to the deposits feeders, feeding on organic particles deposited on and in the sediment of the bottom together with associated microorganisms. This group comprises *Nuculana* sp., *Yoldiella lenticula*, *Y. intermedia*, *Y. fraterna* and *Yoldiella* sp. (Ockelmann, 1958).

The remaining four mollusc species are first and foremost suspension feeders, viz. *Bathyarca glacialis*, *Arctinula greenlandica*, *Thyasira equalis* and *T. gouldi* (Ockelmann, 1958; Bernard, 1979). *T. equalis* is the most frequent species in the samples analysed in the present study. In East Greenland and around Jan Mayen as well as Iceland it lives as a suspension feeder (Ockelmann, 1958), and in the Beaufort Sea it is also a filter feeder, but shows some modifications towards macrophagy (Bernard, 1979). Dando (2001) found it, however, in the UK part of the North Sea to live both as a suspension and deposit feeder. On the continental shelf west of central Norway, Myhrvold et al. (2004) reported it as feeding "on deposits on the sediment surface". Such sediment is more or less water saturated and exchanges water with the water mass above.

The sulphide mining of the species *T. equalis* is noteable. It uses the extensional foot (up to 13.9 cm) to gain sulphur to symbiotic bacteria living on their gills (Defour and Felbeck, 2003). However, this sulphide mining tendency cannot have affected our dates very much, as far as we can see. To test this, six pairs of molluscs from the same or nearly the same level in cores MD992272 and MD992271 were AMS ^{14}C dated, with *T. equalis* forming one half of each pair (Table 3). The $\partial^{13}C$ values measured on *T. equalis* are generally very negative (−7 to −8‰), and possibly some fractionation of carbon isotopes is associated with the bacterial symbiosis. The source of the negative values may relate to the animal's metabolism, or possibly to emanations of gas containing negative $\partial^{13}C$ values. As the values are consistently negative at all coring sites, it is considered less likely that the gas origin is the primary source. The dating results (Table 3, Fig. 4) show overlap in all cases and no significant age difference between *T. equalis* and other species, which include scavengers and deposit feeders. However, the *T. equalis* samples show a tendency towards higher ages in four cases out of six, the exceptions being the deposit feeding *Yoldiella* cf.

Table 3
Radiocarbon dates and calibration results for mollusc pairs with differing feeding habits, dated at the same or nearly the same level in cores MD992271 and MD992272. One of the species in each pair is *Thyasira equalis*, which displays very low $\partial^{13}C$ values. For discussion see text and Table 2

Core	Depth (cm)	Lab no.	Material	^{14}C age BP ± 1 s	Cal. yr BP ± 1 s	Cal. range (s), 1 s	$\partial^{13}C$
MD992272	107.5	AAR-7784	cf. *Dentalium entalis*	1595 ± 45	1162	1101–1214	+1.26
MD992272	107.5	AAR-7783	*Thyasira equalis*	1695 ± 50	1257	1217–1286	−7°
MD992272	111.5	AAR-7785	*Siphonodentalium lobatum*	1602 ± 40	1166	1059–1252	+1.34
MD992272	111.5	AAR-7786	*Thyasira equalis*	1715 ± 50	1269	1233–1298	−7.62
MD992272	186.5	AAR-7788	*Yoldiella* cf. *lenticula*	2560 ± 42	2272	2155–2300	+1.35
MD992272	186.5	AAR-7789	*Thyasira equalis*	2570 ± 36	2282	2179–2302	−6.39
MD992272	300.5	AAR-7792	*Siphonodentalium lobatum*	3791 ± 46	3710	3663–3806	+0.69
MD992272	301	AAR-7793	*Thyasira equalis*	3858 ± 47	3820	3734–3869	−8.13
MD992272	461.5	AAR-7676	*Oenopota* sp.	5987 ± 45	6395	6336–6439	+1.14
MD992272	457.5	AAR-7677	*Thyasira equalis*	6120 ± 80	6542	6442–6641	−7.07
MD992271	200.0	AAR-7770	*Yoldiella* sp.	3356 ± 35	3214	3167–3261	−2.23
MD992271	197.5	AAR-7769	*Thyasira equalis*	3340 ± 55	3202	3167–3261	−7.51

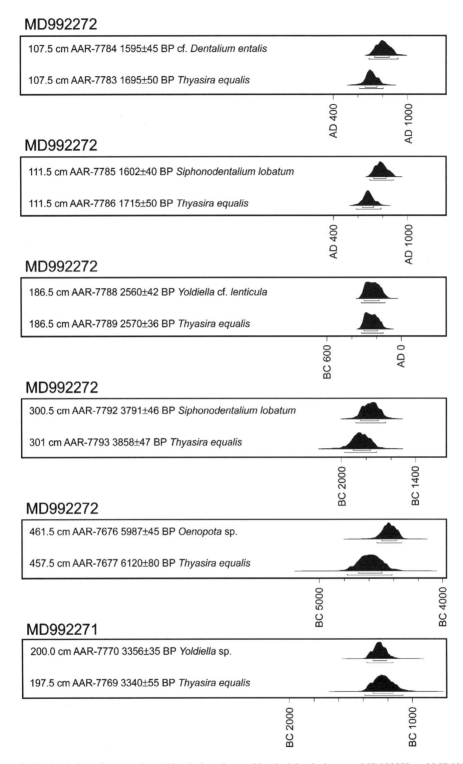

Fig. 4. Calibrated ages of pairwise dating of six samples at identical or close to identical depths in cores MD992272 and MD992271, where one of the mollusc species is *T. equalis,* the other being scavenger or deposit feeder species. Probability plots and one σ as well as two σ error bars are shown. For discussion see text.

lenticula, and *Yoldiella* sp. In high precision dating, this apparent tendency needs to be investigated further to test if the observed age difference tendency is significant or has arisen by chance.

As there appears to be no significant difference in the age of the suspension feeder *Thyasira equalis* and deposit feeders or scavengers when dated pairwise at the same level in the sediment record, it is considered highly unlikely that the so-called *Portlandia* effect (Dyke et al., 2002), i.e. contamination of the shells with old carbon due to deposit feeding, is a significant factor in deviations between tephrochronological age models and AMS ^{14}C based age models in the North Icelandic shelf cores. The deviations are up to 5–10 times larger than the observed (but not significant) species related age difference tendency. All dated occurrences of *Yoldiella* and *Nuculana* (deposit feeders) have been particularly checked, and there is no tendency for anomalous high ages. Total absence of detrital carbonate in the shelf sediments (Knudsen et al., 2004b) reflects the fact that there are no calcareous rocks in Iceland, which is the provenance region for the shelf sediments. Extremely high sedimentation rates in all the studied sediment cores are also considered to reduce the possibility of uptake of old organic carbon by burrowing animals.

3. Results

3.1. The tephra layers

A total of thirteen different tephra layers, spanning the period 10,300 ^{14}C BP to AD 1717 have been identified so far in the five cores presented here (from west to east MD992271, HM107-03, HM107-01, MD992273 and MD992275). Selected intervals of these cores have been processed and the most distinct tephra horizons were sampled and analysed. In core MD992275, eleven tephra layers have been identified (Fig. 3), while three to five tephra layers were identified in the other cores (Figs. 3 and 5–9). Two out of the thirteen tephra layers occur in all five cores, i.e. a tephra layer from the AD 1104 Hekla eruption (Hekla 1104, Thorarinsson, 1967) and Hekla 3, 2879 ± 34 ^{14}C BP (Dugmore et al., 1995), both of silicic composition. The tephra stratigraphy has already been described for two of the cores, i.e. core HM107-03 by Eiríksson et al. (2000a) and core MD992275 by Larsen et al. (2002).

Three tephra layers, previously undetected in the marine sediments off North Iceland, have been identified. Only these are treated in detail here. In core HM107-03, a basaltic tephra layer with poorly vesiculated grains, over 2 mm in longest diameter, is found at 37 cm depth, 16 cm above the Hekla 1104 tephra (Fig. 6 and Table 1B). No matching tephra has been identified on land in North Iceland. This tephra is preliminarily attributed to a submarine eruption northwest of the Grímsey island, described in a contemporary annal for

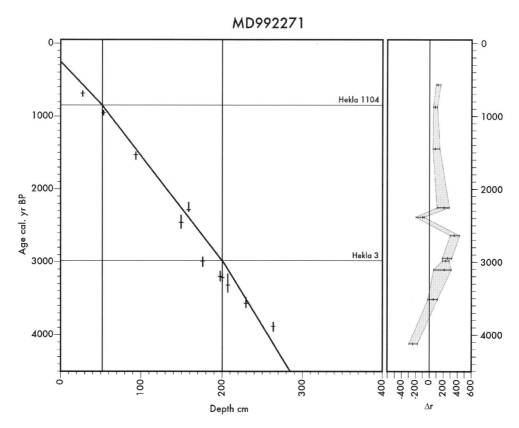

Fig. 5. Age-depth diagram for the last 4500 calibrated years in core MD992271 showing tephra marker horizons as well as calibrated AMS ^{14}C datings of molluscs (shown with ± one standard deviation). A marine reservoir age of 400 years has been applied. The age model is based only on tephra marker horizons (tephrochronological age model). The Hekla 1104 is historically dated, while Hekla 3 is dated on terrestrial material from Iceland. The right-hand diagram shows the deviation (Δr) of each calibrated radiocarbon date from the tephrochronological age of the same level.

HM107-03

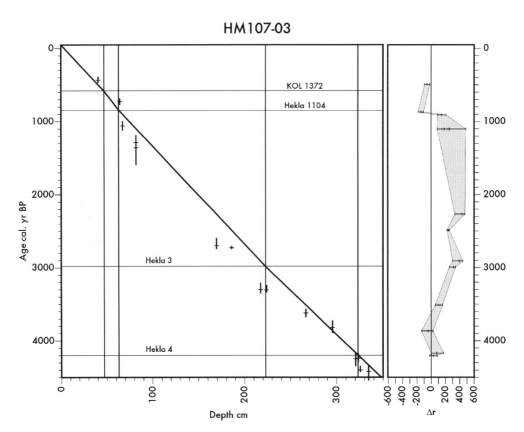

Fig. 6. Age-depth diagram for the last 4500 calibrated years in core HM107-03 (see also Eiríksson et al., 2000a; Knudsen and Eiríksson, 2002) showing tephra marker horizons as well as calibrated AMS ^{14}C datings of molluscs or foraminifera (shown with ± one standard deviation). A marine reservoir age of 400 years has been applied. The age model is based only on tephra marker horizons (tephrochronological age model). The KOL 1372 and Hekla 1104 are historically dated, while Hekla 3 and Hekla 4 are radiocarbon dated on terrestrial material from Iceland. The right-hand diagram shows the deviation (Δr) of each calibrated radiocarbon date from the tephrochronological age of the same level.

the year 1372. This tephra layer, KOL 1372, is also distinct in core HM107-01 (Fig. 7 and Table 1C).

In the easternmost core, MD992275, a basaltic tephra with the chemical characteristics of the Veidivötn volcanic system occurs at a depth of 101 cm (Fig. 9 and Table 1E). It is correlated with a tephra layer found in North Iceland and attributed to an eruption on that volcanic system in AD 1717 (Thorarinsson, 1950; Larsen, 1982). This tephra, V 1717, has also been identified in core MD992273 (Fig. 8 and Table 1D).

At 1552 cm in MD992275, a silicic tephra layer with the chemical signature of the Hekla volcanic system is found (Fig. 9 and Table 1C). Its glass composition matches that of the Hekla 5 tephra (Sigurdsson, 1982). Peat enclosing the Hekla 5 has been dated to 6130±90 and 6070±120 ^{14}C yr BP (Vilmundardóttir and Kaldal, 1982).

The oldest two tephra layers presented here are found in cores MD992271 and MD992275 (Fig. 3). The upper one is 6–8 cm thick and occurs at 602 and 2560 cm, respectively. It has the chemical signature of the Grímsvötn volcanic system (Tables 1A and E) and is correlated to the Skagi-Saksunarvatn tephra that is widely found in North Iceland (Hjort et al., 1985; Björck

et al., 1992; Pétursson and Larsen, 1992; Principato and Geirsdóttir, 2002) and off North Iceland (e.g. Andrews et al., 2002; Geirsdóttir and Ólafsdóttir, 2002). The lower of the two tephra layers is 1–8 cm thick and occurs at 702 cm and 3405 cm depth, respectively. The highly silicic and the basaltic glass (Table 1) match the composition previously published for the Skógar-Vedde glass (Björck et al., 1992; Norddahl and Haflidason, 1992; Lacasse et al., 1995).

In the following sections, we compare the radiocarbon and tephra based age models for the lower part of cores MD992271, MD992273 and MD992275, and we describe the construction of age models for the last 4500 years in all the five cores along an east–west transect and compare the radiocarbon and tephra based age models.

3.2. Age models for the lower parts of cores MD992271, MD992273 and MD992275

The early Holocene segments of the age models for the CALYPSO piston cores MD992271 and MD992275 (Fig. 3) are based on linear interpolation from the lowest marker down to the early Holocene Saksunarvatn ash, and then again on linear interpolation down to the

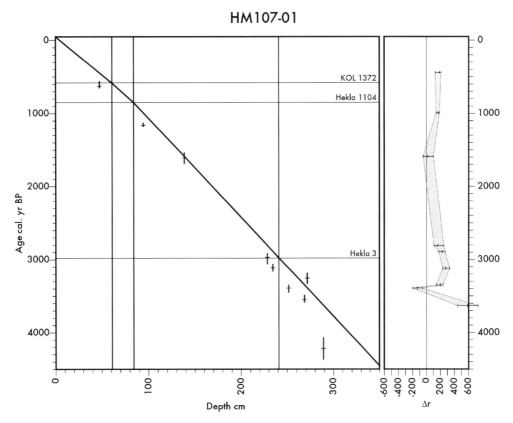

Fig. 7. Age-depth diagram for the last 4500 calibrated years in core HM107-01 showing tephra marker horizons as well as calibrated AMS ^{14}C datings of molluscs and foraminifera (shown with ± one standard deviation). A marine reservoir age of 400 years has been applied. The age model is based only on tephra marker horizons (tephrochronological age model). The KOL 1372 and Hekla 1104 are historically dated, while Hekla 3 is radiocarbon dated on terrestrial material from Iceland. The right-hand diagram shows the deviation (Δr) of each calibrated radiocarbon date from the tephrochronological age of the same level.

Lateglacial Vedde Ash. In core MD992273, a linear interpolation between Hekla 3 and Hekla 4 is used for the lowest relevant segment, and the age model is then extended downcore to a basal radiocarbon date (AA-41841, 6433±93 BP; 6899 (6784–7012) cal. BP).

3.3. Age models for the last 5000 years

Core MD992271—Two tephra markers have been found within the 4500 year record in core MD992271, the Hekla 1104 and the Hekla 3 (Fig. 5). The lowest part of the age model below Hekla 3 is based on a linear interpolation down to the level of the early Holocene Saksunarvatn ash (Fig. 3). The age of the top of the core is estimated to 250 cal. BP. This is based on a comparison of the benthic foraminifera biostratigraphy of core MD992271 with other cores in the region. The mean Holocene sedimentation rate is over 50 cm/ka. Except for two, the calibrated AMS ^{14}C dates yield ages that are higher than the tephrochronological ages of corresponding samples. Increasing deviation is notable from just before 4000 cal. BP up to 2700 cal. BP,

followed by a decreasing trend, which is apparently reversed around 700 cal. BP.

Core HM107-03—The tephrochronological age model for core HM107-03 (Fig. 6) is based on three markers from Hekla: Hekla 4, Hekla 3 and Hekla 1104, and the local tephra correlated with the AD 1372 eruption northwest of Grímsey (KOL 1372). The sedimentation rate averages at over 75 cm/ka. Fifteen AMS ^{14}C datings are available. From the base of the core up to ca. 3700 cal. BP the radiocarbon and tephrochronological age models are similar. The deviation increases sharply close to 3000 cal. BP at the level of Hekla 3 and rises to more than 400 calibrated years. At 1000 cal. BP, the deviation is around 200 years (but with high uncertainty), while negative deviations, observed at 850 and 500 cal. BP, indicate reservoir ages similar to or slightly lower than today.

Core HM107-01—The tephrochronological age model for core HM107-01 is based on Hekla 3, Hekla 1104 and KOL 1372 (Fig. 7). Nine AMS ^{14}C datings are available. Below Hekla 3, a linear extension of the interval between Hekla 1104 and Hekla 3 has been used. High, positive deviations between the radiocarbon and

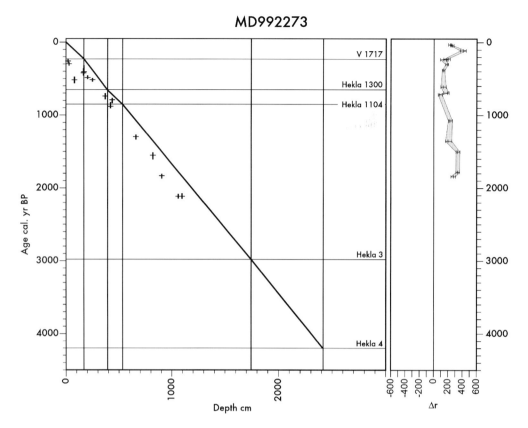

Fig. 8. Age-depth diagram for the last 4500 calibrated years in core MD992273 showing tephra marker horizons as well as calibrated AMS ^{14}C datings of molluscs (shown with ± one standard deviation). A marine reservoir age of 400 years has been applied. The age model is based only on tephra marker horizons (tephrochronological age model). The V 1717, Hekla 1300 and Hekla 1104 are historically dated, while the levels for Hekla 3 and Hekla 4 are based on radiocarbon datings of terrestrial material from Iceland. The right-hand diagram shows the deviation (Δr) of each calibrated radiocarbon date from the tephrochronological age of the same level.

tephra age models are observed immediately above the Hekla 3 tephra. There is no deviation at 1600 cal. BP, while positive values of ca. 200 years are observed again at 1000 and at 450 cal. BP. The 1000 cal. BP value agrees well with the HM107–03 value at that level, while the 450 cal. BP value may indicate either a rapid change between 500 (as registered in HM107-03) and 450 cal. BP (HM107-01), or that the reservoir ages were different at the two sites.

Core MD992273—Numerous tephra layers have been observed in the uppermost 600 cm of core MD992273. In this study, however, we include only three of them, Hekla 1104, Hekla 1300 and V 1717 (Figs. 3 and 8), as they can be correlated with the tephra stratigraphy on land. Further work is needed before reliable correlations can be presented for the other tephras, although preliminary results so far indicate that they fit very well with the predicted age model. A total of 15 AMS ^{14}C dates are available from the Late Holocene interval, covering ca. 2200 cal. yrs. The top of core MD992273 is assumed to date from AD 1950 or 0 cal. BP. The lowest deviation value between the calibrated radiocarbon dates and the tephrochronological age model (Δr) occurs at 700 cal. BP (Medieval

Warm Period), with increasing although slightly fluctuating values up to 100 cal. BP, after which the deviation values decrease. The youngest values are not very reliable due to lack of constraint on the exact age of the top of the core.

Core MD992275—The age model for the Late Holocene part of core MD992275 is based on 6 well-constrained dates of tephra layers younger than 1130 years (Fig. 9), which is the time span since the settlement of Iceland, and on radiocarbon dates of the two older tephra horizons, the Snæfellsjökull 1 and Hekla 3 (for further discussion, see Larsen et al., 2002). A total of 22 AMS ^{14}C dates are available from the interval. There is a relatively high positive deviation between 4200–3500 cal. BP. After a decrease to a low positive value just before 3000 cal. BP, maximum values are reached at 2700–2500 cal. BP. This is followed by a continuous decreasing trend towards a value of close to zero at 1850 cal. BP. Between 1700 and 150 cal. BP, the deviations remain at a level around 100–200 calibrated years, with maximum deviations at 1700–1350, 900–550 and 300–150 cal. BP. A final decrease is observed towards modern time, i.e. to a 400 year reservoir age as found today in this area.

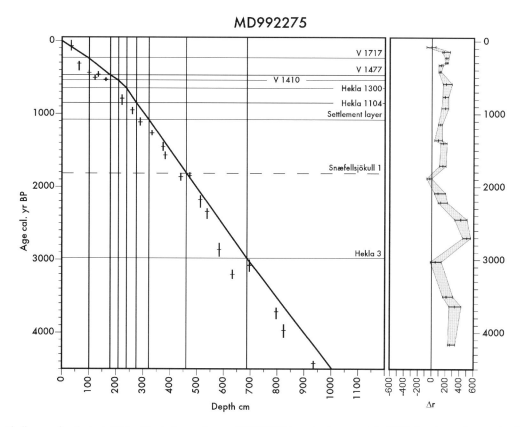

Fig. 9. Age-depth diagram for the last 4500 calibrated years in core MD992275 (see also Larsen et al., 2002) showing tephra marker horizons as well as calibrated AMS ^{14}C datings of molluscs (shown with ± one standard deviation). A marine reservoir age of 400 years has been applied. The age model is based only on tephra marker horizons (tephrochronological age model). The tephra layers V 1717, V 1477 and V 1410, Hekla 1300 and Hekla 1104 are historically dated, the Settlement layer is dated on the basis of correlation with the GRIP ice-core, while Snæfellsjökull 1 and Hekla 3 are radiocarbon dated on terrestrial material from Iceland. The right-hand diagram shows the deviation (Δr) of each calibrated radiocarbon date from the tephrochronological age of the same level.

4. Discussion

If the reservoir age offset, ΔR, in waters off North Iceland had remained constant throughout the Holocene, a tephrochronological age model and a calibrated ^{14}C age model should fit closely, assuming that the marine calibration data set truthfully reflects the oceanic mixing of ^{14}C variability in the atmosphere. The significant discrepancies observed in the area between the two age models may have several causes. In the present study, we examine the possibility that the discrepancies reflect changes in reservoir age caused by reconfiguration of the oceanic Polar Front, separating predominantly Arctic water with a "high" apparent age, and Atlantic water with "normal" apparent age. Other possible causes include deficiencies in the marine calibration data set for the region, or changes in the global thermohaline circulation that might affect the apparent age of the regional water masses. In view of the tests presented in this study and with respect to ecology of the dated organisms, differences in feeding habits cannot explain the large observed deviations.

Fig. 10 shows a summary of the deviations (Δr) between all available calibrated radiocarbon dates from the North Icelandic shelf and the tephrochronological age models, as well as a smoothed curve for the deviations. It is evident that the local reservoir age has generally been higher than 400 years and that we need additional reservoir age correction to fit the tephrochronological age models.

In general, negative Δr values are observed around 4000 cal. BP, followed by an increasing trend that is reversed at ca. 3700 cal. BP. A dramatic deviation peak observed at 2650 cal. BP is the culmination of a trend that started at close to 3500 cal. BP. The next minimum is observed at 2000 cal. BP, and low values are also apparent at 800 cal. BP and perhaps during the last 150 years.

It is appropriate to compare this distinct variability with the available archives of palaeoceanographic variability during the last 4500 years. As an example, the percentage distributions of two selected benthic foraminiferal species from core HM107–03 (Fig. 10) show a clear pattern in the assemblage changes through the last 4500 cal. years. *Cassidulina neoteretis* (Seidenk-

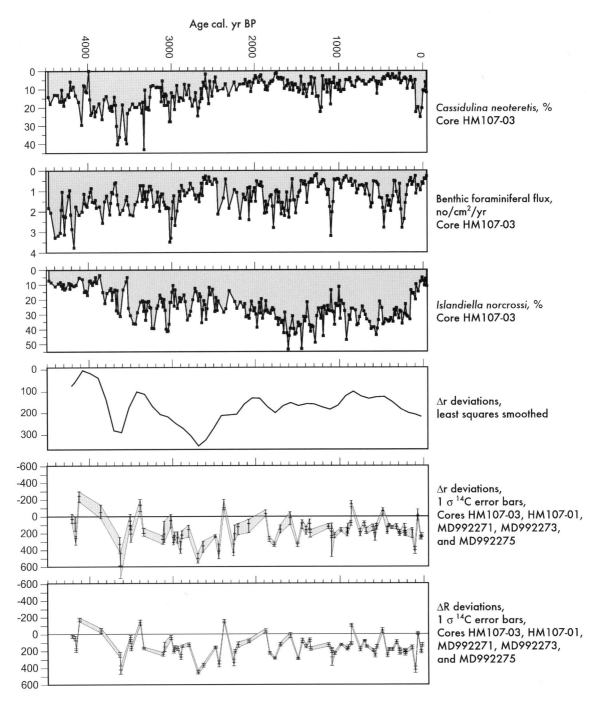

Fig. 10. The percentage distrubution of two benthic foraminiferal species (*Cassidulina neoteretis* and *Islandiella norcrossi*) and the total benthic foraminiferal flux values (no/cm²/year) for core HM107-03 are compared with the deviation of all available calibrated radiocarbon dates (Δr in calibrated years), as well as the calculated additional reservoir age (ΔR in ^{14}C years) of each radiocarbon date that can be compared with constrained tephrocronological age models on the North Icelandic shelf. The Δr and ΔR values are plotted with one σ error bands. A separate smoothed curve for the Δr data was produced by means of distance weighted least squares smoothing with 10% of the data points in each running window.

rantz), which is an indicator of the relatively warm, high-salinity Atlantic water masses of the Irminger Current (see e.g. Jennings and Helgadóttir, 1994; Seidenkrantz, 1995; Hald and Steinsund, 1996), is frequent in the lower part of the core. There is a general decrease after 3000 cal. BP, but with temporary increases again around 1000 cal. BP and in modern

time. The arctic water species *Islandiella norcrossi* (Cushman) (see e.g. Nørvang, 1945; Hald and Steinsund, 1996), however, rises from values around 10% in the lowermost part up to its maximum values of 20–40% that have prevailed since about 3500 cal. BP. Prominent peaks in IRD flux, observed at about 3000 cal. BP, both in HM107–03 and HM107–01, correspond to a period of maximum percentages of *Islandiella norcrossi* (see Knudsen and Eiríksson, 2002). A marked decrease in *Islandiella norcrossi* is recorded during the last about 300 years. The distribution of these two species shows a high degree of variation in the influence of cold water masses from the East Icelandic Current on the sea floor. During cold periods, the Norwegian Sea Deep Water may also have encroached the shelf and replaced the surface water masses at the sea floor of the study area (see e.g. Knudsen and Eiríksson, 2002; Jonsson and Valdimarsson, 2004, Knudsen et al., 2004a).

An interesting feature is that after about 3500 cal. BP, intervals with high total benthic foraminiferal fluxes (number of specimens/cm^2/year) appear to correlate with periods of high influence of the Polar water masses, which are generally less nutritious than the warm Atlantic water masses. In this area, the correlation between high foraminiferal fluxes and cold water masses is presumably related to the proximity to the high primary production zone associated with the Polar Front. On the other hand, the maximum benthic flux values in the lowermost part of the record (Fig. 10) is supposed to be related to the influence of the relatively warm and nutrient rich Atlantic waters of the Irminger Current, i.e. a Polar Front position further to the north in the area. A close correlation between planktonic and benthic fluxes is a common feature, because the primary production in the water column is, with few exceptions, the ultimate food source for benthic communities living below the euphotic zone (Graf, 1992).

Based on a large dataset of sedimentological and biostratigraphical proxies from cores HM107–03 and HM107–01, Knudsen and Eiríksson (2002) identified the first severe cooling event of the Late Holocene at 3700 cal. BP. This was followed by several cooling intervals, i.e. at 3150–2550, 2000–1300, 1200–1000, and 750–100 (Little Ice Age) cal. BP. A warm interval, corresponding in time to the Medieval Warm Period, was identified betwwen 1000 and 750 cal. BP. In a study of the diatom record of core HM107–03, Jiang et al. (2002) recorded a general cooling trend during the last 4600 years, with reconstructed SST values fluctuating from 5.4 to 8.1 °C. Severe cooling events were recorded at around 3700 and 2800 cal. BP. A warm diatom zone, centred at 850 cal. BP was correlated with the Medieval Warm Period. The longest and most severe cooling during the late Holocene was correlated with the Little Ice Age, with an SST drop of 1.5 °C from around 700 cal. BP.

Andrews et al. (2001b) interpreted carbonate fluxes in two gravity cores from the North Icelandic shelf as a measure of productivity. In spite of local variability, a generally decreasing trend was observed for the last 5000 years of the Holocene, with two maxima at 3800 and 2000 cal. BP, and low intervals at around 4300, 3200–2200, 1400 and 300 cal. BP. This was considered

consistent with IRD findings by Bond et al. (Bond et al., 1997) for the Late Holocene interval in the North Atlantic. On the basis of several sedimentological proxies as well as coccolith biostratigraphy, Andrews and Giraudeau (2003) also noted that the last 4–5000 cal. years has been a period of rapid and variable changes in the western part of the North Icelandic shelf. They concluded that a major change occurred at 5000 cal. BP, after which conditions became more variable with a 1.5 kyr cyclicity, including a warm event at ca. 1000 cal. BP, corresponding to the Medieval Warm Period, and a last cold event, corresponding to the Little Ice Age.

Surface circulation changes during the Holocene south of Iceland were demonstrated by Giraudeau et al. (2000), who correlated occurrence minima of the coccolith species *Emiliana huxleyi*, or EH events, with the Holocene record of haematite stained quartz grains (Bond et al., 1997), indicating periods when cool, ice-bearing surface water from the Greenland-Iceland Seas was shifted southward across the Iceland region towards Britain. Three of the EH events fall within the last 4500 cal. BP (at 4500, 2700 and 1500 cal. BP).

Major changes in the on-shore climate in Iceland have been observed and documented in Iceland since the AD 9th century Nordic settlement. The earliest written documents containing climate information stem from the early 12th century. The Icelandic documentary evidence about climate and sea ice has been examined by Ogilvie (1991, 1996, 1997), who concluded that from the settlement of Iceland in the 9th century, climate was generally mild until the late 12th century, after which harsh periods occurred periodically until the 16th century. Particularly cold but variable conditions occurred during the late 17th–19th century. This corresponds to the time period when the sea-ice index reached its maximum around Iceland (Ogilvie and Jónsdóttir, 1996).

There appears to be a clear relationship between the evidence for changes in the influence of Arctic water masses from the East Greenland Current and the East Icelandic Current to the North Icelandic shelf and variation in the deviation of reservoir ages in the area (Fig. 10). The time interval between 4000 and 3700 cal. BP, during which negative Δr values are followed by an increasing trend to relatively high positive values, corresponds to a period with a high influence of warm Atlantic waters followed by a marked cooling due to a change to predominantly Arctic water masses at around 3700 cal. BP (Knudsen and Eiríksson, 2002). The pronounced deviation peak, which is observed at 2650 cal. BP, is the culmination of a cooling trend that started at close to 3500 cal. BP, while the minimum at 2000 cal. BP is preceding the 2000–1300 cal. BP cooling event identified by Knudsen and Eiríksson (2002). Low deviation values, with their minimum centred at 800 cal.

BP, corresponds in time to the Medieval Warm Period, while relatively low deviation at the top of the record may reflect the modern warming at the sea floor after the Little Ice Age (see also Knudsen et al., 2004a).

Although a denser radiocarbon sampling and identification of more tephra markers in some of the analysed cores would probably provide a more coherent pattern, there is a remarkable coupling in the data sets compared above. Palaeoceanographic changes off North Iceland, reflecting the major climate events during the latter part of the Holocene, clearly coincide with variations in the Δr value. The results indicate that the modern value of 400 years as a measure of reservoir age for shelf water masses north of Iceland cannot be extrapolated back in time and that independent control on ^{14}C dating of marine sediment cores in the region is needed.

5. Conclusions

Thirteen different tephra layers have been identified in five marine cores from the North Icelandic shelf (MD992271, HM107-03, HM107-01, MD992273 and MD992275), spanning the period AD 1717 back to 12,000 cal. BP. Three tephra layers are identified in marine sediments for the first time, the V 1717, the KOL 1372 and Hekla 5 tephras.

Tephrochronological age models have been constructed for all five marine cores from the North Icelandic shelf. The deviations between reservoir corrected radiocarbon dates of molluscs or foraminifera and the tephrochronological age models (Δr) for these cores vary from tens to hundreds of years. Sixty-three out of 72 radiocarbon dates show significant positive deviation with a maximum deviation of 500 calibrated years, corresponding to about 450 ^{14}C years, observed at ca. 2700 cal. BP.

A comparison of the deviations with the climatic evidence of a large amount of proxy data from the area, including benthic foraminiferal distributions and productivity, shows that they can be related to incursions of Arctic water masses into this Polar Front boundary area off North Iceland.

The results demonstrate a real need for independent control on ^{14}C dating of marine sediment cores obtained from an oceanographic boundary regions such as the Polar Front separating the Atlantic and Arctic water masses. The results also show the potential of using changes in reservoir ages through time as a geochemical tool for palaeoceanographic reconstructions.

Acknowledgements

The present paper is a contribution to the two European Union 5th framework projects, HOLSMEER (Contract No. EVK2-CT-2000-00060) and PACLIVA (Contract No. ECK2-CT-2002-00143) as well as the PANIS project (Palaeoenvironments on the North Icelandic shelf). Core material was obtained from the BIOICE Cruise in 1995 and the MD9922 IMAGES Cruise in 1999. We are grateful to the Institut Paul-Emile Victor (IPEV) for the IMAGES coring operations onboard the *Marion Dufresne*. Scientific work was supported by grants from the Icelandic Research Council and the Danish Natural Science Research Council. We thank Ole Tumyr at the Geological Institute, University of Bergen, Norway, for assistance and discussions about microprobe analyses presented here.

References

Andersen, G.J., Heinemeier, J., Nielsen, H.L., Rud, N., Johnsen, S., Sveinbjörnsdóttir, Á.E., Hjartarson, Á., 1989. AMS ^{14}C dating on the Fossvogur sediments, Iceland. Radiocarbon 31, 592–600.

Andrews, J.T., Giraudeau, J., 2003. Multi-proxy records showing significant Holocene environmental variability: the inner N. Icelandic shelf (Húnaflói). Quaternary Science Reviews 22, 175–193.

Andrews, J.T., Hardardóttir, J., Helgadóttir, G., Jennings, A., Geirsdóttir, Á., Sveinbjörnsdóttir, Á., Schoolfield, S., Kristjánsdóttir, G.B., Smith, L.M., Thors, K., Syvitski, J.P.M., 2000. The N and W. Iceland shelf: insights into Last Glacial Maximum ice extent and deglaciation based on acoustic stratigraphy and basal radiocarbon AMS dates. Quaternary Science Reviews 19, 619–631.

Andrews, J.T., Caseldine, C., Weiner, N., Hatton, J., 2001a. Late Holocene (ca. 4 ka) marine and terrestrial environmental change in Reykjarfjördur, north Iceland: climate and/or settlement? Journal of Quaternary Science 16 (2), 133–143.

Andrews, J.T., Kristjánsdóttir, G.B., Geirsdóttir, Á., Hardardóttir, J., Helgadóttir, G., Sveinbjörnsdóttir, Á.E., Jennings, A.E., Smith, L.M., 2001b. Late Holocene (~5 cal ka) Trends and Century-scale Variability of N. Iceland Marine Records: Measures of Surface Hydrography, Productivity, and Land/Ocean Interactions. In: Seidov, D., Maslin, M., Haupt, B. (Eds.), The Oceans and Rapid Climate Change: Past, Present and Future, Geophysical Monograph 126. American Geophysical Union, pp. 69–81.

Andrews, J.T., Geirsdóttir, A., Hardardóttir, J., Principato, S., Grönvold, K., Kristjansdóttir, G.B., Helgadóttir, G., Drexler, J., Sveinbjörnsdóttir, Á., 2002. Distribution, sediment magnetism and geochemistry of the Saksunarvatn (10 180±60 cal. yr BP) tephra in marine, lake, and terrestrial sediments, northwest Iceland. Journal of Quaternary Science 17 (8), 731–745.

Austin, W.E.N., Bard, E., Hunt, J.B., Kroon, D., Peacock, J.D., 1994. The ^{14}C age of the Icelandic Vedde Ash: implications for Younger Dryas marine reservoir age corrections. Radiocarbon 37 (1), 53–62.

Bernard, F.R., 1979. Bivalve mollusks of the Western Beaufort Sea. Contribution in Science, Natural History Museum of Los Angeles County 313, 1–80.

Björck, S., Ingólfsson, Ó., Haflidason, H., Hallsdóttir, M., Anderson, N.J., 1992. Lake-Torfadalsvatn: a high resolution record of the North Atlantic ash zone I and the last glacial-interglacial environmental changes in Iceland. Boreas 21, 15–22.

Bond, G., Showers, W., Cheseby, M., Lotti, R., Almasi, P., deMenocal, P., Priore, P., Cullen, H., Hajdas, I., Bonani, G., 1997. A pervasive millennial-scale cycle in North Atlantic Holocene and Glacial Climates. Science 278, 1257–1266.

Dando, P.R., 2001. A review of pockmarks in the UK part of the North Sea, with particular respect to their biology. Strategic Environmental Assessment—SEA2, Technical Report 001, pp. 1–21.

Defour, S.C., Felbeck, H., 2003. Sulphide mining by the superextensile foot of symbiotic thyasirid bivalves. Nature 426, 65–67.

Dugmore, A.J., Cook, G.T., Shore, J.S., Newton, A.J., Edwards, K.J., Larsen, G., 1995. Radiocarbon dating tephra layers in Britain and Iceland. Radiocarbon 37, 379–388.

Dyke, A.S., Andrews, J.T., Clark, P.U., England, J.H., Miller, G.H., Shaw, J., Veillette, J.J., 2002. The Laurentide and Innuitian ice sheet during the Last Glacial Maximum. Quaternary Science Reviews 21, 9–31.

Eiríksson, J., Knudsen, K.L., Haflidason, H., Heinemeier, J., 2000a. Chronology of late Holocene climatic events in the northern North Atlantic based on AMS ^{14}C dates and tephra markers from the volcano Hekla, Iceland. Journal of Quaternary Science 15 (6), 573–580.

Eiríksson, J., Knudsen, K.L., Haflidason, H., Henriksen, P., 2000b. Late-glacial and Holocene palaeoceanography of the North Icelandic shelf. Journal of Quaternary Science 15 (1), 23–42.

Eisma, D., Mook, W.G., Das, H.A., 1976. Shell characteristics, isotopic composition and trace-element contents of some euryhaline mollusks as indicators of salinity. Palaeogeography, Palaeoclimatology, Palaeoecology 19 (1), 39–62.

Feyling-Hanssen, R.W., Jørgensen, J.A., Knudsen, K.L., Andersen, A.-L.L., 1971. Late Quaternary Foraminifera from Vendsyssel, Denmark and Sandnes, Norway. Bulletin of the Geological Society of Denmark 21, 67–317.

Forman, S.L., Polyak, L., 1997. Radiocarbon content of pre-bomb marine mollusks and variations in the ^{14}C reservoir age for coastal areas of the Barent and Kara Seas, Russia. Geophysical Research Letters 24 (8), 885–888.

Geirsdóttir, A., Ólafsdóttir, S., 2002. A 36 Ky record of iceberg rafting and sedimentation from north–west Iceland. Polar Research 21 (2), 291–298.

Giraudeau, J., Cremer, M., Manthe, S., Labeyrie, L., Bond, G., 2000. Coccolith evidence for instabilities in surface circulation south of Iceland during Holocene times. Earth and Planetary Science Letters 179, 257–268.

Graf, G., 1992. Benthic-pelagic coupling: a benthic view. Oceanography and Marine Biology, Annual Review 30, 149–190.

Haflidason, H., Eiríksson, J., Kreveld, S.V., 2000. The tephrochronology of Iceland and the North Atlantic region during the Middle and Late Quaternary: a review. Journal of Quaternary Science 15 (1), 3–22.

Håkansson, S., 1983. A reservoir age for the coastal waters of Iceland. Geologiska Föreningens i Stockholm Förhandlinger 105, 64–67.

Hald, M., Steinsund, P.I., 1996. Benthic foraminifera and carbonate dissolution in the surface sediments of the Barent and Kara Sea. In: Stein, R., Ivanov, G.I., Levitan, M.A., Fahl, K. (Eds.), Surface-sediment composition and sedimentary processes in the central Arctic Ocean and along the Eurasian Continental Margin, Vol. 212. Polarforschung, Berichte, pp. 285–307.

Hansen, B., Østerhus, S., 2000. North Atlantic-Nordic Seas exchanges. Progress in Oceanography 45, 109–208.

Heier-Nielsen, S., Conradsen, K., Heinemeier, J., Knudsen, K.L., Nielsen, H.L., Rud, N., Sveinbjörnsdottir, A.E., 1995a. Radiocarbon dating of shells and foraminifera from the Skagen core, Denmark: evidence of reworking. Radiocarbon 37 (2), 119–130.

Heier-Nielsen, S., Heinemeier, J., Nielsen, H.L., Rud, N., 1995b. Recent Reservoir ages for Danish Fjords and Marine Waters. Radiocarbon 37 (3), 875–882.

Hjort, C., 1973. A sea correction for East Greenland. Geologiska Föreningens i Stockholm Förhandlinger 95, 132–134.

Hjort, C., Ingólfsson, Ó., Norddahl, H., 1985. Late Quaternary geology and glacial history of Hornstrandir, northwest Iceland: a reconnaissance study. Jökull 35, 9–29.

Hurdle, B.G., 1986. The Nordic Seas. Springer, New York 777pp.

Jennings, A.E., Helgadóttir, G., 1994. Foraminiferal assemblages from the fjords and shelf of Eastern Greenland. Journal of Foraminiferal Research 24, 123–144.

Jiang, H., Seidenkrantz, M.-S., Knudsen, K.L., Eiríksson, J., 2002. Late Holocene summer sea-surface temperatures based on a diatom record from the north Icelandic shelf. The Holocene 12 (2), 137–147.

Jonsson, S., Valdimarsson, H., 2004. A new path for the Denmark Strait overflow water from the Iceland Sea to Denmark Strait. Geophysical Research Letters 31(L03305), doi: 10.1029/2003GL019214.

Knudsen, K.L., 1998. Foraminiferer i Kvartær stratigrafi: laboratorie- og fremstillingsteknik samt udvalgte eksempler. Geologisk Tidsskrift 3, 1–25.

Knudsen, K.L., Eiríksson, J., 2002. Application of tephrochronology to the timing and correlation of palaeoceanographic events recorded in Holocene and Late Glacial shelf sediments off North Iceland. Marine Geology 191, 165–188.

Knudsen, K.L., Eiríksson, J., Jansen, E., Jiang, H., Rytter, F., Gudmundsdóttir, E.R., 2004a. Palaeoceanographic changes off North Iceland through the last 1200 years: foraminifera, stable isotopes, diatoms and ice rafted debris. Quaternary Science Reviews, this issue (doi: 10.1016/j.quascirev.2004.08.012).

Knudsen, K.L., Jiang, J., Jansen, E., Eiríksson, J., Heinemeier, J., Seidenkrantz, M.-S., 2004b. Environmental changes off North Iceland during the deglaciation and the Holocene: foraminifera, diatoms and stable isotopes. Marine Micropaleontology 50, 273–305.

Lacasse, C., Sigurdsson, H., Jóhannesson, H., Paterne, M., Carey, S., 1995. Source of Ash Zone 1 in the North Atlantic. Bulletin of Volcanology 57, 18–32.

Larsen, G., 1982. Gjóskutímatal Jökuldals og nágrennis (Tephrochronology of the Jökuldalur area). In: Thorarinsdóttir, H., Óskarsson, Ó.H., Steinthórsson, S., Einarsson, T. (Eds.), Eldur er í nordri, Sögufélag, Reykjavík, pp. 51–65.

Larsen, G., Eiríksson, J., Knudsen, K.L., Heinemeier, J., 2002. Correlation of late Holocene terrestrial and marine tephra markers, north Iceland: implications for reservoir age changes. Polar Research 21 (2), 283–290.

Mook, W.G., 1971. Paleotemperatures and chlorinates from stable carbon and oxygen isotopes in shell carbonate. Palaeogeography, Palaeoclimatology, Palaeoecology 9, 245–263.

Muus, B.J., 1959. Skallus, søtænder, blæksprutter. Danmarks Fauna 65, 1–239.

Myhrvold, A., Hovland, M., Nøland, S.-A., 2004. Baseline and environmental monitoring in deep water—a new approach. Society of Petroleum Engineers SPE 86776, 1–8.

Norddahl, H., Haflidason, H., 1992. The Skógar Tephra, a Younger Dryas marker in North Iceland. Boreas 21, 23–41.

Nørvang, A., 1945. Foraminifera. The Zoology of Iceland II (2), 1–79.

Ockelmann, W.K., 1958. The zoology of East Greenland. Marine lamellibranchiata. Meddelelser om Grønland 122 (4), 1–256.

Ogilvie, A.E.J., 1991. Climatic changes in Iceland A.D. c. 865 to 1598. Acta Archaeologica 61, 9–12.

Ogilvie, A.E.J., 1996. Sea-ice conditions off the coasts of Iceland A.D. 1601–1850 with special reference to part of the Maunder Minimum period (1675–1715). AmS-Varia 25, 9–12.

Ogilvie, A.E.J., 1997. Historical accounts of weather events, sea ice and related matters in Iceland and Greenland, A. D. c. 1250 to 1430. Paläoklimaforschung 23, 25–43.

Ogilvie, A.E.J., Jónsdóttir, I., 1996. Sea-ice incidence off the coasts of Iceland A.D. 1601–1850: Evidence from historical data and early sea-ice maps. In: 26th International Arctic Workshop, Arctic and Alpine Environments, Past and Present. Program with abstracts INSTAAR, 14–16 March, 1996. Boulder, Colorado, pp. 109–110.

Pétursson, H.G., Larsen, G., 1992. An early Holocene basaltic tephra bed in North Iceland, a possible equivalent to the Saksunarvatn ash bed. In: Geirsdóttir, A., Norddahl, H., Helgadóttir, G. (Eds.), Abstracts: 20th Nordic Geological Winter Meeting, 7–10 January 1992. Icelandic Geoscience Society, Reykjavík.

Principato, S.M., Geirsdóttir, A., 2002. Glacial geology of the Strandir and Isafjardardjup coastlines, NW peninsula, Iceland. In: Jónsson, S.S. (Ed.), Abstracts: 20th Nordic Geological Winter Meeting, 6–9 January 2002. Icelandic Geoscience Society, Reykjavík.

Rytter, F., Knudsen, K.L., Seidenkrantz, M.-S., Eiríksson, J., 2002. Modern distribution of benthic foraminifera on the north Icelandic shelf and slope. Journal of Foraminiferal Research 32, 217–244.

Seidenkrantz, M.-S., 1995. *Cassidulina teretis* Tappan and *Cassidulina neoteretis* new species (Foraminifera): stratigraphic markers for deep sea and outer shelf areas. Journal of Micropalaeontology 14, 145–157.

Sigurdsson, H., 1982. Útbreidsla íslenskra gjóskulaga á botni Atlantshafs. In: Thorarinsdóttir, H., Óskarsson, Ó.H., Steinthórsson, S., Einarsson, T. (Eds.), Eldur er í nordri. Sögufélag, Reykjavík, pp. 119–127.

Stefánsson, U., 1962. North Icelandic Waters. Rit Fiskideildar 3, 1–269.

Stuiver, M., Reimer, P.J., Bard, E., Beck, J.W., Burr, G.S., Hughen, K.A., Kromer, B., McCormack, G., van der Plicht, J., Spurk, M., 1998a. INTCAL98 Radiocarbon Age Calibration, 24,000-0 cal BP. Radiocarbon 40 (3), 1041–1083.

Stuiver, M., Reimer, P.J., Braziunas, T.F., 1998b. High-precision radiocarbon age calibration for terrestrial and marine samples. Radiocarbon 40 (3), 1127–1151.

Tauber, H., Funder, S., 1975. ^{14}C content of recent molluscs from Scoresby Sund, central East Greenland. Grønlands Geologiske Undersøgelse, Rapport 75, 95–99.

Thorarinsson, S., 1950. Jökulhlaup og eldgos á jökulvatnasvædi Jökulsár á Fjöllum. Náttúrufrædingurinn 20, 113–133.

Thorarinsson, S., 1967. The eruptions of Hekla in historical times. Societas Scientiarum Islandica The Eruption of Hekla 1947–1948 1, 1–170.

Thorson, G., 1944. Marine gastropoda prosobranchiata. The Zoology of East Greenland. Meddelelser om Grønland 121 (13), 1–181.

Vilmundardóttir, E.G., Kaldal, I., 1982. Holocene sedimentary sequence at Trjávidarlækur basin, Thjórsárdalur, Southern Iceland. Jökull 32, 49–59.

Weiner, N., Helgadóttir, G., Jennings, A.E., 1999. Modern foraminiferal faunas of the western and northern Iceland shelf. Abstract Volume, Geological Society of America 31 (7), A315.